LONG RANGE

RANGE

SHOOTING

Learn from how to shoot from professionals who instruct foreign snipers.

Tell people what you think? Thanks!

Reviews from amazing people like you help other marksmen discover these insights, and make the entire community better and smarter.

Thank you in advance for your help and time!

Join the ranks of the most precise and accurate marksmen on the planet.

Introduction (Learn, Have Fun, and Be Safe)

1. Core Rifle Equipment — 13

Rifles — 13
Ammunition — 17
Scopes — 20

2. Concepts — 20

Long Range — 20
Shooting — 22
Synonyms — 23
Systems of measurement — 23
Angular Measurements — 24

3. Firearm Safety — 24

Treat Every Firearm as if It Were Loaded — 24
Always Point the Gun in a Safe Direction — 25
Safety On and Finger Off — 26
If One Person is Unsafe, Everyone is Unsafe — 26
The Rifle is Only One Part of a Dangerous System — 27
Always Remember! Firearm Safety is a Mindset — 28

Setting Up Your Equipment

4. Setting Up a Scope   31

Reticle (Eyepiece or Ocular) Focusing   32
Target (Objective) Focusing   32
Parallax (Reticle Shift)   34
Adjusting Elevation and Windage Settings with Turrets   36
Getting to Mechanical Center   38
Bore-Sighting   40

5. Holding a Rifle   42

Support Hand   43
Shoulder Pocket   44
Trigger Hand (Firing Hand)   46
Trigger Finger   48
Cheek Weld and Eye Relief   50
Benchrest Shooting   51

Beginners
(0 to 100 Meters or Yards)

6. Loading a Rifle   53
Bolt Manipulation   54

7. Taking Aim at a Target   55
Target Definition   55
Finding a Target   57
Sight Alignment and Iron Sights   58
Sight Picture   60
Holdover and Ballistic Loopholes   63

8. Firing the Rifle   65
Trigger Pull and Flinching   66
Recoil and Muzzle Jump   67
Follow-Through and Reacquiring Sight Picture   71
Cold-Bore Shooting   73

9. Ensuring Accuracy (Grouping and Zeroing)   74
Grouping   74
Picking a Zero Distance   77
Zeroing   79
Resetting Elevation and Windage Turrets to Zero   84

Intermediates (100 to 300 Meters or Yards)

10. Body Control   87
Breathing   87
Natural Point-of-Aim (NPA)   89
Heartbeat   92
Shooting with Both Eyes Open and Eye Dominance   94

11. Shooting Positions   95
Prone   97
Kneeling   100
Standing   102
Using a Sling   102
Using Ground Support   104

12. Finding Distance to a Target   110
Using Rangefinding Tools   110
Using a Standard Reticle (Milling)   113
Using Hands   119
Using Estimation   121
Perception Errors   123

13. Using Scopes for Long-Distance Shooting   126
Magnification   127
Gravity, Drag, and Trajectory   128
Standard Reticles   132
First and Second Focal Plane Reticles   136
Elevation Hold and Dial   137
Precisely Adjusting the Sight Picture   140
Bullet-Drop Compensator Reticles   140
Danger Distance   143

14. Eliminating Rifle Cant   145
Formulas for Cant Offset   147
Holding Cant   150
Reticle Cant   152
Mechanical Adjustment Cant   152

Experts (300 to 600 Meters or Yards)

15. Wind — 157

Wind Value	159
Crosswind Deflection	162
Vertical-Wind Deflection	163
Wind Gradient	164
Multiple Winds over Distance	167
Changing Wind over Time	171
Windage Holds and Dials	173
Converting a Wind to a Windage Hold (Summary)	177

16. Wind Estimation — 178

Flags and Vanes	179
Vegetation	182
Mirage	184

17. Climate — 187

Temperature	188
Air Density	189
Simple Mirage	191
Inferior and Superior Mirage	194
Precipitation	194

18. Inherent Imprecision — 197

Lateral Throwoff	198
Probabilistic Shooting (Mil and MOA-Accuracy)	201
Estimation (Range and Wind) Uncertainty	206

19. Shooting Uphill or Downhill (Inclination) — 207

Perpendicular and Parallel Components of Gravity	208
Change in Air Density with Change in Altitude	212

Information Management

20. Information Gathering   215
Calling Shots   215
Meters   216
Recording   217

21. Calculating   221
Ballistic Calculators   221
Conversion Charts   222

22. Instruction   224
Practice   224
Dry Firing   227
Drills   229
Teaching   230

Appendices

23. Linear Distance   235
Systems of Measure   235
Meters (m) and Yards (yd)   235
Caliber   236

24. Angular Distance   237
Minute of Angle (MOA)   239
Inches Per Hundred Yards (IPHY)   239
Milliradian (Mil a.k.a. MRAD)   240
Comparing and Converting MOA and Mil   240

25. Other Measurements   241
Speed and Velocity   242
Mass and Weight   243
Force and Pressure   244
Work and Energy   245
Momentum   246

26. Functions and Malfunctions   247
Feeding   247
Chambering   248
Locking   249
Firing   249
Unlocking   250
Extraction   250
Ejection   251
Cocking   252
Standard Malfunction Actions   252

27. Glossary   253

28. Credits   261

Introduction

1. Core Rifle Equipment   13

Rifles   13
Ammunition   17
Scopes   20

2. Concepts   20

Long Range   20
Shooting   22
Synonyms   23
Systems of measurement   23
Angular Measurements   24

3. Firearm Safety   24

Treat Every Firearm as if It Were Loaded   24
Always Point the Gun in a Safe Direction   25
Safety On and Finger Off   26
If One Person is Unsafe, Everyone is Unsafe   26
The Rifle is Only One Part of a Dangerous System   27
Always Remember! Firearm Safety is a Mindset   28

Introduction (Learn, Have Fun, and Be Safe)

I have a very strict gun control policy: if there's a gun around, I want to be in control of it.

—Clint Eastwood

This manual uses **easy, step-by-step** explanations to teach you how to shoot and bring out the inner sniper that you were meant to be. After reading this book, you will find that you are way more precise and accurate than ever before, and you will be able to maximize the power of your rifle to the fullest. With enough time, you may even become as proficient as the most elite marksmen in the world: **United States Special Forces**.

Before explaining how to shoot, this introduction quickly covers rifle equipment, useful concepts, and firearm safety. This is to help first-timer marksmen or to refresh more experienced marksmen. If all of this is redundant, then readers can **skip to the next chapter.** (See Setting Up Your Equipment, Pg. 31.)

1. Core Rifle Equipment

This manual assumes that readers possess a **long-range rifle with ammunition and a scope.** These three core components along with the marksman themself comprise a weapon that can propel a bullet to wherever the marksman wants it to go. Additionally, a marksman may incorporate extra equipment that enhances precision and accuracy, such as ballistics calculators and weather meters.

1.a Rifles

A rifle is a weapon with a barrel from which a bullet (i.e., the projectile) is propelled by gunpowder (i.e., the propellant), and with a stock that transfers recoil into the marksman's shoulder through a stock. A rifle consists of five major components: the receiver, the barrel, the action, the trigger group, and the stock. (See Image 1, Pg. 14.)

The **receiver** "receives," or attaches to, all of the other components of the rifle and is the core hub of the rifle. This is not a hard definition, and sometimes it can be difficult to isolate a single receiver. For example, the

Rifle System Parts

Image 1: Remington Model 700 in .30-06 Springfield. This is a **bolt-action rifle**, which means the marksman must themself move the handle behind the scope backwards to eject a fired round and forwards to load a new one after firing.

Image 2: A G82 German Army Barrett M107 variant. This is a **semi-automatic action and rifle**, and so one trigger squeeze only fires the rifle and automatically loads a new cartridge. A fully-automatic rifle would additionally fire that round and keep firing until the magazine were empty or the trigger were released.

U.S. Army's rifle, the M4, has two receivers: an upper and a lower. Both parts attach to other rifle parts and also to each other. (See Image 3, Pg. 15.)

The **barrel** is the metal tube that a bullet travels down. The inside of a barrel is called the "bore." Importantly to be a rifle, a long-gun's bore must have interior grooves, called "rifling," that both grip and spin the bullet. By spinning, bullets can use gyroscopic forces to keep their pointed tip facing forward causing less air resistance and farther, more precise travel. Barrels also keep the exploding gas inside so that it can press against the back of the bullet. The longer time that gas can act on the bullet, the faster a bullet travels, which is why longer barrels create more muzzle velocity than shorter barrels.

The **action** of a firearm refers to the combination of parts that facilitate the loading, firing, ejecting, and general mechanical action of ammunition.

Rifle Receivers

Bolt-Action Firearms

Steyr AUG

Receiver

Receiver
(with bolt inserted)

AR

AK

Receiver

Receiver

Image 3: Rifles come in many forms. The Bureau of Alcohol, Tobacco, and Firearms (ATF) is responsible for regulating, and therefore defining, firearms in the United States. Because a firearm can still discharge a bullet while missing parts such as the scope, the stock, and most of the barrel, the ATF (and thereby the U.S. Government) **defines a "firearm" as its receiver** (i.e., the central connection part). But defining what a receiver is exactly is difficult, so the ATF had to supply the four above images among many others to clarify with specific examples.

What parts of a firearm constitute the action differs depending on the rifle's operating system and is sometimes debated. The location where a bullet is loaded into, ready to be fired, is called the **chamber**. Although the chamber is formed into the rear of the barrel, it is often considered part of the action, since it relates to the manipulation of the ammunition. To hold a round in the chamber, a metal cylinder called a **"bolt"** is placed behind the round. The bolt stays behind the round and locks the chamber while the bullet is fired. The bolt must be able to move and open the chamber up again to allow for the ejection of the spent round and the insertion of a new round. Bolts are also the rifle part that pushes a round into the chamber. (See Image 4, Pg. 16.)

There are two widely used actions in long-range shooting: manually operated actions and automatically operated actions. Manually operated actions require the marksman to manually reload a new round of ammunition after the trigger is pulled, whereas automatic actions automatically load the next round after firing using the force generated by the previous round firing. The most commonly used manual action in long-range shooting is the bolt-

Receiver Area

Image 4: The location of stored cartridges is the **magazine**. If the magazine is detachable, it is inserted into the **magazine well** located in the stock or receiver. The **bolt** is the metal cylinder that inserts, locks in, and ejects cartridges from the rifle. The part of the bolt that touches the back of a cartridge is the **bolt face** (not pictured). The bolt travels along the **bolt channel** to transport rounds from the magazine. The open part of the bolt channel is the **ejection port**. The forward, closed part of the bolt channel is the **breech** (a breech is an entryway for ammunition). After the bolt channel is the **chamber**, where ammunition is locked in, ready to be fired. (As with all vocabulary, definitions vary between sources.)

action. With this mechanism, the shooter pulls the bolt back to open the chamber and eject the spent round, then pushes it forward to load a new round and close the chamber. (See Image 1, Pg. 14.) There are a variety of automatic actions, but since the marksman does not interact with the mechanisms, their operation is not vital to the skill of shooting. The only important distinction to make is between fully automatic actions, which fire bullets as long as the trigger is depressed, and semi-automatics, which do not fire the bullet after loading it. All automatic actions intended for long-range shooting are semi-automatic. (See Image 2, Pg. 14.)

The **trigger** is the part of the rifle that the marksman operates to cause the gunpowder to explode. The trigger does not directly act on the ammunition, and instead is just the first in a chain of parts called the "trigger group" or "trigger assembly." Around the trigger is a thin band of metal called the "trigger guard," which helps to prevent any unintentional depressing of the

Rifle Stock Terms

Image 5: Rifles can be customized to their marksman. The most important customization is the **length-of-pull**, which is the distance from the trigger to the buttstock. A length-of-pull that is too short prevents a marksman from properly placing the buttstock in the shoulder pocket. (See Shoulder Pocket, Pg. 44.), while one that is too long interferes with placing the check on the stock. (See Cheek Weld and Eye Relief, Pg. 50.) A proper length-of-pull leaves **1 cm (0.5 in)** of space between the stock and the biceps when the marksman's finger is in the trigger well and the arm is resting at **90 degrees**.

trigger. The ideal trigger provides a constant level of resistance to being pulled back. A one-stage trigger has a single level of resistance. In contrast, a two-stage trigger suddenly increases the resistance level immediately before the trigger would fire the rifle, which can be a useful safety feature.

The **stock** is the part of the rifle that interfaces with the marksman's body. The shooting hand holds the grip, while the front hand holds the forestock. The rear of the stock, the buttstock, presses into the marksman's torso, and the marksman lays their check on the top of the buttstock, called the "cheekrest" or "comb." A properly fitted rifle stock has a correct **length-of-pull**, which is the distance from the trigger to the end of the buttstock. (See Image 5, Pg. 17.)

Rifles may also come with accessories that are helpful but not strictly necessary. For example, external supports such as bipods and tripods can vastly increase the stability of a rifle.

1.b Ammunition

Ammunition is the expendable part of a rifle setup. It consists of four parts: bullets, gunpowder, primers, and casing. Together, these four parts are called "cartridges" or "rounds." (See Image 6, Pg. 18.)

Bullets are small pieces of metal that are constructed to transfer energy created at the marksman's location to the target while facing the least amount of air resistance possible. To carry more energy, bullets are made of dense materials, such as lead, that have more mass per unit of volume. To be

Cartridge Parts

Image 6: A diagram of .280 / 30 British military sectioned cartridges. All widely used cartridges (a.k.a. rounds) use this structure. Note that identically shaped ammunition may be made with different materials and internal structures.

durable enough to survive being fired and impacting, the lead is coated in a more durable material such as a copper alloy. And to reduce air resistance, bullets are made in a pointed shape that can slice through the air.

Gunpowder, or powder, is a solid material that causes a gas explosion. (Modern "powder" actually comes in pellet form.) The resultant gas explosion is always faster than the bullets it pushes since a slower gas cannot reach and push a faster bullet. However, while it is common to say that gunpowder "burns" or "explodes," the technical term is "deflagrates" since the chemical reaction occurs slower than the speed of sound. (In a controlled environment, inside a firearm cartridge where pressure builds up rapidly, the effective speed of the expanding gases can result in projectile velocities of 500 to 1,200 m/s (1600 to 4000 ft/s) for rifles, but this is not the same as the actual deflagration rate of the powder itself.)

There are various gunpowder compositions that generate gases in a way that balances the power of the gases while minimizing damage that that power does to the barrel. This allows the gas to effectively propel the bullet without causing damage to the rifle.

Primers are small disks that explode when they are struck with a metal pin. That is, they are the interface between the mechanical energy release of the trigger and the chemical energy release of gunpowder. The pin they are struck by is called the "firing pin."

Casings, or cases, are the metal cylinders that hold the bullets, the gunpowder, and the primers in place. Casings are most commonly made of

Scope Turrets, Dials, and Reticles

Image 7: This Schmidt & Bender 5-25x56 riflescope has a parallax turret and an illumination turret on its left side.

Image 8: This Swift 687M riflescope has a parallax dial on the objective bell (the forward end of a scope) and no illumination turret.

Various Reticles

| Basic Crosshairs | Mil-Dot | Modern Rangefinding |

Image 9: The sight picture is the image seen through the back of the scope by the marksman. The reticle is the pattern of hashmarks overlaid on the sight picture.

brass, so they are often called the "brass"; however, casings can also be made of steel. They are made of metal because it is durable, but also because casings serve to dissipate heat. When a hot casing is ejected from a rifle, the heat it carries can no longer be transferred to the rifle nor to subsequent rounds.

All normal, modern rifle store ammunition in the **magazine** (a.k.a. "mag"). Magazines can be an internal part of the rifle, or a detachable container. The location that detachable magazines are inserted into is called the "magazine well." Magazines have internal springs that sequentially push each cartridge into a position where it may be readily loaded into the barrel chamber by the action.

1.c Scopes

A scope (a.k.a. riflescope or telescopic sight) is an optical attachment that is mounted atop a rifle. It is a kind of "**sight**," which is a tool that aids in aiming at a target. Scopes serve two purposes. First, they have internal markings on the lens that show a marksman where a bullet would impact if the rifle were fired. Second, they magnify the target to allow for better sight and more precise aiming. Long-range shooting usually requires a high-power scope ("high-power" means the lowest magnification setting is greater than one).

The internal markings on a scope's lens are called its "reticle." The **reticle** is a pattern of **hashmarks** (i.e., short lines or dots) that show where a bullet would impact if a bullet were fired. There are a variety of reticles that are explained more in-depth later in the book. (See Standard Reticles, Pg. 132.) Reticles can be adjusted up and down with the **elevation turret or dial**, and left and right with the **windage turret or dial**. (See Image 17, Pg. 37.)

Many scopes also have a **parallax turret or dial** on their left side that simultaneously changes the scope's depth-of-field (See Image 41, Pg. 62.) and parallax (See Image 16, Pg. 35.). Depth-of-field is the range of distance that is in focus for the marksman when looking through the scope. Parallax (as far as shooting is concerned) is the ability of the reticle to stay on the same point on a target even if the marksman moves their eye around the scope. Finally to focus the sight picture, a marksman can adjust the scope's eyepiece lens by rotating the eyepiece dial. (See Image 28, Pg. 49.)

2. Concepts

The following concepts are relevant to the rest of the content in this manual. Additionally, there is a glossary of terms included in the appendix at the back of this manual that contains terms which may not be familiar to all readers.

2.a Long Range

Long-range shooting involves using a rifle to hit a target at a significantly long distance. If pressed for a concrete number, a commonly used definition of "long range" is anything 300 m or yd or beyond. If more fineness is desired, "long range" can be defined differently for various rifle sizes. For example, a

Accuracy versus Precision

Low Accuracy   **High** Accuracy   **Low** Accuracy   **High** Accuracy
Low Precision   **Low** Precision   **High** Precision   **High** Precision

Image 10: Accuracy requires the skill of understanding the relationship between where a rifle is pointing and where a bullet would impact. Precision requires the skill of being consistent for every shot taken.

---

small rifle may have a "long range" of 100 m or yd, while a sniper rifle may have one of 500 m or yd.

The problem is that the term "long" can be ambiguous because marksmen generally associate the term with skill level and not actual distance. That is, if a target is difficult to hit, then the target is perceived as being "far." Therefore, a more nuanced and perhaps natural approach to defining "long range" is based on the importance of **accuracy versus precision**. Accuracy and precision are two distinct-but-related concepts, and each requires its own skill set. (See Image 10, Pg. 21.)

Accuracy refers to a marksman's ability to **center the average** of all of their shots on the target. To be accurate, the bullet must be pointed at the target. That is, a marksman must have consistent and reliable equipment setups, such as securely aligned scopes. And the marksman must properly account for wind and gravity.

Precision refers to the ability to repeatedly hit a **small area** no matter where the center of that area is. To be precise, each successive shot must be fired under the same conditions as the previous shot. That is, a marksman must have consistent and accurate shooting techniques, such as consistently pulling the trigger in the same manner.

"Long range" is then the distance at which **properly aiming (accuracy) becomes more challenging than maintaining consistent shooting techniques (precision)**. In practical terms, if a bullet travels relatively flat up to 200 m or yd, aiming is simple because it only requires putting the target in the scope's crosshairs. However, if a bullet were to travel 1,000 m or yd in high winds, it would deviate significantly from a straight line and require calculations to find the correct aiming point and timing.

(Mathematically speaking, this shift occurs because environmental factors such as wind continue to affect a bullet throughout its flight, resulting in a small but exponentially increasing effect on the bullet trajectory. On the other hand, the marksman's influence has a significant effect only while the bullet is in the bore, resulting in a large but linear effect. Over time, the exponential effect caused by environmental factors always surpasses the linear effect caused by the marksman, with an exponent greater than one.)

2.b Shooting

The goal of shooting is to propel a small mass at high speeds toward a specific target. The skill of shooting comprises the numerous acts required to achieve that goal. This manual covers approximately fifty distinct topics related to long-range shooting that range from using the correct posture to hitting targets at vastly different elevations.

It is important to note that not all of these skills are necessary for every marksman. For example, an indoor target marksman and a military sniper both shoot at long range, but the former is far more concerned with equipment while the latter is more concerned with wind. As a result, while the definition of what constitutes the goal of "shooting" is clear, the skills that comprise "shooting" are subjective based on each marksman's particular needs and situation.

In particular, many people object to the idea that one can be a marksman if they do not perform certain acts or skills because they associate the term "shooting" more with the acts and skills rather than the goal. This is relevant to modern technology, which removes much of the work that was previously required to propel a small mass at high speeds toward a specific target. For example, mechanical servos and cameras can allow a rifle to be fired from a location detached from the marksman, removing the need for the marksman to be still and stable.

While the traditionalist view is understandable, today's reality is that electronically assisted shooting systems are becoming increasingly accurate and widespread, with ballistic calculators being almost universally used. Therefore, this manual takes the stance that **"shooting" is based on the goal, which is to propel a small mass at high speeds toward a specific target.** Nevertheless, the information herein assumes that a reader does not always have electronic assistance so as to provide the most comprehensive instructions possible.

2.c Synonyms

If this book uses a term unfamiliar to a reader, it may simply be because the reader uses another term for the same thing, and not because the object or concept itself is something new. Many times **multiple words have come to identify the same concept** because traditionally, rifles and guns have been a very decentralized activity. For example, "dial, turret, or knob" are all interchangeable; the front end of a stock is known as the "forearm," "foreend," "forend," "handguard," "forestock" and more; and the front attachment to a barrel that lessens the sound of firing is called the "silencer," "suppressor," or "sound moderator." This issue is especially prevalent when discussing how bullets move in the air. (See Image 112, Pg. 129.)

2.d Systems of measurement

There are two main systems of measurement used in the world of shooting, the **metric system** (formally known as the "international system of units (SI)") and the **imperial system** (formally known as the "United States Customary System"). Ballistic charts, scopes, and other shooting accessories often use metric units for easier and more precise adjustments. However, it is common for United States civilians to still use the imperial system.

Therefore, this manual presents both systems, usually starting with metric units followed by imperial unit within parentheses (e.g. 100 m (109 yd)). Metric units are first because metric is currently the global standard among militaries, including the United States, and it is slowly becoming the global standard for manufacturers as well. More about these units is explained in the appendices. (See Linear Distance, Pg. 235.)

There are some measures and units that are specific to the world of rifles and shooting that are uncommon elsewhere. For example, a common measure is **caliber (cal)**, which is the internal diameter or bore of a gun barrel. Caliber can be presented in either inches (e.g., 0.45 cal is 0.45 in) or millimeters (e.g., 9 mm). A common imperial unit that is specific to shooting is the **grain**, which is 1/7000 of a pound. Because using such a fine unit of measure was convenient for weighing out gunpowder and bullet metal, grains were the international standard for ammunition-related weights even in metric-centric countries. That is, however, slowly changing as countries move to using grams and milligrams.

2.e Angular Measurements

Understanding angular measurements is vital to understanding how to properly use a scope. In brief, angles are important in shooting because many elements, such as scopes and marksmen, rotate about their axis. The point-of-origin (e.g., a marksman) forms imaginary lines with a starting point (e.g., the point-of-aim) and an ending point (e.g., the point-of-impact). The angle between these two lines can be measured. The two standard systems of measurement for angles are **minutes-of-angle (MOA) and milliradians (mils)**. There are 21,600 MOA in a circle and 6,283 mils in a circle. Approximately 3.5 MOA is equivalent to 1 mil. (The conversion is approximate not only for simplicity but also because the actual definitions of the units are not always clear).

With that said, **if there is any uncertainty on what an angular measurement is, readers may benefit from reviewing that information in the appendix at the end of this manual**. (See Angular Distance, Pg. 237.)

3. Firearm Safety

Firearms are extremely dangerous tools, and if misused, they can cause catastrophic injury or death. Therefore, nothing is more important than practicing safety precautions when handling a firearm of any kind by anyone. That is why every responsible firearm owner must prioritize safety, and why this long-range shooting manual begins with a chapter on firearm safety. (Please note that some of the terms in this section may not be familiar to those who have never shot before, but are clarified throughout the book in their relevant sections.)

3.a Treat Every Firearm as if It Were Loaded

No matter how experienced a marksman may be, it is crucial to handle every firearm as if it were loaded and ready to fire. Specifically, **firearms are never pointed at something that is not intended to be shot**. Disregarding safety precautions with flippant statements such as "Don't worry, the safety is on" or "It's not even loaded" is unacceptable in any setting. Many innocent lives have been lost due to misplaced confidence in the status of a firearm. It is not impolite or unwelcome to ask someone to reclear their firearm (even if they insist it is already clear). Always remember, it's the supposedly "unloaded gun" that causes a negligent discharge.

Image 11: No one is too cool or too skilled to forgo range safety.

To ensure a firearm is unloaded, the marksman must perform a "dry firing" process using the following steps: First, point the firearm in a safe direction where no one can be harmed in case of a discharge. Remember to keep all fingers away from the trigger guard. Next, remove the magazine if it is detachable and place it away from the firearm. Then, manually cycle the action of the firearm a few times to clear any remaining ammunition in the chamber. Visually inspect the chamber. Pull the trigger to attempt to fire the firearm. Ideally, this is a "dry" fire because no ammunition was present in the firearm. Finally, it is recommended to keep the action open and visually inspect the chamber to ensure it is empty. Even after following these steps, the firearm must still be treated with the same extreme caution as if it were loaded.

3.b Always Point the Gun in a Safe Direction

Even if a firearm is loaded, it can only cause harm if it is aimed at someone or if they are close enough to be affected by the hot gases escaping the muzzle. For this reason, a marksman must only aim at things they are willing to shoot, such as a safe target or a location where a negligent or accidental discharge could not cause harm to anyone. Importantly, a safe direction always ends with a target in sight, and is never the air or somewhere far away where there is an unknown impact zone.

Using a Chamber Flag

Image 12: When a rifle is not in use, a yellow or orange piece of plastic called a "safety flag," "breech flag" or "**chamber flag**" is inserted into the chamber to prove that a round cannot be in the chamber.

Examples of safe discharge locations include soft areas of the ground, a hill or berm, a target at a range. Although not ideal, if a marksman needs a safe indoors location, joists at where the wall meets the floor can stop most rounds.

When navigating among firearms and marksmen, people must have the situational awareness to not walk in front of a rifle pointed in a safe direction, thereby rendering the direction unsafe. **To reiterate, awareness of one's surroundings and fellow marksmen is crucial when firearms are present.**

3.c Safety On and Finger Off

Even if a firearm is loaded and pointed at another person, it cannot be fired as long as the safety is on and the marksman's finger is off the trigger and outside the trigger guard. While keeping the index finger on the stock may feel unnatural and using the safety may seem inconvenient at first, these safety precautions eventually become habitual and cannot be disregarded.

Properly using the safety mechanism is one of the best precautions to take, but it must never replace the proper handling of a firearm as all the precautions are redundant, not interchangeable. For example, if there is no potential target, the rifle and magazine must not be loaded in the first place.

3.d If One Person is Unsafe, Everyone is Unsafe

Again, firearm safety is the responsibility of everyone. When firearms are being handed between individuals, trust and responsible handling cannot be

Using a Clearing Chamber

Image 13: A Gunner's Mate 2nd Class instructs a Sonar Technician (Surface) Seaman in clearing procedures for an M16 rifle. USS Jason Dunham (DDG 109), Gulf of Aden, 16 Nov 2012. The muzzle is pointed in a safe direction (into a purpose-built clearing chamber), her fingers are off the trigger, and she has **removed the source of the ammunition feed (i.e., the magazine held in her left hand).**

assumed. During such instances, **both parties** must ensure that the action is open and the firearm is not loaded.

Making negative comments or bullying someone for safely handling their firearm is completely inappropriate. On the other hand, responsible marksmen must be willing to politely but firmly call out any safety violations, even if the other marksman is a stranger. Similarly, if a ceasefire is called for any reason, all marksmen must immediately comply by keeping their firearm pointed downrange, unloading it, and inserting a breech flag, sometimes also called a safety flag or chamber flag.

3.e The Rifle is Only One Part of a Dangerous System

Acting improperly when firing a rifle is just one way to cause damage with rifles. Messing with individual rifle parts is also dangerous. For example, when stripping or disassembling a rifle with a spring, it is crucial to be aware that springs are often under high pressure and that there is a potential for the spring to release its pent-up energy and cause injury.

A marksman must also consider that acts apart from firing can cause issues during a shooting session. For example, tiny rifle parts that are easy to lose can cause problems because marksmen may reassemble a rifle without knowing that particular part is missing. Therefore, **all cleaning and repair activities must be approached with the same level of care as shooting the firearm**. That means always attempting to clear the rifle of any ammunition and keeping the rifle's components orderly and separate from one another. If a piece is lost, the marksman must contact a professional because the accidental interchanging of parts can result in malfunctions or serious harm while operating the weapon.

In addition to the precautions necessary for the firearm itself, ammunition must also be handled and stored with extreme care. A cartridge is a powerful explosive encased in a thin metal cylinder. Each round must be individually inspected for corrosion and defects before being loaded into a firearm and detonated. Good ammunition is clean, clearly labeled, and sealed in its original manufacturer packaging. Because oil-covered rounds can collect abrasives that may damage a rifle's interior or even cause unintentional firing outside the chamber, ammunition must not make contact with lubrication until loaded into the firearm.

Ammunition does not interact well with heat. (See Temperature, Pg. 188.) It must be properly stored in a clean and heat-stable area, away from direct sunlight. Direct sunlight almost never causes ammunition to spontaneously combust; however, extra heat can add power to the gunpowder and cause a more powerful explosion within the rifle. Also, in hot, dry environments the impacted and exploded remnants of bullets have been known to start fires, so a fire safety plan is necessary on every range.

3.f Always Remember! Firearm Safety is a Mindset

Finally, it is vital to understand that **firearm safety is more than just a set of rules**. Safety must be a set of ingrained habits that become second nature that bleed into other situations, rather than a series of conscious decisions.

For example, both the marksman and all other adults in the vicinity of the firearm are responsible for their safety. If any of the safety rules mentioned above are violated, it is everyone's duty to recognize the potential danger and respond accordingly. Children and newcomers to firearms must

Calling for a Cease Fire

Image 14: An Iraqi soldier signals a cease fire during live fire maneuver training. Camp Taji, Iraq, 14 Dec 2016. While different organizations have different signals, **the universally understood signal for cease fire is waving your hands above your head like a madman while shouting "CEASE FIRE!"**

be consistently reminded of these safety rules and closely supervised, or they must be immediately removed from the area where active marksmen are present.

Some additional safety measures to follow when not firing include separately storing firearms and ammunition and keeping them locked and secure. Firearms or ammunition must not be altered or modified without certified training. It is important not to handle firearms within 24 hours of consuming alcohol. Loaded firearms must never be carried while engaging in physical activities such as climbing fences or riding four-wheelers unless they are secured in a proper container that is customized to hold that specific firearm. It is crucial to always have complete situational awareness and not force any part of a firearm to close or open without considering the situation. Lastly, it is also essential to wear proper eye and ear protection at all times while rifles are being fired.

Setting Up Equipment

4. Setting Up a Scope   31

Reticle (Eyepiece or Ocular) Focusing   32
Target (Objective) Focusing   32
Parallax (Reticle Shift)   34
Adjusting Elevation and Windage Settings with Turrets   36
Getting to Mechanical Center   38
Bore-Sighting   40

5. Holding a Rifle   42

Support Hand   43
Shoulder Pocket   44
Trigger Hand (Firing Hand)   46
Trigger Finger   48
Cheek Weld and Eye Relief   50
Benchrest Shooting   51

Setting Up Your Equipment

By failing to prepare, you are preparing to fail.
—*Benjamin Franklin, Founding Father of the United States*

Before shooting starts, you must set up your equipment. While extra equipment may require extra setup, at a bare minimum, a marksman must set up their scope and properly hold their rifle. For every step onwards, you can get the fastest and best results by first consulting your records and the instructions provided by the manufacturer of your equipment.

4. Setting Up a Scope

The first and most crucial step in the setup of a long-range rifle is to ensure that the scope can perform its primary function, which is to **help you as a marksman to predict where your bullets would impact** once they have been shot. Firing a rifle without knowing where the bullets would impact is at best useless, and at worst downright dangerous!

Eventually, a precise alignment of a scope is performed with the zeroing process (See Ensuring Accuracy (Grouping and Zeroing), Pg. 74.); but that requires firing the rifle to see where bullets land on the target. And the rifle cannot be fired until the scope has at least a bit of setup performed.

Therefore before zeroing, marksman must perform a rough alignment to ensure that zeroing can occur in the first place. This alignment process entails focusing the reticle (the internal markings seen on the scope's lenses (See Image 15, Pg. 33.)), focusing the target, and repositioning the reticle to where it can predict the impact location of a bullet (i.e., the point-of-impact).

Some scopes with extra features may also require checking external settings such as battery status and light brightness. Lights are adjusted to their lowest visible setting at first because lights that are significantly brighter than the target may obscure the target.

Scopes must be set up or inspected prior to **every shooting session** because they are held to the rifle by adjustable supports, which are inherently made to adjust. All such adjustable supports (Weaver and Picatinny rail systems and scope mounts, to name a few) can misalign because they are subject to recoil after each gunshot, transported under pressure, knocked around in the field, or just plain dropped. In fact, **a loose scope can explain most of the missed shots** from experienced marksmen.

4.a Reticle (Eyepiece or Ocular) Focusing

The internal markings in a scope that form a grid system are referred to as the "reticle." The reticle can take various forms, appearing as some combination of crosshairs, dots, and dashes. (See Image 115, Pg. 133.) Reticles must be in focus to be useful. **Focusing the reticle before focusing the target is crucial**, because adjusting the reticle-focus affects both the reticle and the target, whereas adjusting the target-focus only affects the target. (See Image 15, Pg. 33.)

The reticle must be focused to the eyes of the specific marksman who plans on using the rifle, as a reticle is focused to the human eye and every person has different eyes. This is also the reason why it is required that marksmen wear any corrective glasses or contacts they plan to use while shooting. In fact, because eyesight can change over time, even the same marksman using the same rifle and scope must periodically recheck the reticle-focus of their weapon.

To focus the reticle, the marksman simply rotates the eyepiece on the scope until the reticle comes into clear focus. (Therefore this step is also known as **"eyepiece"** or **"ocular" focusing.**) (See Image 8, Pg. 19.) Sometimes it may be difficult to see if the reticle has any blur, so it can be helpful to aim the scope towards a bright, blank surface like the sky to create contrast.

Importantly, it is crucial for the marksman to relax their eye when focusing the reticle. This is because the eye also has the ability to refocus, and failing to relax it can lead to mistakenly rotating the eyepiece until the reticle is only in focus when the marksman's eye is unrelaxed. Not only does this cause eye strain, but it also leads to unsteady focus. To relax their eye, a marksman can take quick glances through the scope and blink while adjusting the eyepiece.

Once the optimal focus setting is found, the marksman can lock the eyepiece in place using tape or a built-in mechanism (if either is available).

4.b Target (Objective) Focusing

Once the reticle is in focus, the marksman must bring their specific target or objective into focus. This is because the scope can only bring into focus objects within a range of distance to the marksman (e.g., 100 to 200 m or yd, or 150 to 300 m or yd). This band of focus is called the **"depth-of-field."** (See Image 41, Pg. 62.) This means that the scope cannot bring into focus all of the objects in the sight picture if they are spaced longer than the scope's depth-of-field. While a marksman can focus the target simply by turning

Reticle and Target Focusing

| Start | Step 1: Reticle | Step 2: Target |

| **Out of Focus** | **Reticle is In Focus** | **In Focus** |
| **Out of Focus** | **Target is Out of Focus** | **In Focus** |

Image 15: Seeing **crisp borders and fine details** is very important for a marksman to know exactly where they are aiming and what they are aiming at.

the target-focus dial until their target is focused, it often helps to know the target's specific distance. (See Finding Distance to a Target, Pg. 110.) Also, whenever selecting a new target that falls outside the previous focus range, the marksman must refocus on the new target and its distance.

Notably, some scopes do not have an adjustable target focus to simplify their construction and reduce costs. That means that a target outside such a scope's depth-of-field is always out of focus; however, these scopes have a broader depth-of-field to compensate for this limitation. Non-adjustable scopes are often set with an approximately 250-m or yd depth-of-field that ranges from 50 to 300 m or yd because that is the most common range for hunters (the primary consumer group for these scopes). On the other hand, adjustable scopes may have a depth-of-field range of less than 75 m or yd.

There are two common mechanisms for adjusting the target focus. The first is the **side focus turret** (a.k.a. parallax adjustment turret). This option can be easily reached while shooting, making it more suitable for unknown-distance targets such as game. (See Image 7, Pg. 19.) The other common option is the **adjustable objective housing** (a.k.a. objective bell), which requires rotating the objective housing (i.e., the far end of the scope). Although scopes with adjustable objective housings cannot be operated while shooting, they are lighter and more affordable, making them ideal for targets that area at a known distance. (See Image 8, Pg. 19.)

Once a target is chosen, to focus on it the marksman first sets the focus to infinity and then refocuses backwards. This is done to prevent recoil from pushing forward the internal lenses within inherent looseness in the scope's internal mechanisms. It can sometimes be challenging to remember which direction to turn a side turret when under pressure. To overcome this, some marksmen use a mental trick: if the target seems too distant, the top of the turret is rolled back as if to bring the target closer. Conversely, if the target appears too close, the top of the turret is rolled away to push the target away.

In some cases, cheaper scopes may not be able to focus both the reticle and the target simultaneously due to optical limitations. In these situations, it is always a priority to focus on the reticle since marksmen need to have their attention on the reticle when firing. This is because a clearer reticle is smaller (blur increases width), and a smaller reticle is more precise.

Focus is also affected by the magnification settings. For example, the depth-of-field is inversely proportional to magnification (e.g., higher magnification causes a narrower depth-of-field). This means that when a marksman zooms in on their target with imperfect focus, the blur and parallax are exaggerated. This is typically only a problem with moving targets though, as most experienced marksmen generally can quickly find the distance to stationary targets and focus on them.

4.c Parallax (Reticle Shift)

Parallax (a.k.a. reticle shift) is a phenomenon that occurs when a target and a reticle appear to shift to the side by different increments when the marksman moves their eye. A well-adjusted scope has zero parallax, meaning the reticle appears to perfectly align with the target no matter where the marksman positions their eye in relation to the scope. (See Image 16, Pg. 35.)

Parallax is not a problem if a marksman consistently places their eye in the same location every time. However, if parallax exists and if the marksman does shift their view of the sight picture to see it from a different angle, then the reticle and target would shift differently. In other words, the reticle would become unreliable for aiming, as the point-of-aim with parallax would move all around the target, depending on the location of the marksman's eye.

This phenomenon occurs because when the marksman moves their eye, it moves the same sideways distance relative to all the objects it sees. However, objects at different distances from the marksman (such as the target and the reticle) need to move different linear distances to stay in a straight line

Parallax

No parallax - Target is aligned to focal planes

First Focal Plane   Second Focal Plane

Parallax - Target is too close or far to align to focal planes

No parallax means that the reticle stays on the target no matter where the eye moves.

Parallax means that the reticle moves on the target when the eye moves around the scope.

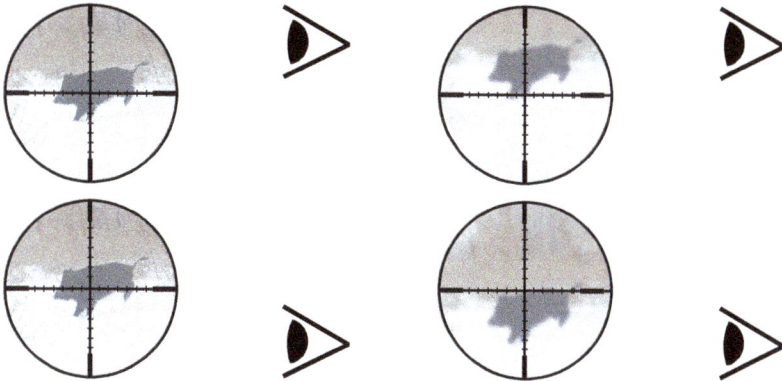

Image 16: Parallax occurs when the light rays of the target do not converge on either the first or second focal planes. (Not aligning to the focal planes also makes the target appear blurry.) This is because the reticle is located on either the first or second focal plane, and so **missing the focal planes means missing the reticle**.

with the marksman's eye as it moves. Therefore, without lenses in the proper locations in the scope, moving the marksman's eye makes the eye, the reticle, and the target no longer form a straight line.

In a quality scope, there is zero parallax when the target is perfectly focused. This is why the target focusing turret or dial is also known as the parallax turret or dial, it serves both functions. (See Image 7, Pg. 19.) That is also why fixed target focus scopes are also known as fixed parallax scopes, as neither can be adjusted.

That means that one way to determine if the target is optimally focused is for the marksman to move their eye around the sight picture. **If there is no parallax, then the focus is correct and vice versa**. Some marksmen believe that removing parallax is the best way to focus on the target because the human eye can accommodate an out-of-focus image by adjusting the eye's lens. In contrast, the human eye cannot fix parallax. Therefore, eliminating parallax always focuses the target with a relaxed eye lens.

With that said, cheaper scopes may be poorly calibrated, and they may not optimize focus while minimizing parallax at the same setting. This has led to much debate on which is more important: perfect parallax or perfect target focus. However, when shooting at distances where this distinction matters (e.g., more than 600 m or yd), the marksman likely has an expensive scope where target focus and parallax align, making the issue irrelevant. That being said, one trick for some cheaper scopes with this issue is to experiment with the eyepiece focus, as there may be an eyepiece setting that allows the scope to achieve an acceptable balance between reticle clarity, target clarity, and minimum parallax simultaneously.

Fixed parallax scopes, with no adjustable parallax, cannot achieve zero parallax except exactly at their set distance. However, they can still be a good option because parallax error is minimal within a significant range For example, a 100-m (109-yd) fixed-parallax scope may have a maximum difference of 10 cm (3.9 in) between the zero-parallax point-of-aim and the shifted parallax point-of-aim at 400 m (438 yd) and 50 m (55 yd), regardless of where the marksman looks through the scope. (The lower limit would be closer to 100 m (109 yd) because parallax increases exponentially as the distance becomes shorter.) Moreover, the 10 cm (3.9 in) difference is rarely noticeable because parallax error is reduced with a consistent and centered eye placement. Big-game hunters particularly benefit from the simplicity of fixed-parallax scopes because they typically engage large targets at distances within 400 m or yd.

4.d Adjusting Elevation and Windage Settings with Turrets

A scope's reticle can be adjusted within the scope without moving the rest of the scope. This adjustment is made using the two dials or turrets that are

Adjusting Scope Turrets

Image 17: A competitor in the USASOC International Sniper Competition adjusts his elevation turret (the top one). Fort Bragg, NC, 22 Mar 2021. He uses his trigger hand (See Image 22, Pg. 42.) so that his support hand can stay in place, allowing the sight picture to remain unchanged. (The right-side turret is the windage turret.)

Image 18: This scope has markings to help operate it. On the top of each dial is written "1 Click = 1/2 Min." ("Min" is "minutes-of-angle"). This means that whenever a turret is rotated, every 2 **audible clicks** indicates the reticle has moved by 1 minute-of-angle. The numbers show how many minutes past 0 the turrets have been rotated. The hex screws indicate that this rifle has removable turret caps, so when the turret is at 0 the rifle is at its zero point. (See Zeroing, Pg. 79.)

present on virtually all scopes: the elevation turret on the top of the scope and the windage dial on the right side of the scope. (See Image 8, Pg. 19.) Rotating the elevation turret moves the reticle vertically within the scope, and rotating the windage turret moves the reticle horizontally. (See Image 17, Pg. 37.) (Many scopes also have a third turret on the left that adjusts target focus and parallax.) (See Target (Objective) Focusing, Pg. 32.)

Because the two dials are mechanically identical, left-handed marksmen sometimes rotate a scope without a parallax turret 90° so that the windage turret is on top and serves as the elevation turret, and the elevation turret is on the left side and serves as the windage turret. This can be a difficult arrangement for complicated reticles, however, because the reticle would also be rotated 90° with the scope.

Turrets on scopes are marked in either minutes-of-angle (MOA) or milliradians (mils). (See Angular Distance, Pg. 237.) When a marksman rotates the turret, they both feel and hear **distinct clicks** indicating the movement of the internal parts that adjust the reticle. (See Image 18, Pg. 37.) Each click represents a specific measurement, which can vary between manufacturers and is always specified in the scope's manual.

For example, a single click on a scope may correspond to a 0.25-mil adjustment, while on another scope, it may signify a 0.1-mil adjustment. Because the click values differ between scopes, communication about adjustments always uses the actual MOA or mil measurement, not the number of clicks. Additionally, some cheaper scopes may not be "**repeatable**," meaning that a click in one direction does not have the same measurement as a click in the opposite direction.

Similarly, the number of rotations a turret can make also varies from scope to scope. Some turrets only rotate for one 360° revolution, while others can complete multiple revolutions. Due to the variety of systems, a marksman must always review and examine the instructions for their specific scope.

4.e Getting to Mechanical Center

The phrase "adjusting a reticle" means to move it within the scope. If the reticle is adjusted so far that it hits the interior of the scope, it can no longer be adjusted in that specific direction. Therefore, to ensure that a scope can always be adjusted up, down, left, and right, the marksman must ensure that the reticle is roughly centered within the scope, and that it also points approximately towards the target.

When turrets are positioned in the middle of their adjustment range, that is called the "**mechanical center**" because it is in the physical center of the

Optically Centering a Reticle

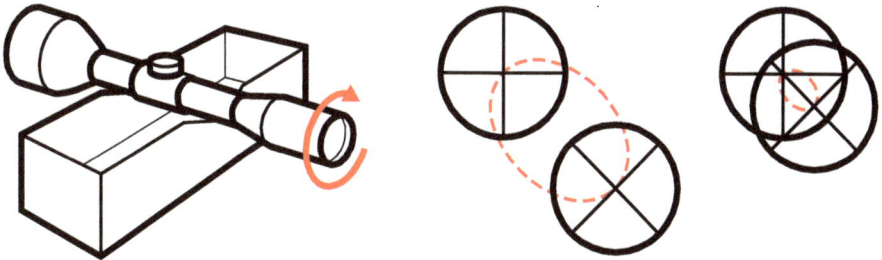

Image 19: Placing a scope in a rest and spinning it rotates the reticle. If the reticle's center rotates, then the reticle is not at the scope's optical center, which is a good approximation of the mechanical center. Therefore to get to the optical center, a marksman can adjust the turrets until **the center rotates on a point**.

scope. This contrasts with the "**optical center**," which is when the reticle is in the center of the scope's sight picture. The two are always close, but may or may not always be the same.

To mechanically center the reticle and still point it at a target, a marksman may need to adjust their scope mounting. Therefore, mechanical centering is done before mounting a scope, and may require removing and remounting an existing scope back onto its rifle.

To find the mechanical center, the marksman adjusts both the windage and elevation turrets all the way in one direction. Then, they rotate the turret completely in the opposite direction and count the number of clicks it takes for a full rotation. (This step can be skipped if the scope's click range is listed in the manual.) Next, the marksman rotates each turret in the opposite direction, but only half the number of clicks it took for the full rotation. This brings the turret to its mechanical center.

Another, faster way to find an approximate mechanical zero is to rotate the scope when it is detached from the rifle. When the reticle is not at the optical center, its center rotates in an elliptical pattern. However, when positioned at the optical center, the reticle's center rotates on a point. (See Image 19, Pg. 39.) Therefore, the scope can be set to an approximate mechanical zero by adjusting the turrets until the center of the reticle rotates on a point.

If the scope cannot be removed from its rifle, the marksman can place a small mirror directly on the end of the scope and look through the scope. Thereafter, the marksmen can see both the reticle as well as its reflection. To optically center the reticle, the marksman adjusts the turrets until both the reticle's center and its reflection's center align (converge) onto the same point.

Bore-Sighting a Bolt-Action

View through Scope

View through Bore

Image 20 et al: A competitor in the USASOC International Sniper Competition bore-sights a rifle by **removing the bolt** and looking down the bore. Fort Bragg, NC, 21 Mar 2023. Something like a stump can be an excellent reference point. Note that everything in the right images is centered.

4.f Bore-Sighting

When installing a scope for the first time or if the scope is significantly misaligned, it is necessary to perform a process called "bore-sighting." Bore-sighting involves aligning the rifle's boreline (i.e., the path a bullet follows as it exits the barrel) with the sightline (i.e., the sightline as the marksman looks through the scope). This initial alignment can aid in the accuracy of subsequent adjustments. (See Ensuring Accuracy (Grouping and Zeroing), Pg. 74.)

There are two commonly used methods for bore-sighting a rifle. The first method involves **removing the bolt of a bolt-action rifle** to have a clear view down the bore. (See Image 20 et al, Pg. 40.) The marksman then aims the rifle at a distinct target such as a paper target, vehicle headlights, or white transformer boxes.

The marksman looks down the rear of the bore to see the target, making sure that the muzzle opening is centered in the bore and the target is centered in the muzzle opening. (A "muzzle" is the end of a barrel.) Once the bore is aligned with the target, the marksman looks through the scope's sight

Bore-Sighting a Semi-Automatic

Image 21: A Gunner's Mate 3rd Class, assigned to Naval Mobile Construction Battalion 74, Headquarters Company, bore-sights an advanced combat optical gunsight (ACOG) **using a laser bore light**. Camp Mitchell, Rota, Spain, 20 Dec 2010. Different laser bore-sighters and sights work slightly differently, so it is vital to read the instruction manual for the specific equipment being used prior to using it.

picture and adjusts the reticle to align with the target. When making these adjustments, it is easy to accidentally bump the rifle. Therefore, the process of physically centering the bore on the target and aligning the scope must be repeated and verified at least a few times to achieve the best results.

The second method of bore-sighting is used for semi-automatic rifles, since their construction prevents a marksman from seeing down the bore from the rear. This method involves using a specialty tool. One such tool is the **bore-sight laser**, which can be placed in the rifle's muzzle. It emits a red dot that the marksman can then align the scope with. The advantage of this method is that even if the bore moves, the dot of the laser moves with it, allowing the rifle to be moved during the bore-sighting process. Another tool is a **bore-sighting collimator**, which is also inserted into the rifle's muzzle. The collimator is aligned to the muzzle, so aligning the scope to the collimator also aligns the scope to the muzzle.

While the reticle can be adjusted via the scope's turrets during the bore-sighting process, the process itself can move the reticle off its mechanical center. Therefore, a better alternative to adjusting the turrets is to adjust the scope's mounts so that the turrets retain their full ranges of motion.

Once the marksman is confident that a bullet shot through the bore would land close enough to their point-of-aim, they can proceed to the next step in calibrating a scope to give precise readings, that is "zeroing." (See Ensuring Accuracy (Grouping and Zeroing), Pg. 74.)

Holding an Unsupported Rifle

Image 22: A Clarksville, Tennessee resident and veteran fires his rifle during the 2018 Fort Benning Multi-Gun Challenge from an **unsupported position**. Fort Benning, GA, 17 Nov 2018.

5. Holding a Rifle

Rifles are designed and built to be as intuitive as possible, so this section may be common sense to some people. However, it is never a mistake to cover the basics. A marksman contacts a rifle in five distinct places: the two hands, the shoulder, the cheek, and the trigger finger. Each location has a best practice for interfacing with and supporting a rifle without interfering with the process of aiming and firing the weapon.

The one rule that is constant throughout this section is that **consistency is the key to accuracy**. To be consistent, first everything must be rigidly held; many marksmen attempt to maintain rigidity through muscle control, but that can certainly lead to fatigue and shaking, resulting in increased inaccuracy over time. This is inevitable. No human being can hold a heavy piece of metal and wood (or polymer) for an extended period of time without experiencing muscle pain and fatigue. Therefore whenever possible, it is important to hold a rifle in a way that minimizes the involvement of muscle tissue, instead **resting the weight on tendons and bones**.

Marksmen must then hold the rifle in this same relaxed position that relies on bones and tendons every time they shoot, allowing it to become automatic. A useful way to test the effectiveness of their habits is for the marksman to close their eyes and attempt to get into the same shooting position multiple

Holding a Supported Rifle

Image 23 et al: These marksmen are firing from a **supported position** because the rifles are supported by (clockwise from top left) a pipe, a piece of wood, a bipod, and another bipod. The support hand is in a different location in every image (at the marksman's discretion) because each object replaces the support hand's support.

---

times. If the same sight picture is consistently achieved, then their habits are effective. (See Natural Point-of-Aim (NPA), Pg. 89.)

5.a Support Hand

The support hand is by definition the hand that does not pull the trigger. (That hand is the "trigger hand" or "firing hand.") Its job is to support and stabilize the front of the rifle. The support hand is most important when the marksman is firing from an "unsupported position," which means there is no support other than the firing hand (e.g., no bench or tripod).

In an unsupported shooting position, the support hand must be placed as far forward on the stock as possible to provide finer control over the rifle. (See Image 22, Pg. 42.) The support hand holds the rifle in the crook of the thumb with the palm facing upwards to prevent the palm from pushing the rifle sideways. The entire palm firmly grips the rifle with the thumb pointing forward in order to help maintain control during recoil, but not so firmly that the hand and rifle start shaking.

The marksman's hand must never touch the barrel during live-firing unless there is a barrel shroud or handguard covering the barrel because barrels get hot enough to burn skin and gloves. The ejection port and action must remain uncovered and clear of any obstructions, as this could hinder both bullet and casing ejection.

When shooting from a supported position, since the rifle is already stable, the support hand may be moved towards the middle of the rifle for easier maneuvering and pivoting, or wherever else the support hand can help maintain the stability of the rifle. (See Image 23 et al, Pg. 43.)

Specifically when using a bipod or other stabilizing forward support, the support hand can be positioned at the rear of the rifle to provide support to the buttstock from below. This positioning creates a stable tripod with two legs at the front and the support hand at the back. The marksman can form their hand into a fist and rest the buttstock (i.e., the rear part of the stock) on it, or hold a sandsock (i.e., a bag filled with fine material like sand or rice) under the buttstock. Making a fist or using a sandsock allows for slight alterations to the sight picture. Squeezing the fist or sandsock makes it taller and moves the sight picture down, while relaxing the squeeze raises the sight picture. (See Precisely Adjusting the Sight Picture, Pg. 140.)

5.b Shoulder Pocket

A support hand or forward support (e.g., a bipod) stops the rifle from moving forward as the torso presses into the rear of the stock (i.e., the "buttstock"). The location on the torso that the rifle presses into is called the "shoulder pocket," which is where **the pectoral muscle ends and the shoulder begins**. (See Image 24, Pg. 45.) Buttstocks and recoil pads (i.e., rubber or plastic pads fixed to the end of stocks) are vertical and thin to allow them to fit into an average person's shoulder pocket.

The shoulder pocket is the lowest point in the local area, so it is the most secure location to hold the rifle during a shooting session. When properly positioned, a rifle can be held with only the support hand holding the front of the rifle and lightly pressing the buttstock into the shoulder pocket.

Some marksmen opt to deepen the shoulder pocket by holding their firing-side elbow away from the body, which exaggerates the front deltoid muscle; however, hunters and military marksmen must be cautious of this technique (called "chicken winging") as it may cause their elbow to catch on something if they are moving.

Once the rifle is in the shoulder pocket, marksmen lean into the rifle during the firing sequence to absorb the recoil into the shoulder. Maintaining

Shoulder Pocket

Shoulder Pocket

Pectoral (Chest) Muscles

Deltoid (Shoulder) Muscles

Buttpad Placement

Image 24: The shoulder pocket is at the meeting of the pectoral and the deltoid.

consistent pressure is crucial, as sudden tensing into the stock in anticipation of recoil is a primary cause of inaccurate shooting.

Thick gear, clothing, or suspenders **may obstruct the shoulder pocket**, leaving no space for the buttstock. Shooting in this manner is uncomfortable at best and unsafe at worst, since the rifle could have uncontrolled movement during recoil. The rifle would also appear to be too long, since length-of-pull measurements are for unobstructed shoulder pockets. (See Image 5, Pg. 17.) Therefore, it is essential to move any gear out of the way to find a secure pocket hold for the most precise shooting positions.

However, sometimes a rifle is supported by equipment such as a tripod or vice so that human support is superfluous. If the rifle is stable enough to stand on its own, **the shoulder pocket becomes unnecessary**, and the marksman can simply rest the buttstock anywhere on their chest. (See Image 87, Pg. 107.)

And when shooting from the prone position (i.e., on one's stomach), there is no true shoulder pocket since the chest is facing the ground. (See Image 64, Pg. 98.) However, while in the prone position it is still a good idea to first place the rifle in the approximate pocket and then lean forward to create a repeatable shooting position and maintain pressure on the buttstock.

Trigger-Hand Grip on a Rifle

Image 25: This marksman is using a rifle with a **military-style grip**. He is holding the grip with his palm, placing his trigger finger outside the trigger well, and operating the safety mechanism with his thumb.

Image 26: This marksman is using a **hunting-style rifle with no grip**. So his thumb goes on top or over the rifle. His trigger finger is in the trigger well because he is ready to fire. J&K, India, 21 Feb 2019.

5.c Trigger Hand (Firing Hand)

Contrary to the support hand, the trigger hand's main role is to pull the trigger. The first step in achieving a good trigger pull is establishing a **firm grip** to ensure consistency and repeatability, and to prevent the trigger finger from shifting due to recoil.

In technical terms, a "firm" grip refers to the strongest hold that does not cause any trembling that affects the scope's point-of-aim. (Therefore, a firm grip is not actually powerful.) Trembling is caused by muscle fatigue, which occurs when stressed muscle fibers are replaced by fresh muscle fibers. Novice marksmen often grip their rifles too tightly, which becomes noticeable when their fingernails turn white due to excessive pressure before firing. (Performing hand exercises can significantly improve baseline hand strength and thereby a rifle grip.)

To properly grip a rifle, the trigger hand's palm firmly presses against the stock while the crook of the thumb rests either in the pistol grip for military-style rifles (See Image 25, Pg. 46.) or on the top of the stock for hunting-style rifles (See Image 26, Pg. 46.). This is because touching the palm to the stock can minimize the impact of the trigger pull on the trigger hand's grip. Then the fingers securely wrap around the stock and grip the rifle. For better grip, some rifles may have features such as palm swells or finger grooves.

However, the trigger (a.k.a., index) finger alone remains outside the trigger well (i.e., the space between the trigger and the trigger guard). It is crucial to **refrain from placing the trigger finger on the trigger** unless there is an intention to fire the rifle.

Trigger Finger Pull

Correct
Finger Pad Method

Correct
First Joint Method

Incorrect

Incorrect

Image 27: Fingers rotate about their knuckles, as seen in the top left diagram. This rotation is difficult to transform into linear force. The arrows indicate the **direction of the force** applied by the finger.

When the trigger is pulled backward, the hand shape changes, making it difficult to maintain a consistent grip. This is why it is essential to press the buttstock firmly into the shoulder pocket. By doing so, for the trigger hand to twist the rifle, it would have to overcome the force of the entire torso pushing into the back of the rifle. To the same effect, many marksmen choose "lighter" (i.e., requiring less force to depress) triggers, reducing the force that the trigger finger can apply to the rifle. In fact, when using a low-power rifle in a supported position, the trigger hand may not need to support the rifle at all, allowing it to avoid gripping the stock or affecting the rest of the rifle.

Besides its primary role of pulling the trigger, the trigger hand is also responsible for operating the safety mechanism on most rifles. And when the support arm cannot provide complete stability, or in instances where the rifle is too powerful, the trigger hand can also help stabilize a rifle; although this may negatively impact accuracy.

5.d Trigger Finger

Proper placement of the trigger finger is crucial to maintaining rifle alignment during the trigger pull. Usually, the trigger finger makes contact with the trigger in the middle of the farthest pad of the finger. (See Image 27, Pg. 47.) This position allows marksmen to efficiently translate their fingers' rotational movement into the trigger's backward linear movement. Importantly, there is no other contact between the trigger finger and the gun to prevent any part of the finger from dragging on the firearm.

However, as long as the trigger is pulled straight backward without any horizontal or vertical movement, any finger position in relation to the trigger is acceptable. Hand placement may vary among individuals due to differences in hand size and shape. Therefore, the natural placement of the finger on the trigger, whether it is in the middle of the finger pad, at the first knuckle, or wrapped around the trigger, **can all be equally effective with practice and muscle memory**. It is never advisable to force the trigger finger into a specific position that compromises a comfortable grip.

Some instructors suggest placing the trigger finger first and then gripping the firearm. This approach allows for the intentional positioning of the trigger finger. However again, intentional positioning assumes that one position is ideal for shooting the weapon which is untrue. It also raises safety concerns as it advises placing a finger on the trigger without the intention to fire, which is a direct violation of basic rules of firearm safety.

One trick for a direct backward pull, is to make the proximal phalanx of the finger (i.e., the bone between the second and third knuckle) remain parallel to the barrel and aligned with the firing arm, without angling up or down. (See Image 25, Pg. 46.) (Adjusting the finger slightly up or down on the trigger can help maximize mechanical advantage, considering that the trigger acts as a lever; however, the entire hand may have to move to accommodate this.) To stop the trigger hand from shifting, some experts recommend focusing on keeping the proximal phalanx and wrist pointing straight at the target while depressing the trigger.

Eye Positioning

Image 28: A good **cheek weld** places the marksman's eye in the eye box. The **eye box** is a three-dimensional shape (i.e., a double cone) made by the scope in which a marksman can place their eye and see the sight picture at full brightness. The maximum diameter of the eye box is called the "**exit pupil**," and it is as large as the scope's front lens diameter divided by the magnification setting (e.g., a 40mm lens at 4X magnification has a 10mm exit pupil). "**Eye relief**" is the distance from the back of the scope to the useful section of the eye box.

Scope Shadow

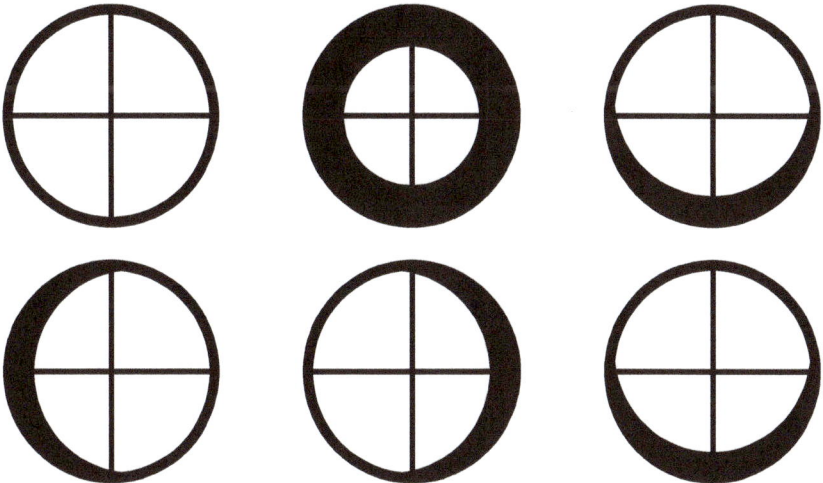

Image 29: Scope shadow is a **thick black border** seen through the scope. When looking through a scope, the only correct sight picture is one with no scope shadow (top left). When scope shadow is present (See Image 101, Pg. 116.) (See Image 102, Pg. 116.), it indicates that the marksman has poor eye relief because they do not have their eye positioned correctly on the stock. This in turn, can be due to a poor check weld or a rifle that does not fit the marksman (e.g., bad length-of-pull (See Image 5, Pg. 17.)).

5.e Cheek Weld and Eye Relief

"Cheek weld" refers to how the marksman's cheekrests on the rifle's stock. The term "weld" can be misleading because it implies forceful pressure. However, in reality, the marksman simply rests their head softly on the stock. Achieving a proper cheek weld is usually straightforward: the marksman turns their head slightly sideways and **rests their cheek on the stock**. Ideally, the head remains mostly vertical. Excessive sideways bending can strain the neck, compromise balance, and affect the marksman's perception of rifle cant. (See Eliminating Rifle Cant, Pg. 145.) When shooting in a supported or prone position, the entire body is adjusted downward to maintain a proper cheek weld, rather than solely relying on neck adjustment.

The purpose of a cheek weld is to position the marksman's dominant eye to view the sight picture through the scope. The sight picture is only visible from specific positions behind the scope. The range of these positions where the eye can see the full sight picture is called the "**eye box**." (See Image 28, Pg. 49.) The eye box becomes smaller as magnification increases; therefore, marksmen need to practice their cheek weld at every magnification setting they plan to use. Larger objective (i.e., front) lenses create larger eye boxes, which is why they are necessary for high-magnification scopes.

The distance known as "**eye relief**" is between the usable section of the eye box (i.e., wider in diameter than the marksman's pupil) and the scope's eyepiece lens, and is typically 5 to 8 cm (2 to 3 in). This is easily measured because before the limit of eye relief, the marksman has an unobstructed field-of-view without the presence of "**scope shadow**." Scope shadow is a circular or crescent-shaped shadow in the sight picture which reduces the sight picture's size. (See Image 29, Pg. 49.) (See Image 101, Pg. 116.) When these shadows are present, a fired bullet would likely not land where the marksman intended due to parallax. (See Image 16, Pg. 35.)

Having a consistent and stable cheek weld is crucial for maintaining a consistent and stable sight picture. If a marksman removes their cheek from the stock while firing, the release of pressure can affect the alignment of the rifle. To check for consistency in their cheek welds, marksmen can close their eyes while obtaining a cheek weld. If the sight picture is thereafter fully visible, they have a good cheek weld. But if any scope shadow is present, marksman must continue to practice creating a consistent cheek weld. To help with consistency, marksmen can **affix a textured material** to the stock to feel the correct cheek position without looking at the stock.

Image 30: **Benchrest shooting** uses a bench, but also holds a rifle stable using equipment such as rifle stands and vices.

The appearance of a stable cheek weld depends on the shape of the rifle and the marksman. If the scope is mounted high and the sight picture is too high, the marksman can attach a cheek piece to the rifle's stock to elevate the cheek weld area (a.k.a., the stock's "comb"). Marksmen who would benefit from a cheek piece are easily identified because they frequently move their heads up and down in search of a comfortable eye position. Marksmen who would benefit from remounting the scope forward or backward are also easily identified because they also move their head to achieve a good sight picture.

5.f Benchrest Shooting

Unlike other shooting styles where the hands or shoulder support the rifle, benchrest marksmen utilize a stable platform, often a bench (hence the name). Typically, both the front and rear of the rifle rest on supports, such as a bipod or sandbag at the front, and an adjustable rest at the rear. This setup provides an incredibly stable shooting platform that still allows the marksman to make adjustments by subtly squeezing or adjusting the support, usually the rear support, rather than the rifle itself.

In benchrest shooting, minimizing physical contact with the rifle helps to substantially reduce human error (making much of this section moot for the discipline). For example, custom triggers with very light pull weights (e.g., 2 oz or 0.5 N of force to depress the trigger) are commonly used to more precisely control the trigger. Furthermore, recoil management is minimal to avoid affecting the rifle during the firing sequence. In fact, some marksmen employ the "free-recoil" technique, only gently depressing the trigger while allowing the rifle to recoil without human restraint.

Beginner Contents

6. Loading a Rifle — 53
Bolt Manipulation — 54

7. Taking Aim at a Target — 55
Target Definition — 55
Finding a Target — 57
Sight Alignment and Iron Sights — 58
Sight Picture — 60
Holdover and Ballistic Loopholes — 63

8. Firing the Rifle — 65
Trigger Pull and Flinching — 66
Recoil and Muzzle Jump — 67
Follow-Through and Reacquiring Sight Picture — 71
Cold-Bore Shooting — 73

9. Ensuring Accuracy (Grouping and Zeroing) — 74
Grouping — 74
Picking a Zero Distance — 77
Zeroing — 79
Resetting Elevation and Windage Turrets to Zero — 84

Beginners
(0 to 100 Meters or Yards)

Being an expert is getting the basics right every time.
—Common saying among U.S. Special Forces

Once setup is complete, a marksman can begin the process of loading and firing ammunition. This process starts, as all do, with the basics. For long-range shooting, that means properly using your sense of sight and being consistent from shot to shot. If you can master those, then you can easily hit any deer-sized target up to at least 100 m or yd.

6. Loading a Rifle

First, the marksman must inspect the ammo to ensure that it is clean and the correct type for the rifle. Sometimes different rounds can appear identical at first glance. For example, the .300 Blackout is visually similar to the 5.56 (.223) caliber round. They can both easily be fed into an AR platform rifle, but mistaking one for the other can have catastrophic consequences.

Second, the marksman must ensure that the rifle is always pointed in a safe direction. Only then can the marksman open or release the bolt to either eject or load a round into the rifle. If the bolt is not spring-loaded, it must be manually closed.

It is vital to never force a cartridge into a rifle. If the bolt does not fully close, the marksman must clear any obstructions and only then attempt to chamber a new round. Firing a round with a partially closed bolt is extremely dangerous and can lead to an explosion that destroys something or someone.

A common situation for occupational marksmen or hunters who may not have the ability to set up a good shooting support is the need to reload a bolt-action rifle in the prone position. (See Image 64, Pg. 98.) This can be a bit more complicated because the rifle is best loaded without looking so that the marksman can keep their eyes on the target and the wind instead. First, repetition is the best answer. However to make loading easier, the marksman keeps ammo in an easily accessible location allowing them to reload without reaching far. For example, a bag-support can both support the rifle as well as hold ammunition.

Bolt Manipulation

Image 31: The bolt handle of a bolt-action rifle can be quickly and efficiently moved by pushing only with the sides of the index finger and thumb. A firm, smooth upward twist and pull, followed by a push and downward twist do the job.

6.a Bolt Manipulation

Bolt-action rifles require manual reloading, and bolts are manipulated aggressively to avoid issues such as the misaligning of the magazine, which causes rounds to not feed smoothly into the action. Beginner marksmen and those using short-movement bolts often grab the bolt between their thumb and index finger. While this is effective, the marksman's fist may interfere with the scope and require more coordination than a different, better technique. (See Image 31, Pg. 54.) To perform this advanced bolt manipulation technique:

1) The top of the firing hand contacts the bolt knob, pushing it directly up, while the thumb maintains its original position for stability. The palm continues to face the stock even as the knob reaches its highest point, and the thumb remains anchored.

2) The side of the index finger rotates around the thumb to begin pushing the bolt knob rearward, with the elbow and shoulder becoming more involved in the movement. When the bolt reaches its farthest backward point, the spent cartridge is extracted and subsequently ejected.

3) The marksman then pushes the knob forward with their palm or thumb, and the thumb drags the knob down to lock the bolt when it is fully forward. Finally, the hand completes the cycle by returning to the firing position.

7. Taking Aim at a Target

Shooting requires the marksman to have a clear understanding of what they are aiming at. While most rifles are built to be "point and shoot," even experienced marksmen can become even more accurate with a deeper understanding of how to precisely get a target into their crosshairs.

7.a Target Definition

Marksmen must define their targets before aiming at them. While this skill may not be as critical when shooting at paper targets, it is crucial for hunters and occupational marksmen to specify what is and is not a target. For example, hunters aim for the vital areas of animals to ensure a quick and clean kill. Different animals require different targeting strategies because their vital areas are located in different parts of their bodies. (See Image 34, Pg. 56.)

Moreover, marksmen must ensure their targets even qualify as targets in the first place. For example, many states require bucks to have antlers with a minimum number of points before they are eligible to be harvested. And in military contexts, snipers must consider factors such as identifying enemy uniforms and determining if the enemy is wearing body armor.

Regardless of the target, marksmen must always aim for the smallest possible target area, which is summarized by the phrase: "**aim small, miss small.**" Redefining the target to something smaller leads to more accurate shooting because marksmen tend to have the same amount of error regardless of the size of their target. Therefore, aiming at a smaller target tends to result in a tighter group of shots.

In practice, this often means **redefining what the target is** to one of the target's vital or central features. For example, aiming at a whole deer is not nearly as useful or effective as aiming for the deer's vital organs, which can be redefined as a hunter's actual target. Aiming at the entire deer and missing by just an inch means missing the entire deer. However, by focusing on a precise point, such as the exact center of the deer, even if the shot misses by an inch, it can still potentially result in a lethal hit. (See Image 32, Pg. 56.)

All that being said, it is essential not to become overly fixated on any specific point of the target, as this may delay the shot. That is because there is a limit to how precisely a marksman can aim due to the slight involuntary

Defining a Target

| Specific feature: "I'll aim at the 8 on the target." | Nebulous outline: "I'll quarter the balloon." | Proxy feature: "My target is button atop his sternum." |

Image 32: **Targets must be as small as possible**. But "small" is only in reference to something else. A marksman creates that reference point in various ways.

| 100 m or yd: Center of Head | 200 m or yd: On Chin | 300 m or yd: Shoulder-Level |

| 400 m or yd: Armpit-Level | 500 m or yd: Center Torso | 600 m or yd: Waistline |

Image 33: **Targets appear different at different distances**. Also, shots become less precise with an increase in distance. Therefore, marksmen must use different points-of-aim at targets of different distances.

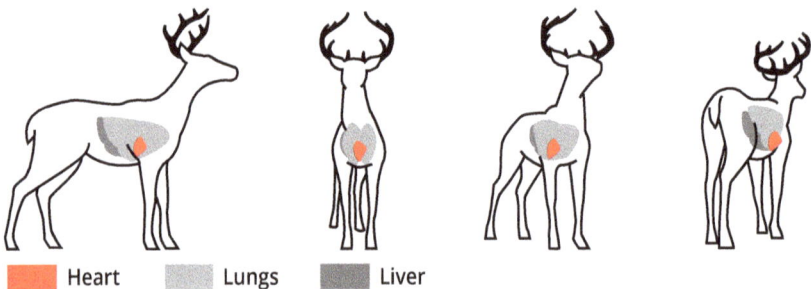

■ Heart   ■ Lungs   ■ Liver

Image 34: Animals are killed faster when vital spots are hit. Therefore, hunters must familiarize themselves with the **proxy locations of vital organs** on the skin.

Search Patterns

Image 35: When searching for a target, a marksman can be more effective by implementing a search pattern. This pattern follows streets and windows. Many patterns **employ a grid** with an X and Y axis.

Image 36: Most locations have **areas of high probability** and areas of low probability of finding a target. For example, animals are most likely to appear at the border of woods and meadows, and around bodies of water. Therefore, that is where a marksman's attention is best allocated. Still use a search pattern.

movement or oscillation of the rifle caused by factors like breathing, heartbeat, muscle tension, and environmental conditions such as wind. Put another way, aiming at targets smaller than one's natural rifle sway is a waste of time.

7.b Finding a Target

Once the marksman has defined their target, they can begin to look for it. For target marksmen, this step barely exists, as the target is right in front of them.

However, for marksmen with living, moving targets, identifying a target can sometimes take hours or days.

While using the naked eye is always a good default, eyes cannot see as far as rifles can shoot. This is of course why marksmen use riflescopes. However, riflescopes are not the best tool for identifying targets. This is because they must have a lot of space between them and the marksman so that the recoil impulse of a rifle doesn't smash the scope into the marksman's eye. (I.e., eye relief, (See Image 28, Pg. 49.)) This distance apart makes the scope's sight picture take up a smaller amount of the marksman's vision than other, closer optics, such as a spotting scope or binoculars. (See Image 40, Pg. 61.) Once a target is identified with one of these tools, only then does the marksman switch to their scope.

When a marksman is using independent optics for target identification, they must have a plan in place to transition to their rifle once the target has been located. For example, they can prepare their rifle by placing it out of a window prior to starting their search. Alternatively, if walking through the woods, a hunter may choose to walk close to tree branches to provide support for their shot. Failing to do so may result in having to take an important shot from an unsupported standing position, which is less stable and more prone to wobbling. (See Standing, Pg. 102.)

Regardless of the devices a marksman may use, searching with **a system is always optimal**. Specifically, the marksman can define a viewing zone with limits and sweep their sight in a back-and-forth pattern, starting from near to far. (See Image 35, Pg. 57.) It is even more effective for the marksman to focus on areas they know the target frequents, such as a border between fields and woods, near water sources, or entrances and exits. (See Image 36, Pg. 57.)

7.c Sight Alignment and Iron Sights

A sight is a tool that a marksman looks at to help aim a weapon. Scopes are a kind of sight, but there are many other kinds of sights too. One such sight system is the iron sights system. These sights are antiquated and never used by modern marksmen for serious shooting because scopes that are properly calibrated for parallax do not need further alignment. (See Parallax (Reticle Shift), Pg. 34.) Iron sights also significantly obstruct the target and also lack magnification.

That being said, many teachers laud iron sights as a teaching tool because they force students to keep their rifles straight while firing. Iron sights also

Shooting with Iron Sights

Image 37: The **front sight post** (in focus) can be seen through the **rear sight aperture** (blurry). Keeping the front sight post in focus while shooting maximizes the alignment.

Image 38: A Soldier peers through the **rear sight aperture** (i.e., the post on the right) to see the **front sight post** (i.e., the post on the left) of an M4 carbine. Fort Jackson, SC, 20 Jul 2023.

Image 39: The front sight post must be aligned in the center of the aperture to ensure that bullets are put on the correct (zeroed) trajectory.

serve as excellent backup sights in case the scope breaks, loses its zero, or runs out of battery power (in the case of electronic aiming devices).

Iron sights consist of a front sight attached to the front of the barrel, and a rear sight attached to the rear of the barrel. The idea behind iron sights is to place two points (the front sight and the rear sight) along the sightline (the imaginary line between the shooter and the target. If the sights are aligned to the sightline, then the barrel would be aligned to the sightline.

In other words, "sight alignment" is to align the eye, the front sight, the rear sight, and the target in one straight line. If any of the four points is misaligned, a bullet cannot travel from the rifle to the target. (See Image 39, Pg. 59.) Any error in sight alignment is proportional to the distance, so a

1 cm (0.39 in) error at 25 m (27 yd) is equivalent to a 10 cm (3.9 in) error at 250 m (274 yd).

All iron sights are designed to allow the marksman to center (referred to as a "center hold") the front sight (e.g., thin metal lines or beads) within the rear sight (e.g., a half or "open" sight, or full circle, "aperture" sight). (See Image 37, Pg. 59.) Some marksmen find it helpful to mark the top of the front sight post with a bright dot for enhanced visibility through the rear sight.

Even for identical iron sights, the proper vertical alignment of the front sight within the rear sight can vary according to the marksman's preferences and how each marksman zeros their weapon. For example, a common issue with iron sights is that the front sight post is opaque, obstructing the target when properly centered within the rear sight. Some marksmen prefer to zero their rifle (See Zeroing, Pg. 79.) so that the bullet impacts slightly above the front sight post, allowing them to aim with a slightly lower front sight post (known as a "six-o'clock hold"), which reveals the point-of-impact during aiming. The downside of the six-o'clock hold is that it can be challenging to consistently achieve the exact same point-of-aim every time, making it far more suitable for hunting large game than for precision target-shooting.

To align iron sights, a marksman performs the same zeroing procedure as they would for any other optic. (See Ensuring Accuracy (Grouping and Zeroing), Pg. 74.) Different iron sights are adjusted in different ways. Many rifles have rear apertures that can be adjusted for windage and elevation, while others (such as the M16) have both front and rear adjustable sights. It is important to read the instructions and understand the specific rifle being used.

7.d Sight Picture

A sight is a tool that a marksman looks at to help aim, and a "sight picture" is what the marksman sees when looking through or at the sight. For example, the sight picture of a scope is the image seen through the back of a scope. (See Image 9, Pg. 19.) For iron sights, the sight picture is simply what the naked eye sees with the aligned sights at the center of the marksman's vision.

The clarity of the sight picture is influenced by the dominance of the marksman's eye (See Shooting with Both Eyes Open and Eye Dominance, Pg. 94.), external conditions, and the ocular focus of their eye.

The most important external condition to consider when shooting is ambient lighting. Shooting a rifle in low-light conditions is extremely dangerous, so marksmen use night-vision tools. Shooting in extremely bright lighting is also a concern because it can cause eye strain and fatigue.

Sight Picture for Various Optics

Naked Eyes

Rifle Scope

Binoculars

Spotting Scope

Image 40: **Marksmen typically use multiple optical devices when shooting.** This is because each device has its own advantages and disadvantages. Hunters use binoculars for the widest field-of-view and binocular vision. Target marksmen use a spotting scope (a small telescope) for very high magnification. Both devices can serve as better optics than riflescopes in their niches because they can be much closer to the marksman's eye than a riflescope. Were a riflescope to be close to the eye, the scope would hit the eye during firing because of recoil.

Specifically, focusing on a target in bright sunlight for extended periods can result in a temporary afterimage of the sight picture imprinting on the marksman's retina. To mitigate the effects of bright light, marksmen must look away every few seconds or between shots to allow their retina to clear the afterimage. They can also partially cover the end of the scope to dim the sight picture. (See Image 165, Pg. 193.)

When looking at things far away, things close up become blurry and vice versa. This is because eyes (like scopes and in fact all optical devices) can only focus at a certain range at any one point in time. This range is called the "depth-of-field." (See Image 41, Pg. 62.) This means that when looking

Depth-of-Field

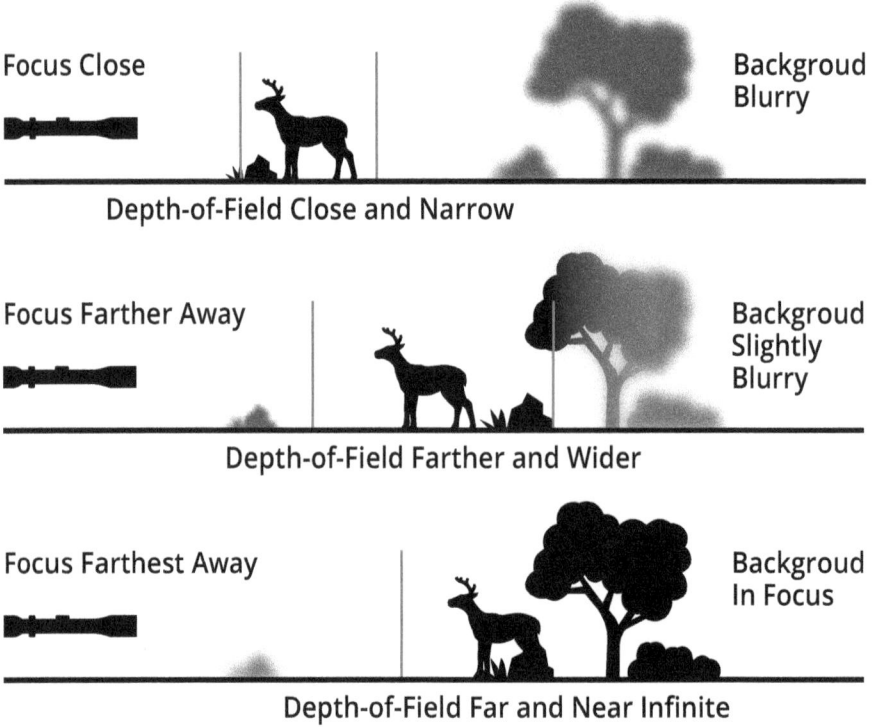

Focus Close

Backgroud Blurry

Depth-of-Field Close and Narrow

Focus Farther Away

Backgroud Slightly Blurry

Depth-of-Field Farther and Wider

Focus Farthest Away

Backgroud In Focus

Depth-of-Field Far and Near Infinite

Image 41: Depth-of-field describes what part of an image is blurry and what part is in focus. If an observer puts a near object in focus, then the depth-of-field is narrow, and therefore, only a narrow depth is in focus. However, **as distance to the observer increases, so does the depth of the in-focus area**. Human eyes also have depth-of-field (as does every optical lens), but the eye's lens constantly adjusts the eye's focus and depth-of-field. It does so very rapidly; so much so that people don't even notice. However, this effect can be observed by putting one hand close to the face and another far away and focusing quickly between them. They cannot both be in focus simultaneously.

through a scope, the marksman's eye can keep the target in focus and the reticle blurry, or the reticle in focus and the target blurry.

When a marksman is searching for targets, they keep the reticle blurry and focus on the target area for two reasons: to better detect any targets that may appear and also to know where to aim. However, when a specific, small target is in a marksman's sight, they switch to the reticle being in focus and make the target blurry. (See Image 37, Pg. 59.) The marksman must actively engage their eye muscles to ensure that their **focus remains on the**

Holdover

Image 42: Scopes are mounted above rifle barrels. Therefore at the muzzle, the sightline (what the marksman sees) is above the bullet's trajectory. **The sightline and the trajectory intersect at the zero point**. However, before the zero point, marksmen must be aware that just because they can see something through the scope does not mean that their bullet has a clear path of travel.

reticle or the front sight post. This is because an in-focus reticle is smaller and therefore more precise than a blurry reticle, and a more precise reticle leads to more precise shooting.

Many novice marksmen are hesitant to shoot at blurry targets and instinctively want to focus on what they are shooting at, rather than the measurement tools. However, shooting with thin reticles is far more accurate than trying to discern the precise details of the target. Furthermore, most targets require hitting the center anyway, and aiming at the center of a blurred object is nearly as easy as aiming at the center of a clear object.

One technique to help marksmen concentrate on their sights is to make the reticle or front sight more visually appealing, such as by buying an illuminated reticle or adding a dot of red paint to the front sight post so that it would stand out more against the background. Then that sight can be contrasted by using blank, uninteresting pieces of paper as targets.

That being said, some marksmen still prefer to focus on a point midway between the sight and the target, which slightly blurs everything. Others choose to focus on the target area during low-recoil shooting so they can observe the impact of their shots in case they miss the target. This is acceptable for practice, but when accuracy is crucial, it is essential to maximize the chances of hitting the intended target by focusing on the reticle or front sight.

7.e Holdover and Ballistic Loopholes

The vertical distance between the center of the bore and the center of the scope is referred to as the "height-above-bore" or "holdover," which is usually

Image 43: A Jordanian Armed Forces' 10th Border Guard Force sniper positions with his barrel inside a **loophole** to avoid hitting the wall. Jordan, 25 Feb 2019.

Image 44: A Romanian sniper aims through **loopholes**. If his aim is just above the bottom, his bullet would hit the wall. Hohenfels, Germany, 04 May 2022.

Image 45: A 10th Group Special Forces Soldier fires through **loopholes** to learn to always consider his height-over-bore. Fort Carson, Colorado, 06 Dec 2018.

no more than 10 cm (4 in). In other words, bullets obviously exit below the scope because the barrel is below the scope. To compensate for this difference, bores are slightly angled upward so that a bullet exiting the barrel intersects the scope's sightline. This point-of-intersection is known as the "zero point." (See Image 42, Pg. 63.)

Until reaching the zero point, bullets can be assumed to travel in a straight line. Therefore, the boreline, the sightline, and the height-above-bore form a right triangle, with the height-above-bore being around 10 cm (3.9 in). This means that in front of the rifle, the bullet requires 10 cm (3.9 in) of clearance below the scope. When halfway to the zero point, the bullet needs 5 cm (2 in) of clearance, and when three-quarters of the way, it requires 2.5 cm (1 in).

A mistake that marksmen sometimes make is looking through their scope, seeing a clear sightline to the target, firing the rifle, and then hitting a wall or branch in front of them instead of the target. They mistakenly believe that if they can see the target, the bullet can reach it. However, bullets exit the bore, not the scope. So, **just because the sightline is unobstructed does not mean the bullet's trajectory is unobstructed**.

To avoid hitting anything, the marksman must ensure that every location along the bullet's trajectory is clear. This is not easy to do in a dense forest with many branches; however, if there is only one obstruction, the marksman may be able to open a ballistic loophole, or "loophole." A **loophole** is a hole in a barrier, such as a wall, that allows the bullet to pass through. (Loopholes only work, however, if the marksman intends to aim at a specific point or area.) (See Image 43, Pg. 64.)

Instead of one large hole for both the sightline and the bullet trajectory, marksmen can create space for shooting by making two small loopholes: a top one for the sightline and a bottom one for the bullet to travel through. The loophole is spaced apart by height-above-bore proportional to the distance from the loophole to the zero point. So for example, if the loophole is halfway to the zero point, the bullet's loophole is spaced half of the height-above-bore below the sightline's loophole.

8. Firing the Rifle

After preparing their rifle, positioning themself, and finding their target, a marksman is ready to fire. Immediately before firing, it is crucial to ensure that the **safety is off**. (The safety is on at all other times!) One of the most common errors is to attempt firing with the safety on. To initiate the firing process, the marksman pulls the trigger, experiences and manages recoil, and prepares for any follow-on shots. Once the firing is complete, the marksman must ensure the rifle is unloaded by inspecting the chamber, magazine, and bolt face to confirm that the rifle is empty of ammunition.

8.a Trigger Pull and Flinching

While pulling the trigger, the marksman's focus remains on the reticle. When the reticle is centered on the target, the marksman applies consistent and intentional pressure on the trigger. **The trigger is not tapped or jerked**, and the trigger finger maintains continuous contact. This is important, as using an excessive pulling force necessitates a greater opposing force from the back of the hand or shoulder. And applying stronger forces all around increases the chances of shaking or rotation, which alters the point-of-aim

During the trigger-pulling process, some marksmen prefer to take their time, spending several seconds gradually pulling the trigger, while others do it in a fraction of a second. Some marksmen use a **two-stage trigger**. This type of trigger is pulled until the trigger pull weight changes; pulling beyond that point discharges the firearm. If using a two-stage trigger, the marksman pauses at the start of the second stage to recheck the sight picture (this pause is commonly known as trigger prep or taking out the slack from the trigger). Regardless, the marksman must automatically release the trigger if anything in their visual field changes unexpectedly.

In contrast, some beginning marksmen repeatedly touch and release the trigger, exerting very little pressure without pulling the trigger. They then may repeat that light pull with the application of greater pressure until the trigger is pulled (or jerked) hard enough to discharge the weapon. Such a lack of consistency and focus greatly increases inaccuracy.

Therefore, because the marksman needs to focus on multiple aspects simultaneously (e.g., the sight picture, trigger pull, and avoiding flinching), the skill of properly **pulling the trigger must become somewhat subconscious** through practice. To practice pulling the trigger, a marksman does not need ammunition or targets and can solely focus on pulling the trigger and resetting the action until it becomes muscle memory. (See Dry Firing, Pg. 227.)

Flinching (i.e., tapping or jerking the trigger) is a common involuntary reflex or muscle movement that occurs just before or during the shot. It can range from a forceful jerking of the entire rifle to losing focus on the target completely (a.k.a., "blacking out"). This reflex can be triggered by the anticipation of recoil, a loud noise, any prior negative shooting experiences the marksman may have had, or simply a fear of the consequences of the shot taken. **All marksmen experience flinching.** It can even resurface in expert marksmen who may be experiencing fatigue, stress, or sudden changes in their surroundings. In that vein, even if beginner marksmen start a session without flinching, it is common for flinching to emerge as their body and mind become fatigued.

That said, avoiding flinching can also be achieved through consistent practice, emphasizing proper trigger control and maintaining a stable shooting position. A marksman can also practice with a light-caliber rifle that produces minimal recoil, helping to condition the mind to not associate pulling the trigger with a significant recoil.

To counteract fatigue-related flinching, it is essential to during shooting sessions and ensure proper rest and hydration. Additionally, to prevent mental panic, a marksman can audibly talk themself through the process. By hearing their own words, the marksman compels themself to focus on the present task rather than the future consequences.

Marksmen may be unaware of any flinching that may be occurring and ultimately affecting the shots they have taken. This might be because they experience no flinching during practice where they know they are not firing live rounds, or because the recoil of the rifle masks any flinching when live rounds are present. It is possible to determine the extent of a marksman's pre and post-firing flinching with the help of a useful technique. That is to practice firing with a random mix of live rounds and dummy rounds. Most helpful would be if these rounds are loaded by an assistant so that the marksman does not know the sequence of the rounds present in the chamber. When "firing" dummy rounds, the marksman can then determine if they flinch without the interference of recoil. (See Image 196, Pg. 231.) Even better, the assistant can video-record the shooter in slow motion to determine if there is even a minuscule level of flinching.

8.b Recoil and Muzzle Jump

Recoil (a.k.a., kickback) refers to the forceful, backward movement of the buttstock that occurs after firing a rifle or any other firearm. **Muzzle jump** (a.k.a. muzzle rise) is the forceful, upwards movement of the muzzle after firing. It occurs because the bullet travels through the barrel, above the stock, causing the rifle to rotate about the marksman's grips which serves as a fulcrum (i.e., a center of rotation). (See Image 47, Pg. 69.)

Both recoil and muzzle jump are direct consequences of laws of physics (which are vital to ballistics). Newton's third law of motion states that for every action, there is an equal and opposite reaction. In the context of firearms and long-range shooting, the action is the expulsion of the bullet from the barrel. The reaction is the backward thrust of the rifle into the marksman's shoulder pocket and the rotation of the rifle about the center-of-mass. Any adjustment to the action affects the reaction; an increase in the amount of powder causes a corresponding increase in recoil and muzzle jump.

If strong enough, **recoil can cause discomfort or injury** to the marksman, resulting in fatigue, bruising, or even ineffectiveness. This is common after firing multiple rounds of high-powered ammunition. Additionally, excessive muzzle jump affects the sight picture of the scope between shots. Consequently, various techniques and equipment have been developed to mitigate the effects of recoil and muzzle jump.

There are two types of recoil reduction methods. The first type, **true recoil** reduction, involves creating a counteracting force. For example, muzzle brakes redirect the gas exiting the muzzle to the side. This redirection pushes the rifle forward upon impact with the muzzle brake, counteracting the forces pushing the rifle backward. Similarly, compensators redirect gas to exit from the top-side in order to counteract muzzle jump. (See Image 49, Pg. 71.)

Muzzle brakes and compensators neutralize force equal to the percentage of gas redirected sideways, and double the percentage redirected backward. To illustrate, a .50 Browning Machine Gun (BMG) fires a bullet weighing approximately 700 grains with a powder charge weighing 300 grains. If all of the gases produced by the gunpowder were redirected sideways, recoil would be reduced by 30%. If the gases made a U-turn and were redirected backward, recoil would be reduced by 60%.

Similarly, true recoil reduction can also be achieved by redistributing the force of recoil to rifle supports such as bipods, tripods, or sandbags. These supports absorb and distribute the force to the ground, minimizing the impact on the marksman. When choosing a support, it is crucial to consider its ability to transmit recoil effectively and remain stationary. Any movement of the support between shots indicates improper usage. For example, if bipod legs creep forward, it signifies that the marksman is applying excessive pressure on the rear of the bipod.

The second type of recoil reduction is known as simulated reduction or "**felt-recoil**" reduction. When a bullet is fired from a rifle, Newton's third law dictates that a counteracting force is generated. However, the intensity of this recoil force can be lessened by making the force present over a longer period of time. For example, instead of experiencing a sudden recoil of 1000 newtons of force (~100 kg or ~45 lb of weight on earth) in 0.1 s, the body might perceive a 500-newton force spread over 0.2 s.

The Effects of Recoil

Image 46: A Soldier assigned to 40th Cavalry Regiment (Airborne), 4th Infantry Brigade Combat Team (Airborne), 25th Infantry Division, U.S. Army Alaska, absorbs the **recoil** from an M2010 Enhanced Sniper Rifle. Statler Range, Joint Base Elmendorf-Richardson, Alaska, 06 Apr 2018. The muzzle is slightly elevated due to **muzzle jump**. The cheek is being dragged back due to recoil. And the marksman's eyes are closed because of an instinctual reaction to the force.

Rifle Barrel

Recoil Force

Rifle's Center-of-Mass
(Rotation Point)

Image 47: The force of recoil presses the barrel backwards. However, the center-of-mass is below the rifle. Any force that does not act through the center-of-mass causes rotation. Because the barrel is above the center-of-mass, it forces the muzzle to rotate up (i.e., muzzle jump) and the buttstock to rotate down.

For example, adding springs between the marksman and the explosion can extend the duration of the recoil force. Many weapon systems utilize metal spring mechanisms, such as the AR-15 platform which has a buffer and a spring in its stock. Advanced firearm designs may incorporate hydraulic or pneumatic shock absorbers, or even a combination of springs, cams, and levers, to modify, dampen, or dissipate the rearward impulse created when the bullet is fired. Each mechanism has its own advantages. For example, gas springs are more expensive but do not have bending metal parts such as coil-type springs that can break.

Interestingly, the primary "spring" used to distribute recoil over time in most rifles is the buttstock pad, also known as a recoil pad. In the context of long-range shooting, the buttstock pad functions as a spring in that when the rifle is fired, the padding compresses and gradually decompresses over a more extended period, extending the shock-of-force over time. Other cushioning materials, such as rubber or even a common sponge, can also be used as recoil padding. These materials can be worn over clothing or integrated into jackets.

In fact, the human body itself also acts as a spring. When the body is positioned in a way that allows it to naturally move with the recoil, the force is spread over a greater distance, giving the force more time to dissipate. Therefore, marksmen must relax their bodies as they shoot and avoid tensing their shoulders. Similarly, if the marksman does not firmly press the firearm into their shoulder pocket, the rifle can hit the shoulder with more force. This happens because the rifle accumulates backward momentum instead of dissipating that momentum into the shoulder from the moment of firing. (See Shoulder Pocket, Pg. 44.) (See Momentum, Pg. 246.)

In addition to extending force over time, pads also decrease force over area. That is, they reduce pressure. Pressure is determined by the combination of force and area, so pressure can be minimized by distributing the force over a larger area. That is one reason why stock dimensions are so important. Specifically, stocks with a low, sharp comb (where the cheekrests) or a small, hard buttstock-end-plate (i.e., buttplate) concentrate the recoil into a smaller area, thereby increasing the pressure and making the recoil feel more intense. Some marksmen dissipate the force by wearing thick jackets or draping towels over their shoulder pockets. However, they risk filling in the shoulder pocket and making holding the rifle somewhat more difficult.

An alternative to springs that reduces felt-recoil is to increase the mass of the rifle. An object with more mass moves slower when subjected to the same force, while an object with less mass moves faster. This is called "the principle of inertia." Therefore, increasing the mass of the rifle decreases its movement speed for the same force, giving the body more time to adapt and absorb the incoming force and thereby reduce the felt recoil, despite the fact that the true recoil remains unchanged. A simple way to add mass to a rifle is by incorporating a lead weight in the buttstock. The reduction in felt recoil is directly proportional to the mass added; therefore, to halve felt recoil, the marksman would need to double their rifle's mass. That being said, most marksmen prefer lighter rifles because they are less expensive and easier to carry.

Regardless of the type of recoil or type of mitigation, shooting with and without mitigation requires slightly different techniques. Therefore, marksmen must either mitigate recoil both in practice and in execution or in

Muzzle Brakes and Compensators

Image 48: An Iraqi security forces soldier fires a Steyr HS .50 anti-materiel rifle. The rifle has a muzzle brake (back-directing) to reduce the intense recoil of a .50 round by forcing gases to exit from both sides to cancel out each other's force. Camp Manion, Iraq, 29 Mar 2017.

Muzzle Brake, Back-Directing (View from Top)	Muzzle Brake, Side-Directing (View from Top)	Compensator, Up-Directing (View from Side)

Image 49: Muzzle brakes that **redirect gas** backwards counteract recoil by pushing gas in the opposite direction of the bullet. Muzzle brakes direct the gas sideways to neutralize the force of gas by pushing it to either side equally. Compensators counteract muzzle jump by **forcing gas upwards**. These devices do not reduce a bullet's velocity because they only affect gas that has already left the barrel.

neither of the two. For example, a target marksman's tournament performance is diminished if they practice with a padded vest but compete without one.

8.c Follow-Through and Reacquiring Sight Picture

"Following through" refers to the act of maintaining a shooting position after firing a weapon until after the recoil pulse has subsided. This includes

Image 50: A U.S. Marine with 3d Reconnaissance Battalion, 3d Marine Division, holds the trigger to the rear after firing. Camp Hansen, Okinawa, Japan, 27 Jan 2021.

holding the trigger to the rear and delaying a trigger reset. (See Image 50, Pg. 72.) Follow-through serves a few important purposes:

1) It reinforces the mental habit of trigger control, allowing the marksman to focus on the present shot rather than any future shots.
2) It enables them to pay more attention to detecting any inadvertent flinching that may be occurring.
3) It ensures that the shooting process ends with a still position rather than a moving one (i.e., reengaging). If a marksman plans to move after the shot, they are more likely to move during the shot, whereas planning to remain still after the shot increases the likelihood of staying still during the shot.
4) It allows the marksman to quickly reacquire their sight picture. Recoil from the firing sequence may cause the sight picture to move off the target, but reducing the number of inputs (such as resetting the trigger) can lessen the movement of the sight picture. Marksmen must aim to see their target again as soon as possible. (Even if the marksman does not intend to take follow-on shots after any one particular shot, it is always necessary to reacquire the sight picture after every shot in order to develop a consistent habit of following-on.)

The possibility of the fourth point, reacquiring the sight picture, depends on the power of the ammunition. On a low-recoil firearm (such as a .22 LR rifle), the marksman can maintain the same sight picture throughout the shooting process with good follow-through. However, if the recoil is medium-to-high, it causes marksmen to lose their sight picture after firing no matter what. Therefore, marksmen prepare by remembering their current sight picture and matching what they see to the previous one. They may also lower the magnification setting to get a larger sight picture if they have to shoot at the same target multiple times. Finally, marksmen rely on their proprioception (i.e., the sense of knowing where body parts are) to learn and remember their body position. This allows them to accurately reposition their body after firing, returning to the same stance they recall. These techniques to reacquire the sight picture require full focus and are rarely exact.

Once the marksman has consciously executed a follow-through, the trigger is smoothly reset to the firing position and the next round is chambered. Again, performing a smooth reset is faster overall compared to rushing through the whole firing sequence.

8.d Cold-Bore Shooting

A "cold-bore shot" refers to the initial shot fired from a firearm during a shooting session. Because the act of firing a firearm burns gunpowder, the first shot as well as all subsequent shots heat the bore. This means that the first shot always occurs when the bore is at its coldest, hence the name.

Historically, professionals and other long-range shooting specialists have justified the existence of the cold-bore shot with the concept of thermal expansion. Thermal expansion occurs when materials expand due to temperature changes. In the case of steel barrel rifles, the barrel expands and changes shape as it heats up due to the energy and heat expended by exploding gunpowder (propellant). In sum, a cold barrel is physically a slightly different shape than a hot barrel.

However, the expansion does not continue forever. Air surrounding the heated barrel cools and contracts it back down to the ambient temperature over time. Therefore, through a consistent cycle of shooting, the burn of each round and the subsequent air cooling theoretically reaches a quasi-equilibrium, in which the barrel remains within a range of temperatures and sizes. The logic is that the cold-bore shot is performed with a barrel that is the farthest outside this range of barrel size and expansion, and is therefore most affected by it.

This logic is not completely incorrect. Thermal expansion does occur; however, it does not significantly contribute to the inaccuracy of only the initial shot. In simple terms, each round fired from a rifle increases the barrel's temperature by a similar amount. When shooting rapidly, the barrel continues to heat up because the air does not have enough time to cool it. If temperature were the primary cause of the initial inaccuracy, it would continually affect the point-of-impact for every shot. Furthermore, while the predictable and repeatable nature of metal's thermal expansion theoretically suggests that it would cause consistent cold-bore shots, many marksmen find that their cold-bore shots deviate away from their point-of-aim in different directions each time, to varying degrees.

Another speculated factor that may affect the trajectory of the first bullet fired from a cold bore is the residue left in the barrel from previous shooting sessions. However, although this phenomenon is also genuine, it does not have as significant an impact on the point-of-impact as marksmen often report when experiencing a cold-bore shot.

The only plausible cause of the infamously inaccurate cold-bore shot is that **marksmen tend to flinch in anticipation of recoil** more on their first shot of the day than with subsequent shots in their session. In other words, the first shot of many is typically the least comfortable. A test to confirm this can be done by having a marksman fire two rifles, one after the other.

Tested marksmen generally find that the cold-bore shot only occurs with the first rifle and not when they switch to the second, which is also technically being shot cold. The best solution to overcoming the "cold-bore" shot is dry-firing practice prior to moving to live rounds. (See Dry Firing, Pg. 227.)

9. Ensuring Accuracy (Grouping and Zeroing)

Rifles can be highly precise instruments. But they must be calibrated so that scopes can accurately predict where bullets impact. To align where the bullet goes (i.e., the point-of-impact) to what the marksman sees (i.e., the point-of-aim), marksmen perform a process known as "grouping and zeroing."

In short, grouping is to fire a series of rounds together as a group at the same point-of-aim. A "grouping" on a target is simply a group of shots. "Zeroing" is to calibrate the rifle off of a group by adjusting the scope so that following shots are more accurate. By iteratively grouping and zeroing, a marksman can bring together their point-of-aim and point-of-impact.

Before grouping and zeroing, a marksman must always verify that all parts of their weapon system, particularly the scope, **are securely fastened** to make the rifle as repeatable as possible. Furthermore, when using a rifle for the first time, a marksman may benefit from firing a few times to become familiar with their rifle's recoil and grouping patterns.

9.a Grouping

Grouping is both a verb and a noun. The act of "grouping" refers to firing several shots at a target as closely together as possible. The result is a tight "group," "shot group," or "grouping" of impacts. The number of shots in a group is up to the marksman. While some marksmen prefer to use three-shot groups to save ammunition, others prefer five-shot groups for a more accurate average.

The tightness of a group depends on many factors. Just one example of that is that some rifles are more accurate with certain brands and types of ammunition. The properties of ammunition that can influence group-tightness include: the material of the projectile (copper jacketed or lead), the shape of the projectile (wad cutters, semi wad cutters, round headed), the weight (grains), the amount of propellant, and the amount of variability from one cartridge to the next. This is partly why 'match grade' ammo exists; it is designed to be more accurate than regular ammo.

The purpose of grouping is to find the center of the group, which is the average point-of-impact. The average is the best estimate a marksman has of where a rifle would shoot without variation or external influence.

"Average" can have two different meanings. When measuring the performance of a rifle, the "average" is the distance between the centers of the two farthest shot holes in a group (hereafter, referred to as the "**diameter**"). On the other hand, when trying to find the expected point-of-impact without random or environmental influences, the "average," or "**center average,**" is the central point of the group of shot holes. (See Image 52, Pg. 76.)

The center average of a group provides an estimate for the distance between the marksman's point-of-aim and the rifle's point-of-impact. Due to the randomness of individual shots and the influence of the marksman, the actual difference between these two points remains somewhat uncertain. However, a tight group improves the accuracy of the estimated distance and direction.

When shooting multiple groups on the same target, the marksman can differentiate between them by using a different color marker to connect all the bullet holes in each group. Alternatively if one marker color is available, the marksman can use different kinds of lines: solid, dotted, dashed, etc. A dot or mark is then made to indicate the group's center, and a number is written alongside each group. (See Image 51, Pg. 76.)

As a general guideline, groups larger than 0.5 mils (2 MOA) in diameter (See Angular Distance, Pg. 237.) using a minimum of three rounds are not precise enough to base adjustments on, and are therefore invalid. While there is no specific lower limit for acceptable group size, groups between 0.25 and 0.5 mil's (1 and 2 MOA) are typically sufficient as most marksmen and rifles simply cannot achieve groups tighter than 0.25 mil's (1 MOA, or a "sub-MOA rifle") even in ideal conditions.

Grouping Bullet Impacts

Image 51: This marksman made a **group of 5 shots**, and marked them in three distinct ways. First, they circled each hole. Second, they connected the holes. Third, they found the center average of the group and wrote down the number of the grouping (in this case "1"). They can also use a different colored marker for each group to distinguish them from each other.

The **diameter** is the distance between the farthest two points in the group.

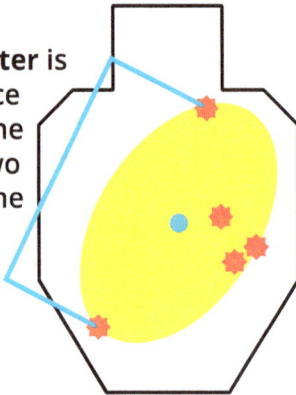

The **center average** is the most central point of the grouping (usually estimated), represented here as the center of an ellipse fit to the grouping.

Image 52: There are two common measurements of a group: the diameter and the average (a.k.a., center average).

Considering that groups have a limit to how small they can get regardless of the marksman's skill level, one common mistake marksmen make is fussing too much over their aim. The inherent inaccuracy in equipment is precisely why marksmen must rely on the average of a group to make sight adjustments in the first place. (See Inherent Imprecision, Pg. 197.) Another mistake is to adjust the point-of-aim from the first shot in the group to subsequent shots. **Every shot in a grouping must have an identical point-of-aim for the group to be valid.** A third mistake is to believe that an exceptionally tight group represents the true precision of a rifle. It is not uncommon for a few shots to randomly deviate in the same direction.

It is okay to disregard problematic shots (i.e., outliers or "fliers") at any point during the zeroing process since the point of grouping is to assess the equipment, not the marksman. If a shot feels off during shooting, the marksman can fire an additional round so long as the marksman knows which bullet hole is being excluded when they assess their grouping. When shooting with a spotter, the marksman must promptly notify the spotter of any bad shots, and the spotter must inform the marksman of any inaccuracies that they may see on the target.

9.b Picking a Zero Distance

While light travels in a straight line, bullets follow a curved path. As a result, the trajectory of a bullet intersects with the sightline of a scope at two specific points (i.e., "**zero points**"). In other words, a zero point is where the point-of-aim (as seen through the center of the reticle) aligns perfectly with the point-of-impact (where the bullet hits its target). (See Image 53, Pg. 78.)

(When shooting at distances greater than 500 m or yd, it may be beneficial for the marksman to not base their zero point off of the center of the reticle, but instead slightly above the center to allow for a longer vertical distance below their point-of-aim. This extra vertical distance allows a bullet to fall farther before exiting the sight picture.)

There is one zero point where the bullet is rising and nearer to the marksman, the "**near-zero point**," and another when the bullet is falling and farther from the marksman, the "**far-zero point**." (If the apex of the bullet trajectory is exactly tangential to the sightline, there could in theory be one zero point, but that is very rare.)

While a zero point is the location of the intersection between the point-of-aim from the center of a reticle and a bullet trajectory, a **zero distance** is the distance between the marksman and a zero point. The center of a reticle does not match the point-of-impact except at the two zero distances, and therefore the marksman must use another location on the reticle as their point-of-aim.

Zero distances, just like zero points, come in pairs: one zero distance for each zero point. In theory, using the far zero distance is preferable since the bullet only falls after the far zero, instead of rising and falling as with the near zero distance, and so calculating the total vertical difference between the sightline and the bullet's impact location (i.e., bullet-drop) is easier. However, in practice, the difference is accounted for when bullet-drop tables or ballistic calculators are used anyway. (See Ballistic Calculators, Pg. 221.)

In theory, a rifle could be calibrated to have a zero point at any distance as long as it can shoot that far. But most marksmen choose a zero distance

Zero Distance

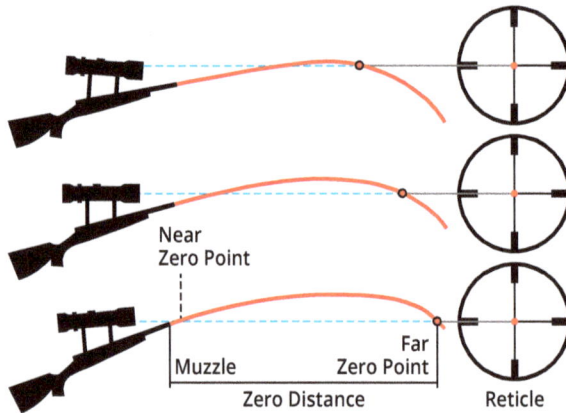

Image 53: The sightline (dotted line) intersects with the boreline (curved line) at two points. The far intersection is known as the "far zero point," or usually just the "zero point." In all three diagrams, the bullet's trajectory is about the same; however by adjusting the reticle up or down, the marksman can extend out the far zero point because the bullet has to travel further to meet the sightline again. The distance from the marksman to the zero point is the zero distance. In sum, a marksman can **adjust their reticle elevation** to attain their desired zero distance.

based on the distance at which they are most likely to be shooting. Thereby, **shooting is easier because the marksman can often use the center of their reticle to predict where their bullets impact on the target.**

Typical zero distances can range from 25 to 300 m or yd. However, the exact distance depends on the individual marksman's specific needs. For example, a marksman in a dense forest may not shoot at targets beyond 100 m or yd, while a plains hunter may frequently engage targets over 300 m or yd away. An extreme-long-range marksman could theoretically use a zero distance of 1000 m or yd, although they are much more likely to use something closer.

That said, zero distances of 50 and 100 m or yd are popular choices for three reasons: 1) they are convenient round numbers; 2) they are far enough to make any inaccuracies noticeable; and 3), they are not so far that external factors (e.g., wind) can significantly affect a bullet's trajectory during its flight to the target.

If a marksman selects a zero distance beyond 100 m or yd, such as 200 m or yd, they must consider that the bullet travels a significant distance (i.e., multiple cm or in) above the sightline at the apex of its trajectory. If a marksman has a faraway zero distance, and the target is at the halfway point, this is the only situation in which the point-of-aim is actually higher than the center of the reticle. However, this concern is irrelevant if the marksman does not plan to shoot before their zero distance.

9.c Zeroing

The zeroing process involves adjusting the sights so that the point-of-aim of a sight aligns with the point-of-impact of a bullet. Specifically, "zeroing" is a five-step process that is repeated as many times as necessary. (See Image 54, Pg. 83.)

1) Select a zero distance (See Picking a Zero Distance, Pg. 77.) and place there a sturdy target (i.e., one unaffected by the wind). Large, blank sheets of paper work well as they are readily available and inexpensive. The marksman can use a straightedge to draw a large vertical and a horizontal line on the target as a reference. Commercially-sold zeroing-targets are also a very good option.

    (Optional step 1.5) If a rifle or ammunition is suspected to be very inaccurate, the marksman begins with a very close target (e.g., 25 m or yd) to ensure that the impacts are at least known to be on the target. This process is called "**getting on paper**," because once the impacts are known to be on a paper target and zeroed a little bit, the marksman can move the target to the actual zero distance to get a precise zero. Good bore-sighting usually makes getting on paper unnecessary up to 100 m or yd with a man-sized meter or yard target. (See Bore-Sighting, Pg. 40.)

2) Shoot a group (See Grouping, Pg. 74.) using the center of the reticle as the point-of-aim. It is important to secure the rifle in place as rigidly as possible, ideally using a rigid support. For example, a shooting benchrest vice is an excellent option for securing. The smaller the groups, the better.

3) Find the center average of the group. (See Image 52, Pg. 76.) If a marksman does not want to walk to the target to examine a group, they can estimate the center average from a distance through the rifle's scope.

4) Determine the angular distance on the target from the center average to the point-of-aim. (See Angular Distance, Pg. 237.) The marksman does not determine the diagonal distance. They instead separately determine the vertical distance for use on the elevation turret, and the horizontal distance for use on the windage turret.

    The angular distance can be directly determined through the scope by reading the hashmarks in the reticle if there are any. For example, if the reticle indicates 1 mil left and 2 mils up, the adjustment to the turret would be the same. The marksman must consult the scope manual to determine how many clicks on their turrets correspond to a mil or an MOA on their scope. (See Image 18, Pg. 37.)

However, at the target, a marksman must use a ruler or the target's grid squares to calculate the linear horizontal and vertical distances. But since this step is complicated and involves math, **many marksmen prefer to use trial and error, or buy zeroing targets that come with printed-on instructions.** To convert this linear distance on the target to an angular distance on the reticle, the marksman uses the following:

Variables
- **Linear Distance on Target** *(Linear D on T)* (usually meters or yards) is the distance measured on the target itself.
- **Marksman Distance to Target** *(Marksman D to T)* (usually meters or yards) is the linear distance from the marksman to the target.
- **Mil Distance on Target** *(Mil D on T)* is the angular distance on the target in mils.
- **MOA Distance on Target** *(MOA D on T)* is the angular distance on the target in MOA.
- The *(Linear D on T)* and *(Marksman D to T)* must use the **same units** for the formulas to work (usually meters or yards). This is easy in metric and somewhat more difficult in imperial, as it often requires converting inches into fractional yards or vice versa.
- **3,438 MOA** is a constant from using MOA and does not hold significance.
- **1,000 mils** is a constant from using mils and does not hold significance.

Formula for mils
- *(Mil D on T) = (Linear D on T) x (1,000 mils) ÷ (Marksman D to T)*

Formula for MOA
- *(MOA D on T) = (Linear D on T) x (3,438 MOA) ÷ (Marksman D to T)*

For example, at a target distance of 100 m, one mil is equivalent to 10 cm of distance on a target. So, if a bullet misses the point-of-aim by 10 cm, the marksman must adjust their point-of-aim by one mil. That is:
- *(Mil D on T) = (Linear D on T)*   *x (1,000 mils) ÷ (Marksman D to T)*
- *(Mil D on T) = (Linear D on T)*   *x (1,000 mils) ÷ (100 m)*
- *(Mil D on T) = (0.1 m)*   *x (1,000 mils) ÷ (100 m)*
- *(Mil D on T) = (100 m)*   *x (1 mil)*   *÷ (100 m)*
- *(Mil D on T) = 1 mil*

Because using this formula for MOA and yards can get complicated, **imperial-system users have a shortcut**, which is to use a zero distance of 100 yd and approximate. That is, there are 3600 inches in 100 yards, and 3600 is pretty close to 3,438, so the MOA formula is simplified:

▸ *(MOA D on T) = (Linear D on T) x (3,438 MOA) ÷ (Marksman D to T)*
▸ *(MOA D on T) = (Linear D on T in inches) x (3,438 MOA) ÷ (3600 in)*
▸ *(MOA D on T) = (Linear D on T in inches) x (~3,600 MOA) ÷ (3600 in)*
▸ *(MOA D on T) = (Linear D on T in inches)*

Thereby, each inch of distance on the target **at a distance of 100 yd** (or 1/2 in at 50 yd, or 2 in at 200 yd, etc.) is corrected with 1 MOA of adjustment on a scope. This technically overcorrects, but because zeroing is an iterative process, any overcorrection can just be fixed with another iteration of zeroing.

5) Adjust the reticle by the angular distances found in step 4. This involves inputting the vertical angular distance into the elevation turret and the horizontal angular distance into the windage turret. (See Adjusting Elevation and Windage Settings with Turrets, Pg. 36.) Moving the turret marked "up" or "down" on a scope shifts the point-of-impact of the bullet up or down relative to the point-of-aim in the sight picture. It is said, "**The direction is for the bullet**." If the turrets do not have enough rotation to zero the rifle, the scope must be mechanically centered and remounted. (See Getting to Mechanical Center, Pg. 38.)

If done correctly, a marksman can align the center of the reticle to create a zero point at the zero point in one iteration. However, many marksmen over and under-correct and so find that they must repeat the process a few times before finding their zero point.

That being said, all rifle equipment has inherent imprecision (e.g., usually shooting within a .25 mil or 1 MOA area). (See Inherent Imprecision, Pg. 197.) Zeroing is done not when the point-of-aim exactly matches the point-of-impact, but when the two are closer together than the inherent imprecision of the equipment. Inherent imprecision is measured by the smallest grouping diameter that can be achieved by the equipment (See Image 52, Pg. 76.), and this metric is usually advertised by rifle and ammunition manufacturers. Attempting to attain grouping diameters smaller than the equipment itself can attain is a futile exercise, and such a mistake is called "**chasing rounds**," since marksmen constantly make small, erratic adjustments to their point-of-aim, making it seem like they are chasing something.

In some cases, a rifle may have multiple sights, such as a scope and iron sights, or a scope and a red dot. Typically, the primary sight is more precise while the secondary sight is more durable and is used if the primary breaks. Once the first sight is zeroed, the second sight can be "co-witnessed" off the first. Co-witnessing is the act of aligning one sight's point-of-aim to another's without verifying the point-of-aim by firing a bullet. This is fine for hunting or target shooting; however, if the backup sight's accuracy is crucial for the marksman's safety, the second sight must also be properly zeroed independently of other sights. Not co-witnessing is also important when using an infrared laser and a night vision device for shooting in the dark. Such systems differ significantly from scopes, and so co-witnessing can introduce error in the secondary sight.

After zeroing is complete and the reticle is adjusted, marksmen must know **how much more adjustment the reticle still has in all directions**. For example, if an elevation turret allows for a total adjustment range (a.k.a., "travel") of 25 mils, at the mechanical center the reticle can be adjusted 12.5 mils up and down. If the zeroing process required the reticle to be adjusted 10 mils up, then the scope would only have 2.5 additional mils of upward adjustment range. This is important for long-range marksmen because if they are unaware of their remaining range and then need that range, the scope would not be able to physically adjust the reticle enough to aim at a target.

If the turrets do not have enough remaining range for the marksman's purposes, the scope must be mechanically centered and remounted. (See Getting to Mechanical Center, Pg. 38.) Because upwards reticle adjustment is much more useful than downwards adjustment (because bullets mostly fall), some marksmen choose to mount their scopes with an inclined shooting block, known as a riser, to tilt the scope slightly upward and purposefully misbalance the range of the elevation turret. If neither of those solutions gives the marksman enough vertical adjustment range, a marksman must buy a scope with more adjustment range.

The process of zeroing is done in a safe environment and is a good time to learn. A marksman can receive feedback from a helper or instructor observing the process. (See Teaching, Pg. 230.) A marksman can also test their equipment by firing many rounds to attempt to identify subtle issues with their rifle, such as a loose scope or warping stock. This extended familiarization and checkup can be performed each time any rifle component changes, such as the installation of a new scope or a change in ammunition.

Zeroing a Rifle

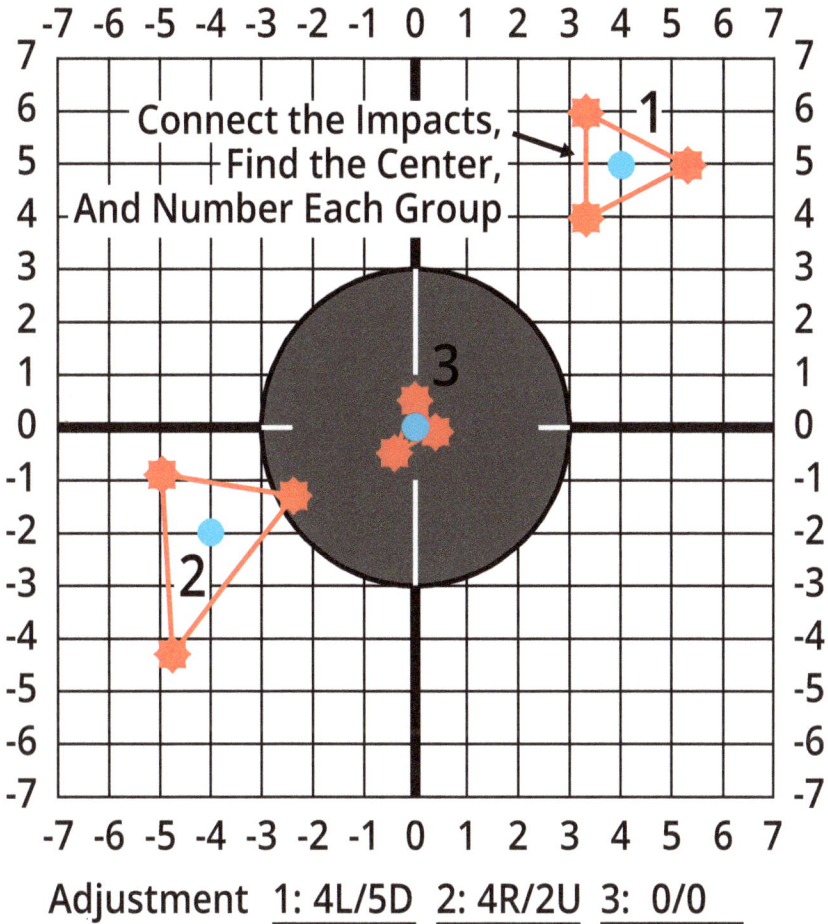

Image 54: This diagram shows the results of the step-by-step process. The first group, labeled "1" was shot too high and too far to the right. After using a marker to draw lines between the impact holes, the marksman determined the center of the group was at the coordinates 4, 5. Therefore, the marksman needed to adjust their scope (or rather the reticle inside the scope) to the left by 4 units (4L) and down by 5 units (5D). Unfortunately in this example, the marksman overdid their corrections (for whatever reasons) and ended up on the other side of the target center. Performing the same steps, the marksman determined that the center was 4 to the left and 2 down from the center. Therefore, they adjusted their point-of-aim 4 to the right (4R) and 2 up (2U). Afterwards, the third group's center was at the center of the target, and the rifle and scope were successfully zeroed.

9.d Resetting Elevation and Windage Turrets to Zero

The elevation and windage turrets on a scope respectively control the vertical and horizontal movement of a reticle inside a scope. When the reticle needs to be moved up or down (i.e., elevation), or left or right (i.e., windage), the marksman turns these turrets. (See Adjusting Elevation and Windage Settings with Turrets, Pg. 36.) However, after being rotated, the numbers on the turrets are set randomly; whereas instead, they must have the number "0" be represent the zero point and zero windage.

Ensuring that the turrets correctly represent the reticle's deviation from its zero is critical. Resetting the numbers on the turrets allows the marksman to be able to get back to their zero setting consistently. (In theory, any digits on the turrets could represent the zero setting, but shooting is stressful and makes marksmen easily forget or confuse basic facts, especially digits.)

So that turrets can be set to zero without changing their internal setting, most modern scopes have removable turret caps on top of the actual adjustment mechanism. The cap is screwed off, usually with an allen wrench (a.k.a. hex key), and rotated so that the number "0" on the turret faces directly backwards. (See Image 55, Pg. 85.) The cap is then screwed back on to set it in place. (It is crucial to avoid over-tightening and potentially stripping screws.) These caps are useful because to zero a rifle a marksman may have to turn the turrets a random amount, and remembering that the zero is at "7" for elevation and "3" for windage can be arduous. It is far easier if the zero is at "0" on both dials.

If a click is heard or felt while rotating the turret, those clicks must be reversed by the equivalent number of clicks in the opposite direction. That is because a click represents an adjustment to the reticle; and therefore, failure to properly reverse the adjustment would require the marksman to re-zero the scope.

If a scope does not have removable caps, or if the marksman wants an additional marking, they can **draw a line down each turret** onto the scope. Then if a turret were to be rotated, each half of the line would become misaligned. The marksman can reset the scope to zero by rotating the turret until the two halves of the line are connected again.

A third, extreme, method for recording the zero setting is for a marksman to record how far the zero setting is into the entire adjustment range for each turret. For example, a marksman can take a zeroed rifle and determine how

Image 55: A Marine pre-scout sniper course instructor prepares to remove a turret cap on a Leupold TS-30A2 Mark 4 scope by removing its set screws.

many clicks it takes to reach the end of the adjustment range from the zero setting in all four directions. With this information, the marksman can turn a turret all the way to the end of its adjustment range in any direction and then adjust back by the number of clicks they recorded to reach the zero setting. This method is the most time-consuming, but also the most reliable as it relies on the physical limitations of a scope's internal parts.

Some scopes come equipped with a **"zero-stop"** feature on the elevation turret. It prevents the marksman from going into negative numbers on the elevation dial. In other words, it is an artificial wall that blocks off part of an elevation turret's adjustment range. With a zero stop, a marksman doesn't have to look at their elevation dial to reset it to zero; they can just rotate it until it stops rotating. Zero stops are useful because needing elevation above the zero point is very rare. However, if a marksman does want to prepare for that, they can set their rotation stopping feature to a negative number instead of zero. For example, they can set their stop to negative four clicks, and always adjust four clicks backwards after blindly rotating the zero stop to its limit.

Intermediate Contents

10. Body Control   87

Breathing   87
Natural Point-of-Aim (NPA)   89
Heartbeat   92
Shooting with Both Eyes Open and Eye Dominance   94

11. Shooting Positions   95

Prone   97
Kneeling   100
Standing   102
Using a Sling   102
Using Ground Support   104

12. Finding Distance to a Target   110

Using Rangefinding Tools   110
Using a Standard Reticle (Milling)   113
Using Hands   119
Using Estimation   121
Perception Errors   123

13. Using Scopes for Long-Distance Shooting   126

Magnification   127
Gravity, Drag, and Trajectory   128
Standard Reticles   132
First and Second Focal Plane Reticles   136
Elevation Hold and Dial   137
Precisely Adjusting the Sight Picture   140
Bullet-Drop Compensator Reticles   140
Danger Distance   143

14. Eliminating Rifle Cant   145

Formulas for Cant Offset   147
Holding Cant   150
Reticle Cant   152
Mechanical Adjustment Cant   152

Intermediates
(100 to 300 Meters or Yards)

The truth is that any good modern rifle is good enough. The determining factor is the man behind the gun.
　　　—*Theodore Roosevelt, 26th U.S. President and Army Colonel*

At distances under 100 m or yd, hitting a standard-sized target is generally straightforward. Simply aiming and firing are usually all that is required for you to hit the target. Therefore, the art of shooting truly starts when targets are situated 100 m or yd away or beyond because you can no longer solely rely on your intuition to improve.

Most marksmen at this intermediate level are skilled enough to know how to control and position their bodies. They can also find the distance to their target without a rangefinder, fully utilize their scopes, and keep their rifles perfectly vertical without any cant.

10. Body Control

The human body **continues to move** even as a marksman remains as "still" as they possibly can. Many muscle movements are only partially voluntary, such as breathing and blinking, while others are completely out of a person's control, such as the beating of the heart.

These muscle movements translate to the rifle and can slightly move it, making aiming difficult. Therefore, marksmen can shoot more precisely by exerting conscious control over their semi-voluntary body movements to calm them. Additionally, a marksman can increase their situational awareness and comfort if they learn to skillfully control how they look through a scope.

10.a Breathing

Breathing is a simple topic: moving the chest up and down while shooting moves the point-of-aim up and down as well. To stay the point-of-aim, a marksman must suspend their breathing. That is, to maintain a consistent chest height, the marksman must control their exhaling and inhaling. To do this, a marksman **waits for their natural exhale to complete**. Then the marksman does not inhale until they fire their rifle to ensure that their lungs

Holding Breath to Shoot

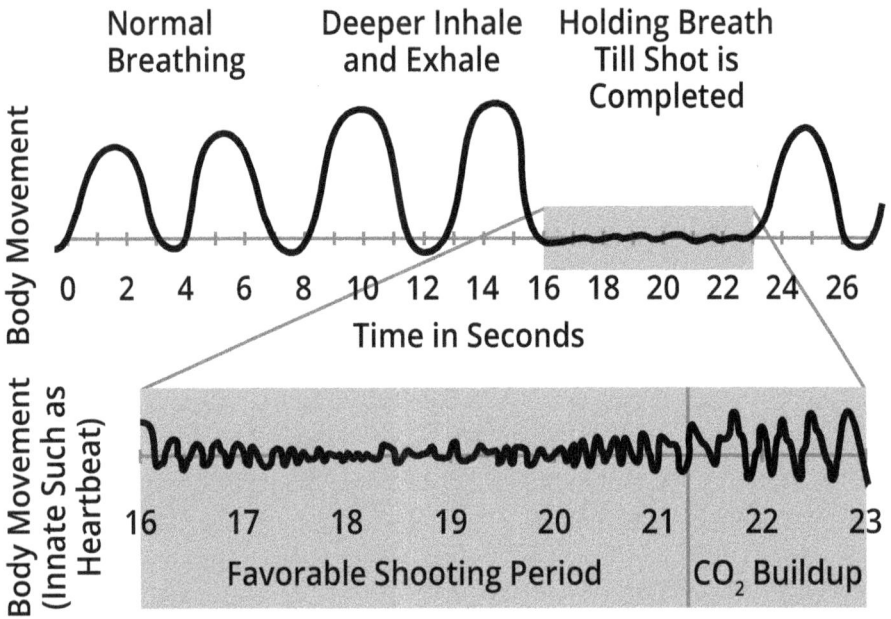

Image 56: Breathing moves a marksman's chest up and down. So to keep their body as still as possible, marksmen hold their breath while shooting. To prepare for this, marksmen prepare with deep inhales. Holding breath for too long creates **carbon dioxide buildup**, which negatively affects precision. This diagram illustrates just one marksman's experience. Some marksmen can effectively hold their breath for longer or shorter periods than other marksmen can.

are not moving (and therefore cannot alter their point-of-aim) while shooting. (See Image 56, Pg. 88.)

Not breathing for a moment before shooting is perfectly fine, but any pause longer than about eight seconds can lead to imperceivable muscle shaking and issues related to carbon dioxide buildup and oxygen deprivation. If eight seconds pass and the marksman has not shot, they simply need to breathe in and out again to reset the process.

The completion of a natural exhalation is the ideal part of the breathing cycle to pause at, since it is the most repeatable. In contrast if a marksman chooses to partially exhale, they may exhale a different amount of air each time they aim their rifle, meaning their chest would be at different heights and their points-of-aim would therefore become inconsistent. That being said, it is important for the exhale to be natural, and not a complete emptying of the lungs. The goal is repeatability, and an unnatural exhale is neither repeatable nor comfortable.

A useful acronym for breathing control is **BRASS**:

Breathe – The marksman takes a deep breath and aligns their sights during the natural exhale. They continue to refine their point-of-aim during subsequent natural exhales.

Relax – While maintaining a calm and regular breathing pattern, the marksman actively tries to relax their muscles and mind.

Aim – The marksman finalizes their point-of-aim.

Slack – The marksman pulls the trigger to the point just before it fires (this is called "removing slack.").

Squeeze– They then squeeze the trigger when they have both reached the next natural exhale and confirmed that their point-of-aim is still accurate. The trigger is calmly pulled and never jerked.

All that being said, the ability or need to control one's breathing is often over-exaggerated because the concept of timing one's breathing cycle is easier to explain than many other concepts. In reality, the movement of the chest during breathing is relatively small and uninfluential. Moreover, in military and hunting scenarios, marksmen sometimes are not able to wait to enter the perfect breathing pattern before taking the shot but must shoot anyway before the target moves. In fact, military marksmen regularly participate in "stress-shoots," where they practice marksmanship after intense exercise to learn to remain accurate and precise in high-stress situations during which their breathing is not fully controllable.

10.b Natural Point-of-Aim (NPA)

The natural point-of-aim (NPA) refers to the location that a held rifle points when the marksman's body is relaxed. Therefore, shooting with an NPA is almost synonymous with shooting while as relaxed as possible (i.e., a marksman must feel so relaxed that they could almost fall asleep).

There are two reasons why a marksman would want to shoot from a relaxed position with an NPA. First, **it is far easier for a marksman to consistently use a single shooting position**; full relaxation is a single, known state. In contrast, a marksman can be partially relaxed in any number of ways and body configurations. This is the same reason that marksmen shoot at their full, natural exhale, instead of only partially emptying their lungs.

This repeatability also allows a marksman to take more shots in faster succession at a target because it is easier to fully relax than to partially relax any particular part of their body. This is important because recoil always relaxes the body anyway, so if a marksman is already fully relaxed then recoil cannot relax the marksman's body significantly more.

The second reason for using an NPA is to reduce muscular tremors, which are caused by muscles attempting to maintain a fixed position. These tremors can transfer to a marksman's rifle, causing both the sight picture and point-of-aim to waver. Shooting from a relaxed position minimizes this instability.

That is not to say that all wavering can be completely eliminated. That is simply not humanly possible. A common mistake made among new marksmen is trying to overly perfect each shot they take. Instead, a marksman must act smoothly and fire when the crosshairs have sufficient, not perfect, stability.

To further relax their muscles, many marksmen employ stability tools such as slings and supports. (See Using a Sling, Pg. 102.) (See Using Ground Support, Pg. 104.) For example, by employing a proper kneeling position and a loop sling, a marksman can minimize errors to less than 1 mil (or 4 MOA) for extended periods.

The ideal shooting position is **just relaxed enough** so that a marksman can aim their rifle from that position without any additional force. That is, a marksman must control their body in such a way that it interferes the least possible amount with shooting. This is achieved by being slow and calculated with each movement and shot.

To determine an NPA, a marksman can assume any shooting position and close their eyes. Once comfortable in that position, they must focus on completely relaxing their body and take a few deep breaths. This breathing work and subsequent relaxation allow the rifle to settle and naturally shift into its NPA. The marksman can then open their eyes. Wherever the rifle points at that time would be the marksman's NPA in that specific shooting position. (It is normal for the NPA to be off-target at that point; it is rare for a marksman to close their eyes and be able to maintain their target.) An alternative method to finding an NPA is to fire the rifle to force a relaxation. Wherever the rifle settles after recoil is its NPA.

Every shooting position for every marksman has its own NPA. However, it is important to note that the NPA of a shooting position may not necessarily be aligned with the marksman's target. Therefore, **expert marksmen can predict which positions can lead to an NPA that is aligned to the target** before even getting into position, which helps them avoid readjusting their aim and rifle. To develop this predictive skill, marksmen must consistently use the same shooting positions, so that they can associate each position with its corresponding NPA.

Adjusting the Natural Point-of-Aim

Vertical Adjustment

As the body moves forwards, the rifle and shoulders rotate around the support elbow and the boreline and sightline lower.

Rifle and Shoulders Rotate

Body Moves Forward

Muzzle Moves Down

Rotation Point

Horizontal Adjustment

Rotation Point

Image 57: Because the natural point-of-aim is present when the body is in a relaxed position, to grossly adjust the point-of-aim, the body needs to move to an entirely new position. This is done **through rotation**. Adjusting the point-of-aim up or down is done by moving the lower body forward or backward to rotate the upper body about the support elbow. Adjusting horizontally requires the marksman to rotate left or right about their hips.

Having said that, even the most advanced marksmen still need to make adjustments to their shooting position once they are in it. These adjustments align the NPA with the target, and can be either gross or minor.

Gross adjustments to the NPA require pivoting the entire body and any supports used. In most cases, these bodily adjustments are made using the lower body. In the prone position, for instance, a marksman would use their legs to rotate around their supporting elbow. In the kneeling position, the entire body would pivot on the forward foot. Lastly, in a standing position, the marksman would rotate around their support.

Minor adjustments, on the other hand, involve adjusting one's limbs independently. For example, in the prone position, the NPA can be adjusted horizontally by slightly moving the support elbow to the left or right. And the NPA can be adjusted vertically by nudging the elbow on the side of the trigger, which raises or lowers the shoulder and consequently the muzzle of the rifle. Other minor adjustments include: moving the support hand forward or backward on the handguards; positioning the stock higher or lower on the shoulder; or finding a stable footing by digging toes into the ground and either pulling their body forward or pushing it back.

10.c Heartbeat

The heart pumps blood with pressure waves (i.e., beats), and not in a consistent flow. A beat of the heart may have enough force to transfer through a marksman and into the rifle, shifting the point-of-aim. Therefore, many depictions of snipers show them timing their shooting to a specific, consistent step in the beating process, just like skilled marksmen time their shots to their breathing. (See Breathing, Pg. 87.)

However, any such depictions are inaccurate because skilled marksmen do not time their shots to their heartbeat and disregard their pumping cycle entirely. Simply put, the heartbeat is relatively unimportant compared to other priorities during shooting, such as breathing and keeping the target in the crosshairs. Timing a heartbeat would also be more difficult than timing a breathe because, unlike breathing which can be consciously paused, the pumping cycle is very short and cannot be paused.

That is not to say that the heartbeat is irrelevant to shooting. If a marksman claims to see the reticle "beat" alongside their heartbeat, they may be gripping the rifle too tightly. Alternatively, they could be so overly excited that their blood is being pumped more strongly than normal. Either way, the best and only way to control a strong, fast heartbeat is to calm down. If a marksman's heart is racing, they can **engage in breathing exercises to become more physically calm**. Specifically, a forceful exhale not only ensures a consistent chest height, but also aids in relaxation and naturally slows down a rapidly beating heart.

Eye Dominance Testing

Image 58: A Marine checks for his dominant eye using the Miles Test. Joint Base Andrews, MD, 03 May 2013

Image 59: The **Miles Test**: Step 1 - Make a triangle with your hands, centering an object. Step 2 - Fully extend the arms. Step 3 - With both eyes open, look through the triangle to the object. Close the left eye. If the object remains in view, you are right-eye dominant, and if the object disappears you are left-eye dominant.

Image 60: The **Porta Test**: Step 1 - Fully extend the arm, and point at an object. Step 2 - Keeping the finger in place, alternate closing each eye. Step 3 - Keeping both eyes open, move the finger towards the face while keeping the finger on the object. The eye which the hand moves towards is the dominant eye.

Shooting with Both Eyes Open

Image 61: An infantryman with a Scout Sniper Platoon, in the 31st Marine Expeditionary Unit shoots targets with **both eyes open** using an M110 Sniper System during a static live fire range. Camp Hansen, Okinawa, Japan, 08 Jul 2020.

Image 62: The Peruvian marksman on the left favors the right side of his body, while the marksman on the right favors the left side of his body. Both marksmen keep both of their eyes open to observe their surroundings. Cerro Tigre, Panama, 15 May 2024.

10.d Shooting with Both Eyes Open and Eye Dominance

Almost all long-range rifles are designed to have the marksman's right eye look through the scope. Most marksmen, therefore, close their left eye to increase the brain's focus on the sight picture that is seen by the right eye. And that is a good technique to avoid fatigue if the marksman fires quickly and they can be unaware of their surroundings (e.g., when target shooting). However, many marksmen must maintain awareness of their surroundings at all times (e.g., military marksmen) and so cannot afford to close one of their eyes. These marksmen must learn to shoot with both eyes open.

The process of learning to shoot with both eyes open begins with determining which of the marksman's eyes is dominant. The majority of people have a brain that naturally favors the input from one eye over the other, just as the vast majority of people favor one hand over the other (although being ambi-eyed is far more common than being ambidextrous). The image from the scope is always more important than the surroundings, so **marksmen with both eyes open always shoot better when they look through the scope with their dominant eye.**

To determine eye dominance, there are a couple of tests available. The Miles Test (See Image 59, Pg. 93.) and the Porta Test (See Image 60, Pg. 93.) are both good options. It is not uncommon, however, to have equal dominance in both the left and right eyes; therefore, any eye-dominance test has a risk of generating a false positive.

It is crucial to understand that the dominant eye does not necessarily have clearer vision. Instead, it is the brain that prioritizes visual information from one eye. Notably, some individuals may have variable eye dominance, where the brain prioritizes one eye over the other based on external factors such as brightness or distance.

Cross-dominant marksmen, who are right-handed and left-eye-dominant or vice versa, must make a choice regarding which side to shoot from. While shooting with the dominant eye always makes processing the sight picture easier, most cross-dominant marksmen opt to use standard rifles designed for right-sided shooting. They therefore sacrifice better focusing on the sight picture for an easier experience using their equipment.

Learning to shoot with both eyes open involves training oneself to instinctively **ignore, rather than eliminate**, the visual input from the non-dominant eye. This is not intuitive at all, and a marksman must master shooting with only one eye open before advancing to two eyes open. At the beginning of practicing drills with both eyes open, seeing double or experiencing blurry vision may be a common issue. However, this diminishes with time and practice.

One drill that can help train the eyes and brain to favor one eye when both are open involves holding a pen upright at arm's length. With the non-preferred eye closed, aim the pen's tip at an object located at least 1 m or yd away. Then, without moving any part of the body, open and close the non-favored eye. The goal is to maintain the sight picture of the favored eye. This exercise allows marksmen to recognize the appropriate visual input for the favored eye when shooting with both eyes open. This drill is essentially doing the Porta test repeatedly while attempting to force a result for either the right or the left eye.

11. Shooting Positions

A shooting position is the set position that a marksman places their body into to fire a rifle. It is distinct from body control because while body control is ongoing, a marksman only enters their position once before shooting and only needs to change their position if they want to grossly alter their point-of-aim.

Shooting positions must optimize for four qualities: recoil management, consistency, visibility, and stability. Differences between shooting positions mostly entail how marksmen position their lower body, as the upper body is mostly preoccupied with holding the rifle. (See Holding a Rifle, Pg. 42.)

Comparison of Shooting Positions

	Speed to Assume Height above Ground Field-of-View Mobility	Stability Protection Recoil Management Consistency
Prone	Worst	Best
Kneeling	Middle	Middle
Standing	Best	Worst

Image 63: There are different strengths and weaknesses to different shooting positions. Often, the best position is dictated by one's surroundings.

Recoil management is the ability of the marksman to return to their point-of-aim after they have fired a rifle. The ability to hold a rifle is described in an earlier section. (See Holding a Rifle, Pg. 42.) However, even if the force of shooting is properly transfered to the marksman's body, the marksman must be able to transfer that force to the ground, or else they might fall over. This holds particularly true for large-bore centerfire rifles designed to shoot powerful military and hunting calibers. To prepare for the recoil impulse, a marksman always orients their body in a way that allows for the maximum recoil absorption in a straight line. That is, a marksman properly directs the recoil force from the shoulder pocket backwards to the lower back, backwards to the hips, backwards to the feet, into the ground. The marksman also places the majority of their body weight forward of their center-of-gravity. This is already done when in the prone position (i.e., lying on one's stomach); however, when kneeling or standing, a marksman must lean forward slightly.

Consistency is the ability of the marksman to position their body in the same way every time they shoot. A consistent shooting position is crucial because it minimizes any deviations and variables that a marksman has to account for. That is, if a marksman can enter a great position automatically, they have more time and ability to consider other optimizations. Such consistency is gained through lots of practice, which creates muscle memory. With muscle memory, a marksman can perform complex actions automatically and precisely without any conscious effort.

Consistency with oneself is far more important than conforming to someone else's standards. The sports world is full of highly successful athletes who used unorthodox methods, but had simply practiced those methods so much that their expertise in that method allowed them to win. Moreover, everyone's body is different and has different proportions, so overly specific prescriptions would only apply to the "average" person anyway.

Rifle height above the ground is good in that it allows a marksman greater **visibility** above obstacles that may be present between the marksman and their target. However, it can also be dangerous because height allows a marksman to be better seen. Additionally, the lower the rifle is, the more stably a marksman can hold it. **Stability** is always good because it allows the marksman to have a more consistent natural point-of-aim (See Natural Point-of-Aim (NPA), Pg. 89.) and better recoil management.

Given all of the above, a marksman can contort their body in infinite ways while holding a rifle. And in fact, if a marksman is faced with imminent danger, they may have to assume uncomfortable, suboptimal positions to properly utilize their surroundings. That being said, there are three basic shooting positions from which all other positions can be derived: prone (i.e., on one's stomach), kneeling, and standing. All three positions have tradeoffs; for example, prone is the lowest and most stable, and standing is the highest and least stable, with kneeling in the middle. (See Image 63, Pg. 96.) (Benchrest "position" was not included here, as marksmen in this position simply sit down without having to orient their legs in any particular way.)

Before diving into the bulk of the text on shooting positions, it is important to note that an effective shooting position relies on **bone-to-muscle** support rather than bone-to-bone or muscle-to-muscle support. For example, placing the triceps on the knee (muscle-to-bone) is superior to resting the elbow on the knee (bone-to-bone). This is because two bones in contact slip off of each other, and two relaxed muscles in contact are not rigid enough to stay in consistent contact for extended periods. Therefore, the most rigid, stable connection is muscle-to-bone.

Of course, due to the limitation of the human body, using artificial supports greatly helps accuracy and precision. All positions can also be subdivided into "unsupported" and "supported." An unsupported position relies solely on the marksman to hold and support the rifle, while a supported position involves using techniques similar to those in unsupported positions but incorporates additional tools such as bags and sticks under the rifle to support and stabilize it better. (See Using Ground Support, Pg. 104.)

11.a Prone

The prone position is the most stable unsupported position because it minimizes the number of joints between the rifle and the ground (i.e., the force of recoil goes into the wrist joint, into the elbow joint, and then into the

Prone

Image 64: A Marine rifleman with Hotel Company, Weapons Training Battalion demonstrates a proper straight-leg **prone** firing position. This marksman is **maximizing contact with the ground**. Camp Lejeune, NC, 10 Feb 2015.

Image 65: An Iraqi soldier lays in the **prone** unsupported position. Al Taqaddum, Iraq, 07 Feb 2018. This marksman is **more comfortable** because lifting a leg allows the head to point forward without bending the neck as far back.

ground). **It is commonly used when maximum precision is required**, such as when zeroing a rifle. However, it is also the least mobile shooting position, requiring more time and effort to both assume and exit out of it. Additionally, objects such as tall grasses or small walls can obstruct the marksman's sightline. As a result, the prone position is not commonly used outside of controlled environments such as zeroing a rifle, shooting in a tournament, or a prepared sniper site.

To assume any prone position, a marksman first lies stomach down on the ground. In the **traditional prone position**, marksmen have their legs straight and spread wide apart. Many marksmen then dig their toes into the ground; however, the most flexible marksmen opt to turn their toes outward to touch the ground with their heels, using the inside of their feet and knees to create more contact with the ground. (See Image 64, Pg. 98.)

Marksmen who find it difficult to comfortably arch their back in the traditional prone position for extended periods can use a variation called the **frog-leg** or **sling prone position**. This alternative position requires less flexibility and strength, allowing shooters to maintain stability and accuracy without straining their backs. In the frog-leg prone variation, the marksman bends and raises their shooting-side knee. (See Image 65, Pg. 98.) Bending the knee requires less flexibility and raises the stomach off the ground, making breathing easier. Moreover, the frog-leg knee position allows for better utilization of a sling. (See Using a Sling, Pg. 102.) However, slings are less necessary in the prone position due to the availability of bipods and other supports such as backpacks and rocks.

In the prone position, the upper body must have both elbows on the ground for support. It is important to adjust the elbow spread according to the marksman's body proportions, as having the elbows too close together can cause a rifle to shift left or right, and too far apart can lower the rifle too far down. Additionally, small adjustments to the distance between the elbows are used to adjust the rifle up or down and thereby its point-of-aim.

Moving on to the support arm, the support bicep pushes the rifle into the marksman's shoulder to more effectively manage recoil. The rifle is held in such a way that the marksman's body is **positioned directly behind the gun**, not off to the side. That is, there is a straight line through the barrel, through the marksman's back, through their firing-side hip, ending in their feet. Some marksmen prefer to align the rifle more inward, with the barrel pointing backwards towards their belly button. Either way, ensuring that the barrel points parallel or into the spine and legs ensures that the force of recoil is evenly dispersed throughout their body, reducing the felt impact and allowing for faster follow-on shooting.

The support hand's main role here is to prevent the rifle from rotating and developing a cant. (See Eliminating Rifle Cant, Pg. 145.) Also, moving the support hand and support elbow allows for minor adjustments to the point-of-aim. Moving the support hand up and down the stock adjusts the point-of-aim down and up respectively, while moving the elbow left and right moves the point-of-aim left and right.

If there is a support attached to the fore-stock, such as a front sling swivel attachment point or bipod legs, a marksman can grab it and make a fist under the fore-stock. Placing that vertical fist on the ground with the fore-stock on top is called a Hawkins position. Since fists are not very tall, this position often requires additional support beneath the fist.

If a marksman would prefer not to use their support hand as a monopod, they could alternatively rest their rifle on its magazine (if it has one). However, **using the magazine as a monopod can be unstable or cause jamming** (failure to feed issues) due to putting pressure on and moving the magazine into an unintended position in the magazine well. It is crucial to test each magazine with the firearm during training to ensure its compatibility and effectiveness as a standalone rifle support.

11.b Kneeling

Kneeling serves as a midpoint between the lower and more stable prone position, and the higher and more mobile standing position. (See Standing, Pg. 102.) With three points-of-contact with the ground (i.e., two feet and one knee), **kneeling is ideal for quick shots** where some minor stability is sufficient or when brush, grass, or other vegetation would otherwise obstruct the marksman's sightline in a lower position.

The proper, unsupported kneeling position of a right-handed marksman involves planting their left foot firmly on the ground with their left leg bent at the knee at an approximate 90° angle. The top of the left foot is bent inward to increase side-to-side stability. The marksman then sits on their right foot with the right side of their buttock, with their right knee on the ground. (See Image 68, Pg. 101.) In combat situations where shooting around the sides of cover is necessary, marksmen must keep their right knee much closer to their left knee to keep it behind whatever cover the marksman is using.

Once in the kneeling position, the marksman's torso must be comfortably erect and relaxed, avoiding a slouched posture, with their body rotated 30 to 45 degrees towards the rifle. The marksman then rests their left triceps (muscle) on their knee (bone) by positioning the elbow (bone) just before (See Image 66, Pg. 101.), or just before (See Image 67, Pg. 101.), the knee, avoiding bone-to-bone contact and thereby any potential slipping. The support arm is directly under the rifle when the marksman looks down. If the arm is not properly aligned, the marksman may be overexerting their bicep muscles and not using their natural point-of-aim. (See Natural Point-of-Aim (NPA), Pg. 89.) (That is, overuse of supporting muscles can lead to muscle fatigue and trembling over time. This, in turn could potentially lead to wobbling of the barrel and thereby decrease the accuracy of the shot.)

Once the marksman has established a comfortable kneeling position and their natural point of aim, they can adjust the scope's vertical sight picture by moving their forward foot closer to or farther from their body. This allows for precise alignment without compromising stability.

Kneeling

Image 66: A Soldier fires from a hasty **kneeling** position. This marksman places his **elbow behind his knee**. Camp Adazi, Latvia, 16 Nov 2021.

Image 67: A Marine kneels with a sling. (See Using a Sling, Pg. 102.) He places his **elbow in front of his knee**. Camp Lejeune, NC, 10 Feb 2015.

Toes Flexed Up

Front of Toes on Ground

Side of Foot on Ground

Ankle Supported by a Shooting Bag

Top of Foot on Ground

Image 68: A lower rear foot is more stable, but slower to rise from and less comfortable. Inflexible marksmen may not even be able to enter a low position.

Image 69: A U.S. Marine **kneels** with his back leg vertical and his rear toes pointing outwards. This shooting position is **marginally higher but less stable** than a standard position. Drasi, Italy, 02 May 2018.

Image 70: A U.S. Marine **sits** cross-legged and digs his elbows into his legs. This shooting position is **marginally lower but more stable** than a standard position. Camp Hansen, Okinawa, Japan, 09 Jul 2018.

11.c Standing

The standing position (a.k.a., offhand position) **offers the marksman the best view of their surroundings** since the eyes and the rifle are higher than in any other position. However, standing is also very unstable. There are only two points of contact with the ground, the feet, and there are also many joints between the rifle and the ground that are prone to shifting (i.e., the vertebrae, the hips, the knees, the ankles, etc.). Therefore, standing is not viable beyond 300 m or yd.

A marksman can maximize their stability first by placing their feet shoulder-width apart, with the trigger-side foot slightly behind the support-side foot. The marksman's body must be relatively rigid to minimize shaking. Part of that rigidity comes from keeping the knees mostly straight, but not necessarily locked. (See Image 71, Pg. 103.)

In this position, the marksman shifts their weight so that most of their weight is on their front foot. This is done to prepare for recoil pushing them back. A slight lean forward into the buttstock can be helpful. (See Image 72, Pg. 103.) Then, the support hand grasps the rifle at the midway point of the fore-stock. If possible, a marksman braces their elbows against their ribs.

Although rigidity is important, relaxation is also important because the **inherent instability of standing amplifies the effect of any muscle shaking**. As a test of relaxation, a properly relaxed marksman is able to wiggle their toes. Both rigidity and relaxation can be best achieved by keeping the center-of-gravity over one's feet and by visualizing the path that recoil follows (i.e., through the shoulder, through the spine, through the hips, and into the ground).

To hold the rifle, a marksman uses the standard technique. (See Holding a Rifle, Pg. 42.) Beyond the standard, the support hand grips the handguard or rail system as far forward as it can while still remaining comfortable and safe. To ensure stability and accuracy while aiming, the support elbow is kept tucked under the rifle stock.

11.d Using a Sling

While some people use the terms "strap" and "sling" interchangeably, they are different items. A rifle strap is a tool that goes around the marksman's body and is attached to the rifle to allow the rifle to be carried. In contrast, a sling is a piece of equipment used to support and stabilize a rifle for more accurate shooting (although it can also be used to carry.) The difference in construction is that **a sling wraps around the back of the marksman's**

Standing

Image 71: A 101st Airborne Division sniper squad leader fires at a downrange target while **standing**. Tactical Base Gamberi, Eastern Afghanistan, 05 Jun 2015. His legs are wide, and an observer can follow the path that the recoil force follows: from the barrel, to the stock, to the shoulder, down the body, through the back leg, into the ground.

Image 72: A U.S. Air Force Staff Sergeant of the 615th Contingency Response Wing shoots while **standing**. Joint Base Lewis-McChord, WA, 26 Jul 2011. He leans into the stock while maintaining an upright position with his center-of-gravity over his legs. He holds a bipod leg. His legs are more forward-facing than the marksman to the left.

support-arm bicep or shoulder, pulling the rifle into the shoulder pocket when the marksman pulls their arm backwards. (See Image 73, Pg. 105.)

Because the sling forces the rifle into the shoulder pocket, the support forearm is freed from using pure muscle strength to hold the rifle up and back. Therefore, the support forearm can move into a different position that allows for more stability or fine movements. In contrast, a strap attaches around the body, does not pull the rifle backwards, and does not free the support arm. Although some marksmen claim that straps improve their accuracy, **there is scant evidence of straps increasing precision since any extra pressure that a strap applies to the rifle risks canting the rifle.** (See Eliminating Rifle Cant, Pg. 145.)

When using a sling, the support arm moves to a position that is as vertically under the rifle as possible. In the prone position, that means moving the support elbow closer to the chin. (See Image 74, Pg. 105.) In the kneeling position, that means moving the support elbow more inside the knee. (See Image 75, Pg. 105.) And in a standing position, that means placing the support elbow on the hip, and the support palm under the fore-stock. (See Image 76, Pg. 105.)

11.e Using Ground Support

In both unsupported and sling shooting, the marksman's body is between the rifle and the ground. Replacing that human support with a rigid, inanimate support that directly connects a rifle to the ground can increase stability by **bypassing the human element entirely**. Although not always possible, an ideal support allows a marksman to only direct and fire a rifle without using their hands to support it.

When a rifle is supported by something other than the marksman, the marksman is much more free to position their body in whatever way they want to, so long as the principles of shooting are maintained (i.e., stability and relaxation). This flexibility makes it difficult to define a standard supported shooting position. Of course, the starting point is whatever the equivalent unsupported shooting position would be. However to a large degree, the marksman can freely adjust from there. And just as often, they have no choice and are forced into whatever contortion allows them to form a stable support.

That being said, some generalizations can be made. First, ground supports usually **support rifles in the middle** so that they are balanced on the support. This has the side effect of increasing the chance of experiencing muzzle jump, as the rigid fulcrum causes more energy into the rotation about the support. (See Recoil and Muzzle Jump, Pg. 67.) To avoid exacerbating this effect and also avoid bending the rifle components generally, it is bad practice to depress a rifle down into its support with great force. It is also a good idea to support a rifle from the front and the back.

Second, the **best supports can hold a rifle on the target independently** of human intervention. For example, sandbags can conform to the rifle to hold it in position. And most tripods can be adjusted in a way that enables the rifle to be positioned in a particular way. That is not to say that non-independent supports are useless. A rifle using a branch, shelf, or bipod for support is much more stable than a rifle not using any support. However, non-independent supports are still not as rigid as independent supports.

Third, **supports must always be tested before use**. For example, shooting with an impromptu support from any position, such as a rock or a branch, requires actively searching for a spot on that support that allows the

Image 73: A Marine of the 2nd Recruit Training Battalion, tightens his **sling** around his support-arm biceps. Parris Island, SC, 26 Jun 2014. Slings are most often used by combat marksmen and are particularly common among Marines. This is because they are a lightweight way to add precision to shooting. Some hunters also use slings. However, for non-military competitions and hobby marksmen, more secure ground supports are used (if any support is used at all).

Image 74: Marines fire in the prone position while using **slings**. Ewa Beach, HI, 10 Feb 2014. Using a sling in the prone position is more stable than shooting unsupported, but less stable than resting the rifle on a sandbag. Slings can also severely restrict movement, which is important when shooting moving targets.

Image 75: A Marine with Weapons Training Battalion, Camp Lejeune, demonstrates a kneeling position with a web **sling**. Camp Lejeune, NC, 10 Feb 2015. The sling allows his elbow to be more interior and support the rifle more vertically from the bottom.

Image 76: A Marine fires his 8 kg (17 lb) M16A4 rifle with a non-slip leather rifle **sling**. Camp Lejeune, NC, 14 Apr 2014. His support arm is directly under the rifle. Some marksmen place their support elbow on their hip. This marksman uses his fist to support the rifle; although other marksmen overturn their fingers to the outside to support with their palm.

Image 77: The most basic supported position is to **put the fore-stock** of the rifle on local objects, such as dirt, rocks, or in this case, sandbags.

Image 78: There are commercial supports called "**sandsocks**," which are lightweight, portable cushions. They can be foam, pellets, or the like.

Image 79: Sandsocks are a staple of shooting because they are always useful for **interfacing the rifle** to a hard surface, no matter the position.

Image 80: Many rifles come with **built-in bipod legs** at the end of the fore-stock. Unlike sandsocks, bipods restrict rotation on uneven surfaces.

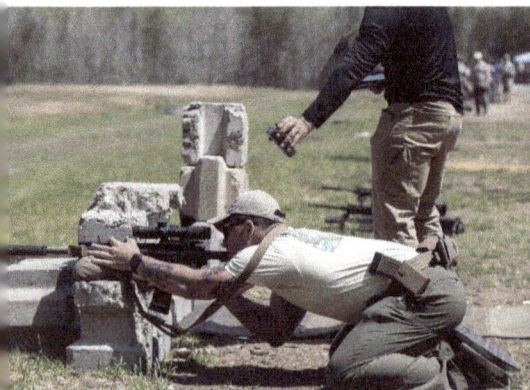

Image 81: If this marksman weren't being timed, he would be better off finding a location where his **center-of-gravity would be over his feet**.

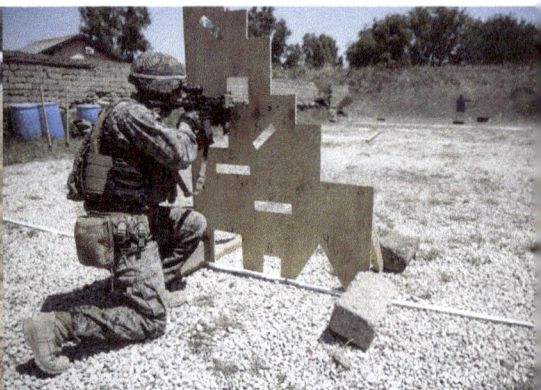

Image 82: This plywood sheet has been cut to allow marksmen to practice using platforms of different heights to **see what works for them**.

Image 83: A marksman uses a **vertical support** by pushing forward his support hand and using his thumb as a catch. This also works for trees.

Image 84: This marksman grabs the pallet not to stabilize the rifle, but to **take his weight off the rifle**. Forcing down a rifle decreases its precision.

Image 85: Literally **anything can be used as a support** if it is rigid and stable. This marksman uses the ladder he is standing on and a small bag.

Image 86: Supports are not always about precision. This marksman just wants to be **more comfortable**. All he is missing is a beverage!

Image 87: A rifle that is set on a tripod can **stand on its own**. Of course, the more stable the rifle, the more difficult it is to adjust.

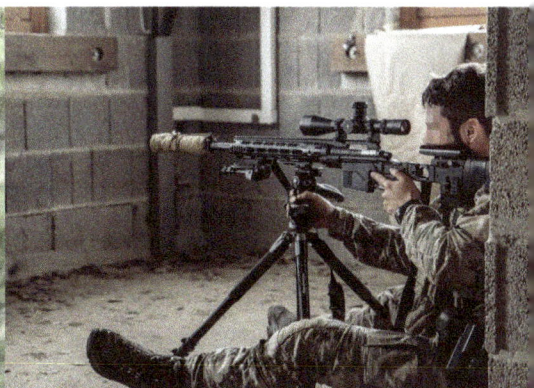

Image 88: Tripods are so stable that **marksmen can take up any position**. This marksman is comfortably sitting. All he needs to do is pull the trigger.

Image 89: The marksman on the right is aggressively leaning forward to **prepare for recoil** and quickly regain their sight picture.

Image 90: The marksman on the left is relatively upright because they are **confident in the stability** of their tripod. They are also wearing a ruck.

Image 91: There is a lot going on here. Sometimes, marksmen must **use whatever is available** to make their shot successful.

Image 92: Marksmen have always used sticks they find as support. The commercial equivalent is called **"shooting sticks,"** as seen here.

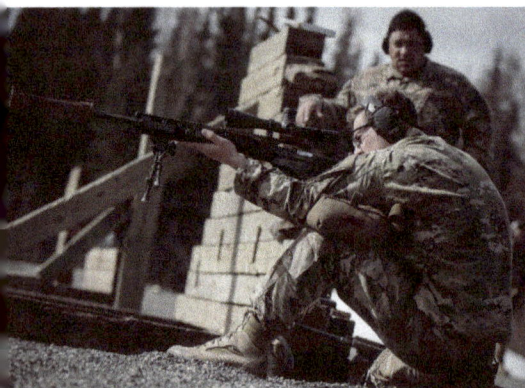

Image 93: Marksmen can **support their "unsupported" shooting positions**. This marksman is using a cushion to stabilize his kneeling.

Image 94: This marksman is using a tripod to stabilize his trigger-arm instead of the rifle directly. **This is less stable, but more adjustable**.

rifle to rest without falling. In contrast for example, if the rifle is accidentally put on something slippery, it could slip at any point in the shooting process. This is not always obvious; for example, rotted bark can look stable, but in fact it can slide right off the tree. Even manmade supports require calibration. Bipod and tripod legs must be the correct height for the marksman, and all the legs must be secured in their position, or else they could randomly slip and make the rifle fall over.

Using a support in the prone position (See Prone, Pg. 97.) is relatively straightforward: a marksman rests the rifle on the support and then uses their hands to operate the rifle in the same way as if the rifle were unsupported. This means that anything, even concrete blocks, can be a prone-position rifle support. That being said, there are two qualities that make for the ideal support: low weight and the ability to conform to the rifle. Both of these qualities are met by a "**sandsock.**" (It is so-called because marksmen would carry a sock with them to fill with sand or dirt from around their shooting position and then use the sock to support their rifle.) Nowadays, a rifle support cushion is always called a "sandsock" even though the fabric need not be a sock and the fill need not be sand. Sandsocks come in every conceivable shape and size and it is up to the marksman to decide what to use. For example, more expensive sandsocks are lighter and larger, but many marksmen don't care and simply place their smaller sandsock on local rocks.

In the supported kneeling position, marksmen can use impromptu supports such as a branch or low wall; or they can use manmade supports, such as shooting sticks, a bipod, or a tripod. Again, this means that **a marksman can position themself whichever way allows them to best take advantage of the support to aim the rifle.** For example, the marksman may not need to place their support elbow on their knee to stabilize the weapon. In fact, many marksmen prefer to keep their trigger-side knee instead to support their trigger hand. All this is to say that there are so many supports, body types, and shooting heights that not many generalizations can be made for supported kneeling shooting. For that reason, practicing likely scenarios is the most paramount for this shooting position.

Standing supported is much more stable than standing unsupported, and standing gives a marksman a much better view of their surroundings, so **shooting from a standing supported position is fairly common** when shooting at long ranges. There are rarely natural supports that are exactly the height of the marksman, so most standing supports are manmade, with the most common example being the tripod. Tripods can be adjusted to put the rifle at whatever height the marksman requires. Sometimes rifles have attachment points that connect the tripod directly to the rifle; other times, a marksman may need to place a sandsock on top of the tripod as a platform for the rifle to interface with.

Marksmen can choose their position behind a tripod or other support based on their own comfort and personal stability. The lower body can either stand straight up (See Image 90, Pg. 108.) or bend forward at the waist while leaning into the support (not leaning into the rifle itself!) (See Image 89, Pg. 108.) to better manage recoil. The body can face the support with each leg spread equally in either direction, creating a tripod-like formation between the support and each of their spread legs, or the body can be rotated with the trigger-side leg staggered behind the support-side leg. In either case, **marksmen usually position their feet slightly wider than shoulder-width apart.** And their legs are straight yet relaxed, with their toes pointed towards the target or outwards.

There are a few unorthodox ways to use support. If a marksman has a rifle with an attached strap or sling, they can attach the strap or sling to a support above them, such as a branch, a doorframe, or a window frame, and hang the rifle from above. Also, if there is a wall or side barrier (i.e., a side-support), a marksman can use their support arm to press the rifle into the side-support. This technique is easier if the side-support is on the trigger side; then the marksman can use their support arm to press the rifle into the side support and use their support thumb to support the rifle from the bottom. If the side-support is on the support side of the rifle, the marksman can grab the rifle in their hand as they would in an unsupported position. They then pull the rifle into the side-support, while simultaneously leaning the rifle into the side-support. (See Image 83, Pg. 107.)

12. Finding Distance to a Target

The distance from the marksman to the target is vital information for a marksman to know. It is an important factor in many formulas used in long-range shooting, such as determining how far a bullet travels, and thereby how much time gravity has to pull the bullet down towards the ground. (See Gravity, Drag, and Trajectory, Pg. 128.)

12.a Using Rangefinding Tools

Finding the distance between two objects is a common enough problem that special tools have been invented to do it, such as laser rangefinders and maps.

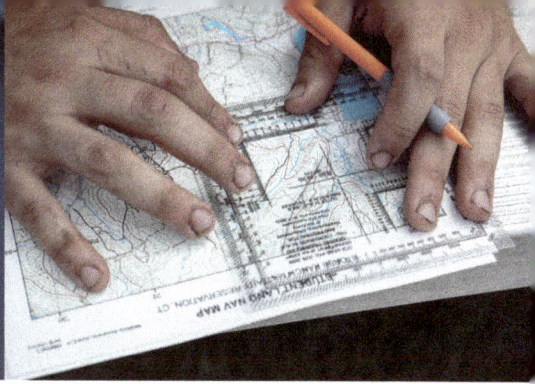

Rangefinding Tools

Image 95: Modern laser rangefinders are **small and light** enough to be carried and used as a part of any setup.

Image 96: While they have recently fallen out of use, **physical maps** were used for hundreds of years to find the distance between different locations.

Rangefinder Errors

Reflective Properties

Size of the Target

Oblique Surfaces

Atmospheric Conditions

Vibration

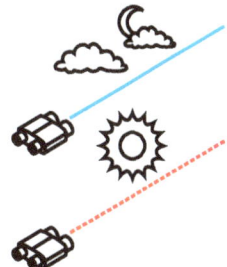

Lightning Conditions

Image 97: Laser rangefinders are very accurate, but can give poor readings in certain conditions. This means that **marksmen must take multiple readings,** and then ensure they are all consistent with each other to ensure precision.

Laser rangefinders are handheld devices that emit a laser beam that bounces off the target and then calculates the time it takes for the beam to return. The longer it takes for the beam to return, the farther away the target is. Because the hardware and software of laser rangefinders are very precise, even less expensive devices can provide distance measurements with errors significantly less than 1% at ranges of 1000 m or yd or less. More expensive rangefinders are of course more precise at farther distances. As with any equipment, marksmen must familiarize themselves with their specific device's instructions and limitations, and practice using it before using it for important events.

Despite their pervasiveness, ease-of-use, and relatively low cost, there are situations where using any kind of laser is inappropriate. The most obvious instance for which a laser rangefinder is inappropriate is during certain military operations. Military operations prohibit the emitting of any signal (including lasers) to the enemy, or else the enemy could detect the origin of the signal and shoot at that location. For these marksmen, specialty technologies such as stereo cameras can determine distance with input light alone. However, no matter the primary means of determining distance, it is recommended to have a rangefinder as a backup (if the marksman can afford one) because being unable to find an exact range is severely detrimental to shooting accuracy.

The more common scenario for not using a laser rangefinder is that some marksmen simply do not like relying on electronics. Such marksmen can use maps instead. **The map method** involves the marksman determining on any map: their own position, the target's position, and the distance between the two. A marksman can pull up Google Maps, measure the distance between them and their target, and use that distance to shoot accurately.

To derive the most accurate measurements from a map, a marksman must exactly know their own position and that of their target. They can determine their own location using a GPS device. Most modern smartphones can output GPS coordinates. If a marksman is able to visit their target, they can store the GPS coordinates of their target to use it as well. If both GPS coordinates are known, the marksman may not even need a map, and can instead use a GPS difference calculator.

Milling Circle

1 Circumference = 360 Degrees
1 Degree = 60 Minutes-of-Angle (MOA)
1 Circumference = 21,600 MOA (360*60)
1 Circumference ≈ 6,283 Mils (Milliradians)
1 Mil ≈ 3.5 Minutes
1 Degree ≈ 17 Mils

Circle Radius = Distance to Target

1 Circumference = Radius * (2π ≈ 6.283)
1 Radius = 1,000 Mils ≈ 3,438 MOA
Target Height ≈ 8° ≈ 136 Mils ≈ 480 MOA
Radius ≈ 7.2 Target Heights
Distance to Target ≈ 7.2 Target Heights

Image 98: For milling, the marksman is always at the center of a giant imaginary circle. The radius of the circle is the distance to the target. The target occupies a certain percentage of the circle's circumference, which is denoted in either MOA or mils. For example, **in this diagram, the deer occupies 8 degrees, 480 MOA, and 136 mils.** A deer that is 1 m tall would have to be 7.2 m away to appear this large.

12.b Using a Standard Reticle (Milling)

The traditional way that snipers determined distance was through a method called "milling." Milling is not as precise as using a laser rangefinder nor reading a map, but it requires no electronics and is much faster than using a map. Before continuing, to mil (derived from the word "milliradian"), a marksman must be familiar with angular measurements. (See Angular

Object Height Examples

Object:	Height:
Door	2.0 m (6 ft, 7 in)
House Wall, External	2.6 m (8 ft, 4 in)
High House Fence	1.8 m (5 ft, 11 in)
Picket Fence	0.9 m (2 ft, 11 in)
Farm Fence and Gate	1.1 m (3 ft, 7 in)
House Window Bedroom	1.3 m (4 ft, 3 in)
House Window Bathroom	1.0 m (3 ft, 3 in)
Toyota Hilux	1.8 m (5 ft, 11 in)
Average Car	1.4 m (4 ft, 7 in)
Concrete Block	0.2 m (0 ft, 8 in)
Brick	0.1 m (0 ft, 4 in)
Garage Door, Single	2.0 m (6 ft, 7 in)
Road Sign	0.9 m (2 ft, 9 in)
Stop Sign from Ground to top	2.1 m (7 ft, 0 in)

Image 99: Observers who mil are familiar with the height of common objects. Objects with global standard height, such as concrete blocks, Hiluxes, or oil barrels, are more useful than customizable objects, such as fences.

Distance, Pg. 237.) and how to use a standard reticle (See Standard Reticles, Pg. 132.).

Milling works by using a formula with three variables. Any two variables out of the three can be used to determine the third. The three variables are:

1) Angular distance that an object occupies in a sight picture. This is measured (a.k.a., milled) with a reticle. For example, if a deer is between 9 hashmarks (i.e., occupies 8 subtensions) in a reticle that uses mils, its angular distance is 8 mils. (See Image 98, Pg. 113.)

   Angular distance is always an observed variable and can be used with target height to determine target distance, or with target distance to determine target height.

   Understanding that mils can be used to measure angles is confusing because most students are only taught to measure angles in degrees; that is, a circle has 360°. However, minutes-of-angle (MOA) (1 degree has

Deer Height Examples

Species:	Average height at shoulder:
Roe Deer	0.70 m (2 ft, 4 in)
Sika Deer	0.83 m (2 ft, 9 in)
Fallow Deer	0.90 m (2 ft, 11 in)
White-tailed Deer	0.93 m (3 ft, 1 in)
Mule Deer	0.93 m (3 ft, 1 in)
Reindeer (Caribou)	1.18 m (3 ft, 10 in)
Red Deer	1.22 m (4 ft, 0 in)
Elk (Wapiti)	1.35 m (4 ft, 5 in)
Moose	1.75 m (5 ft, 9 in)

Image 100: Hunters who mil are familiar with the height of their prey.

60 minutes) and mils can also be used to measure angles, just like meters and yards are both different measures of linear distance. It may help to know that a mil equals about 1/17 of a degree or about 3.5 MOA.

2) Target linear height (a.k.a., "height") is how tall the target is in meters or yards. Height is almost always used because reticles have vertical hashmarks and heights are easy to find. However, any observable dimension can be used, including width. (See Image 99, Pg. 114.) (See Image 100, Pg. 115.)

3) Target linear distance (a.k.a., "distance") is how far the target is to the marksman in meters or yards. (The process of milling is easier in meters because height is usually measured in feet and inches, which is a pain to convert to yards.)

The concept of milling is best explained using the geometric concept of a giant circle, with the marksman in the middle. The distance to the target is the circle's radius. The target height occupies an arc of the circle (an arc is a length on a circle's perimeter between two points). The angular distance of this arc can be measured in mils or MOA. (See Image 98, Pg. 113.)

Milling exploits a geometric fact of circles: **the linear relationship between a specified circle arc (i.e., the milled angular distance) and the circle's radius is fixed**. That is, they expand and shrink at the same rate, no matter how they are scaled. (See Image 98, Pg. 113.)

Image 101: This target occupies 2.6 mils on the reticle. From the center, it extends a little more than 2 dots up, and a little more than 0.5 dots down. E-type targets (i.e., the targets in the picture) are standardized to 101.6 cm (40 in) tall. Using the formula:

(D to T) = (T Height) ÷ (T Angle in Mils) × 1,000

(D to T) = 101.6 cm ÷ 2.6 mils × 1,000

(Distance to Target) = 39076 cm ≈ 391 m (428 yd)

Image 102: This shed occupies 2.5 mils on the reticle. Assume the shed is found to be 1200 m (1312 yd) away with a laser rangefinder. To find the shed height using the formula:

(T Height) = (D to T) × (T Angle in Mils) ÷ 1,000

(T Height) = 1200 m × 2.5 mils ÷ 1,000

(Target Height) = 3 m (3.3 yd)

This phenomenon can also be explained by the concept that as objects move farther away, they appear smaller because they occupy less angular distance in the marksman's field-of-vision. For example, if the distance to the target is doubled, the apparent size is halved accordingly, and vice versa. For example, a target at 100 m would appear twice as large if moved to 50 m, and half as large if moved to 200 m.

Person Height to Distance Table

Height of person in mils	Distance of a standing, 168 cm (5 ft, 6 in) person	Distance of a standing, 183 cm (6 ft, 0 in) person	Distance of a kneeling, 168 cm (5 ft, 6 in) person	Distance of a kneeling, 183 cm (6 ft, 0 in) person
1.0	1646 m (1800 yd)	1829 m (2000 yd)	823 m (900 yd)	914 m (1000 yd)
1.5	1097 m (1200 yd)	1219 m (1333 yd)	549 m (600 yd)	609 m (666 yd)
2.0	823 m (900 yd)	914 m (1000 yd)	411 m (450 yd)	457 m (500 yd)
2.5	686 m (750 yd)	732 m (800 yd)	343 m (375 yd)	366 m (400 yd)
3.0	549 m (600 yd)	609 m (666 yd)	274 m (300 yd)	304 m (333 yd)
3.5	470 m (514 yd)	522 m (571 yd)	235 m (257 yd)	262 m (286 yd)
4.0	411 m (450 yd)	457 m (500 yd)	206 m (225 yd)	229 m (250 yd)
4.5	366 m (400 yd)	406 m (444 yd)	183 m (200 yd)	203 m (222 yd)
5.0	329 m (360 yd)	366 m (400 yd)	165 m (180 yd)	183 m (200 yd)
5.5	299 m (327 yd)	333 m (364 yd)	151 m (165 yd)	166 m (182 yd)
6.0	274 m (300 yd)	304 m (333 yd)	137 m (150 yd)	153 m (167 yd)
6.5	253 m (277 yd)	282 m (308 yd)	127 m (139 yd)	141 m (154 yd)

Image 103: When milling, getting the height of a target wrong can lead to a significant error in the estimate of the distance to the target. However, these averages can be used if there is an acceptable ranging error of 20%, since almost all human heights are within 20% of one of the two categories.

To calculate the distance to the target using a mil reticle requires the following:

Variables:

▸ **Mil constant** = *1,000*
▸ **MOA constant** = *3,438*
▸ **Target Angle** *(T Angle)* = The angle of the target as measured in a reticle in mils or MOA.
▸ **Distance to Target** *(D to T)* = Distance from the marksman to the target. Both *(D to T)* and *(T Height)* must use the same linear units, such as meters or yards.
▸ **Target Height** *(T Height)* = The height of the target. (Again, while nominally height, this can actually be any dimension that is perpendicular to the marksman, and is measurable or known.)

Formulas for mils and MOA to **calculate the distance to the target**, *(D to T)*, with a known *(T Height)* and an observed *(T Angle)* are:

- *(D to T)* = *(T Height)* ÷ *(T Angle in Mils)* × *1,000*
- *(D to T)* = *(T Height)* ÷ *(T Angle in MOA)* × *3,438*
- (See Image 101, Pg. 116.)

The milling formula can be reversed to determine an object's dimensions, an ability that a rangefinder does not have. With a laser rangefinder and accurate milling, the dimensions found can be extremely precise. The formulas for mils and MOA to **calculate an object's dimensions**, *(T Height)*, with a known *(D to T)* and an observed *(T Angle)* are:

- *(T Height)* = *(D to T)* × *(T Angle in Mils)* ÷ *1,000*
- *(T Height)* = *(D to T)* × *(T Angle in MOA)* ÷ *3,438*
- (See Image 102, Pg. 116.)

Observing a target's angular distance (i.e., its height in mils or MOA) requires considerable precision while holding the reticle as steadily as possible. An observer must use decimal-point fractions, and consider, for example, whether a height is precisely 1.0 or 1.1 mils. This is because being off by 10% in the measurement of the target's angle translates to being 10% off in the target's distance. Such precision is vital because at far distances, a bullet can drop fast due to gravity. For example, a bullet may drop over a meter from 900 to 1000 m or yd. That means a 10% error would cause a marksman to aim over a meter higher or lower than their desired point-of-impact.

This is especially relevant when the linear height of the target is an estimate, such as with human height. **While "standard" height is a single number, actual humans have a range of heights.** (See Image 103, Pg. 117.)

To mitigate the difficulty of measuring a target's height with a reticle, observers use the edges of hashmarks rather than their centers. This is particularly useful when using mil-dot reticles, where it is easier to align a specific edge rather than trying to keep the edge of the target precisely in the middle of the dot. (See Image 115, Pg. 133.) Another method to improve estimates is to calculate from multiple dimensions, such as measuring height and width instead of height alone. Similarly, if the heights of multiple objects at the same distance are known, measuring all of the objects in the area and calculating their average provides a more accurate estimation.

Targets are not always aligned perpendicularly with the marksman's sightline and can appear at an angle. Sometimes a marksman may be looking down at a target from a higher vantage point; other times they may observe a target from the side. Angled targets appear shorter or narrower than their actual size. If the angle of the target is known, trigonometry can be used to determine the actual angle. However, since the actual angle is rarely known, estimations are usually more practical. For example, if a perpendicular target appears to be 100% of its true width, a target turned to 30° appears as 87% of its true width, 45° appears as 71%, and 60° appears as 50%. For example, if a target appears to be 0.7 mils wide but is angled at 45 degrees, its true width is approximately 1 mil (i.e., 0.7 ÷ 71% ≈ 1). This relationship is determined by the cosine of the angle. (See Image 178, Pg. 210.)

In stressful situations, it is inadvisable to perform any calculations. To avoid doing algebra under stress, marksmen can carry a small calculator and premade conversion cards. For example, the Mildot Master (a cheap, brand-name product) is available in metric (height in m ÷ distance in m), imperial (height in ft ÷ distance in yd), and mixed (height in ft ÷ distance in m). (See Image 177, Pg. 209.)

12.c Using Hands

To measure angular distance. An observer can use their **fully outstretched** hands and fingers. When a body part is fully outstretched it always takes up a set amount of angular distance (i.e., mils or MOA) in the observer's vision. With one eye closed, the marksman then aligns the bottom of their body part with the base of the target.

While each body part can only be used to measure one angular distance, an observer may use their multiple body parts, such as their thumbnail, the height of all five fingers, a few finger joints, and even parts of a pen, to determine the distance to a standing target. If the target is larger than the body part, the marksman can simply use a different, larger body part. Or if they are more comfortable with one body part, often the thumb, they can stack thumbnail heights together by moving their thumbnail up in intervals. Conversely if the target is smaller, the marksman can use a fractional percent of the body part, such as one-half or one-third.

To use the thumbnail as an example, an average, fully-outstretched thumb occupies approximately 30 mils or about 120 MOA (**these numbers are just examples, and heavily depend on an individual's personal body proportions**). Therefore, if an object appears to be the same width of a thumbnail, it would be 30 mils wide. Angular distance can be used as a relationship between the distance to the target and the height or width of the target. In this case, a 30-mil-tall object would be 30 m tall at 1000 m away,

Using Hands to Measure Angles

30 mils   15 mils   10 mils

30 mils   70 mils   100 mils   125 mils

180 mils   300 mils

Image 104: Body parts can measure angular distance in either mils or MOA. First, the hand is always outstretched to ensure every measure is consistent. Then the width of the body part (e.g., thumb or finger(s)) is compared against a reticle. Once the conversion factor is known (e.g., 30 mils to one thumbnail), it never changes. Thereafter, a marksman can use that outstretched body part as a ruler. A marksman can get the measure of multiple body parts for quicker measurements. **Every marksman's body parts are different, and the above are only examples.** (The top right shows thumbs, while the middle left shows an index finger.)

15 m tall at 500 m away, or 3 m tall at 100 m away. That is, given the same angular measure, an object at half the distance would be half the height, and one at double the distance would be double the height.

This method uses the same formulas as milling, but with a different constant at the end. That is, formulas to calculate *(D to T)* with a known *(T Height)* and an observed *(T Angle)* may include:

- ▸ For mils: *(D to T) = (T Height) ÷ (T Angle in Mils) × 1,000*
- ▸ For MOA: *(D to T) = (T Height) ÷ (T Angle in MOA) × 3,438*
- ▸ For pinkies: *(D to T) = (T Height) ÷ (T Angle in pinkies) × 70*
- ▸ For thumbs: *(D to T) = (T Height) ÷ (T Angle in thumbs) × 33*

The constant for thumbs (33), for example, was found by dividing the constant for mils (1000) by the assumed mils-per-thumb ratio (30) so that $1000 ÷ 30 ≈ 33$. The same math is done to find the constant for any object.

For an observer to most effectively use their hands to mil, they must:

1) know and be able to use the angular distance of many body parts of their hands, such as each finger and fingernail width;
2) be able to accurately estimate midpoints for each body part for small measurements, such as half of a pinky finger; and
3) use a measure that is as repeatable as possible, for example fully outstretching the hand every time so that the body part always occupies the same space relative to the marksman.

12.d Using Estimation

Sometimes a tool is simply not available; however more often, marksmen simply enjoy learning the skill of estimation. Being able to quickly and accurately estimate distance to a target can even a fun skill to practice in everyday life.

Known-Distance Method – This technique involves comparing the space the marksman is trying to measure to a space they already know the distance of and are familiar with. A common example for Americans is comparing a distance to a gridiron football field. People from other countries may use a soccer pitch. Visual references such as fence posts, power pylons, and telephone poles are especially helpful because they are often spaced at regular intervals that are standard to the local area. For example, a marksman might determine that the distance between their position and the target is equivalent to two football fields or eight power poles away.

Known-distance estimates can be fairly accurate up to around 400 m or yd. However, because a marksman must visualize what their reference distance looks like and also how many of those reference distances it

Using Estimation

Step 1: Estimate the halfway point to target.

600m (656 yd)

300m (328 yd)

200m (219 yd)

100 m (109 yd)

Step 2: Estimate the number of fields to the halfway point.
Step 3: Add the lengths of the fields and double it.

Image 105: A marksman can estimate the distance to the target by combining multiple methods. In this diagram, the marksman estimates the halfway point to the target by using the mid-distance method (Step 1 and Step 3) and the known-distance method (Step 2). To get an even more accurate estimate, the marksman can estimate the distance using the appearance-of-things method (Step 4), and average the results from Step 3 and Step 4 using the averaging method (Step 5).

would take to occupy the space, the accuracy of this method strongly decreases over longer distances. This method used to be much more common; however, with the prevalence of GPS today, fewer people have memorized various distances in their life to draw upon.

Mid-Distance Method – One way of extending the distance that an estimate can be is to estimate the halfway point between the marksman and the target and then estimate the distance to that. Alternatively, the estimator can determine the distance from the midpoint to the target. Then the estimator doubles their result to find the total distance from themselves to the target. (See Image 105, Pg. 122.)

This method is the most effective when only part of the distance to the target can be easily estimated. For example, a hunter may know the distance from themselves to a pond. If a target is across the pond, they can estimate the partial distance to the pond, and then estimate how many of those partial distances comprise the whole distance.

Appearance-of-Things Method – An observer using this method determines a distance based on how the characteristics of something (e.g., a person, weapon, or vehicle) appear to them. (See Image 106, Pg. 123.) To use

Appearance-of-Things Example

Distance to target:	Common appearance for 20/20 vision:
50 m or yd	Whites of the eyes disappear.
100 m or yd	Blurry eyebrows, eye color, or small wrinkles.
200 m or yd	Blurry hair texture, facial expression, and posture.
300 m or yd	Blurry facial detail, but face color is visible.
400 m or yd	Blurry body detail, but body outline is visible.
500 m or yd	Body shape is vaguely present, but neck disappears.
600 m or yd	Body shape appears wedge-like.

Image 106: How things appear varies significantly from one person to another, so marksmen must experiment at each distance to find what works for them.

this method accurately, marksmen must be very familiar with their own visual observations. Since visibility can vary between individuals due to factors such as biology and experience, each marksman should develop their own visual references instead of relying on others'. Marksmen wanting to develop an expert skill set must create a mental catalog of associations between the appearances of objects at varying distances.

Averaging Method – With the averaging method (a.k.a., "bracketing method") the marksman estimates the minimum and maximum distance away the target could be and then averages them together. (I.e., Average Distance = (Min Distance + Max Distance) ÷ 2) For example, if the marksman estimates a target to be between 440 and 560 m or yd away, then the average distance would be 500 m or yd. (I.e., 500 = (440 + 560) ÷ 2.)

Averaging is most effective when using estimates from different sources, such as different people. A common example is averaging the distance between the estimates from a marksman and a spotter on a team. Averaging can also be done by combining estimates from different techniques, such as averaging a known-distance estimate with an appearance-of-things estimate.

12.e Perception Errors

The various methods of estimating distances rely heavily on the marksman's perception of their surroundings. Therefore, if the surroundings change then so do the marksman's perceptions and estimations as well, even if the true

Vertical Error from Bad Ranging

Approximate Distance to Target	Approximate Bullet-Drop over 10 m or yd of Horizontal Travel for an Average Rifle System
200 m or yd	0.2 cm (0.1 in)
300 m or yd	1.6 cm (0.6 in)
400 m or yd	3.6 cm (1.4 in)
500 m or yd	6.1 cm (2.4 in)
600 m or yd	8.9 cm (3.5 in)
700 m or yd	12 cm (4.8 in)
800 m or yd	16 cm (6.4 in)
900 m or yd	21 cm (8.2 in)
1000 m or yd	26 cm (10 in)

Image 107: Bullets fall faster as they move farther from the marksman. While the amount of drop may be easy to predict if a precise distance to target is known, even being 10 m or yd off of the actual distance can lead to severe vertical inaccuracy at long distances. For example, this table shows that a bullet would impact about 26 cm (10 in) higher on a target at 990 m or yd compared to a target at 1000 m or yd.

distances do not change. In other words, **perception can influence whether an object appears to be closer or farther away than it actually is**.

There are two distinct types of perception errors that a marksman may experience. The first occurs when the target blends in with its surroundings, and there are other surrounding points-of-reference around the target. In these instances, the target may **appear farther away** (See Image 108, Pg. 125.):

▸ There is an observable depression with visible ground between the observer and the target.
▸ The surrounding environment is large and dominating, such as near large buildings, cliff faces, or large trees.
▸ The marksman is lower than the target.
▸ Targets blend into the background or have irregular outlines, such as a deer in the grass.
▸ The sun is behind the target.
▸ Obscurity from low light (dusk or dawn) or weather (smoke, fog, or rain).
▸ Moving objects in the environment are scarce, slow, or both.

Image 108: The windows in these two images have been roughly standardized to the same size. Despite that, the building in this image **appears farther away** because: the ground is clearly visible all the way to the target; the target is large and imposing; the viewer has a low vantage point; the building outline is irregular, and the sun is behind the target at dusk, thereby lowering the contrast.

Image 109: These buildings **appear closer** for all the opposite reasons of the above image.

Conversely, the second type of perception error can occur when the target is highly visible and the surrounding environment is largely vacant. Then, the target may **appear closer** (See Image 109, Pg. 125.):

▸ There is an unobservable depression between the marksman and the target.
▸ The surrounding environment is flat and uniform, like a desert plain.
▸ The marksman is elevated in relation to the target.
▸ The target contrasts with its background and has a regular outline, such as a house against the sky.
▸ The sun is behind the marksman.
▸ There is full sunlight.
▸ Surrounding objects appear to move fast, such as cars or leaves in the wind.

Perception errors can strongly affect the marksman's accuracy: estimates that are too short cause bullets to hit low, and estimations that are too long cause bullets to hit high. (See Image 107, Pg. 124.)

Overcoming perception errors in shooting requires practice. One way to start learning is by becoming familiar with how various objects appear at a standard distance (e.g., at 100 m or yd) under the same conditions. Alternatively, one can overcome perception error by becoming familiar with the same object under various conditions (e.g., raining versus sunny). Natural markers found in the shooting environment, such as distinctive trees in open areas, clusters of rocks in the distance, or patches of grass on open terrain, make for excellent practice objects.

Marksmen begin their practice by estimating distances to random objects in their surroundings and then confirming the distance with a laser rangefinder or GPS device. Once marksmen feel comfortable identifying objects at the standard distance they chose (e.g., 100 m or yd), they can progress to range distances that are progressively farther away from this standard distance. Estimating distances can even become a friendly competition with a partner to see who can come closest to the correct distance.

13. Using Scopes for Long-Distance Shooting

Scopes are essential tools for long-range marksmen. (See Scopes, Pg. 20.) They are a type of sight, which is any device used to assist in precise visual alignment (i.e., aiming) of weapons. To aim with a scope, a marksman uses the

Image 110: A Marine radio operator with Battalion Landing Team 1st Bn., 4th Marines, 11th Marine Expeditionary Unit, looks through his M8541A optic. This "3-12x50" scope has a magnification range of 3 times to 12 times the naked eye, and has an objective lens 50 mm in diameter. USS Somerset, At Sea 12 Jan 2017.

scope's internal markings called the "reticle." Because aiming is the primary purpose of a sight, having that reticle is the primary purpose of a scope.

However, what sets scopes apart from other sights is their ability to allow marksmen to overcome two challenges specific to long-range shooting: small targets and gravity. For the former, scopes magnify the image that the marksman sees. Almost all long-range scopes have the ability to make an image at least eight times larger.

For the latter, gravity drops bullets down to Earth. In fact, the distance that a bullet falls as it flies is called its "bullet-drop." To compensate for bullet-drop from gravity, scopes have special features on their reticles and turrets that allow marksmen to readjust their point-of-aim upwards to compensate.

13.a Magnification

Humans can make finer adjustments to their aim than what the naked eye can perceive; therefore, shooting accuracy is limited by the human ability to see. A key advantage of using a scope is that it magnifies the target, allowing the marksman to see targets in greater detail and therefore utilize more precise aiming adjustments onto smaller targets. (See Target Definition, Pg. 55.)

However, **the drawback to using magnification is that it restricts the marksman's field-of-view**. That is because the ability of the eye to receive information is limited, and any enlargement of one image necessarily covers another part of a person's overall sight. The restricted field-of-view causes various problems, each with their own solution.

First, limiting a marksman's sight can cause the marksman to lose awareness of the situation around them (a.k.a., situational awareness or SA) and make it difficult to track moving targets. These external problems can be mitigated by keeping the second eye open while using the scope. This will help create greater situational awareness since the marksman would be able to discern threats via their peripheral vision. (See Shooting with Both Eyes Open and Eye Dominance, Pg. 94.) However, this requires a lot of practice since the mind tends to concentrate on one image at one time. That's why many marksmen prefer to keep one eye closed while long-range shooting.

Second, a magnified image by definition enlarges an area that is small relative to one's whole vision. Finding a target area or interpreting a mirage pattern (optical distortions of air) (See Simple Mirage, Pg. 191.) with only a small, bounded area takes more time than if a marksman were to use their naked vision. Enlarging a small area also has the side effect of amplifying any trembling of the sight picture. Psychologically, a shakier image can lead to increased trigger jerking and excessive focus on the shaking, even if the actual shaking remains consistent across different levels of magnification. Therefore, scopes are typically carried and used to search at low-magnification settings and adjusted as needed. (See Finding a Target, Pg. 57.)

Third, limiting or altering the image seen by one eye makes it difficult for a marksman to perceive and interpret details that are to the side of the target. For example, if a marksman wants to know the wind speed, they may have to look away from their scope to check a wind indicator; whereas they could see both the target and the indicator at the same time without a scope. There is no solution here but practice, as eventually marksmen learn to be as accurate as possible after they have acquired the necessary information to shoot.

Finally, high magnification can result in mechanical or optical limitations. For example, some manufacturers design scopes with first focal plane reticles that can hide the bottom of the reticle when at maximum magnification settings. (See Image 119, Pg. 137.) Scopes with a small objective lens may show a dimmer sight picture in low-light conditions. And magnification shrinks the eye box. (See Image 28, Pg. 49.) The solution for these issues is easy, although expensive. Marksmen must buy a better scope with a larger objective lens. Although this has a second-order effect of making the scope heavier, so **marksmen always have to balance their preferences for picture-quality, cost, and weight when choosing a scope.** All too often, marksmen overestimate how much magnification they need to purchase.

13.b Gravity, Drag, and Trajectory

Gravity is a pseudo-force that effectively pulls objects together. From the moment a bullet exits the muzzle, it reacts to the Earth's gravity, which accelerates the bullet downward at a rate of 9.81 m/s² (10.7 yd/s²). That is, gravity causes bullets to move vertically as they travel, no matter how straight

Actual Trajectory of Bullets

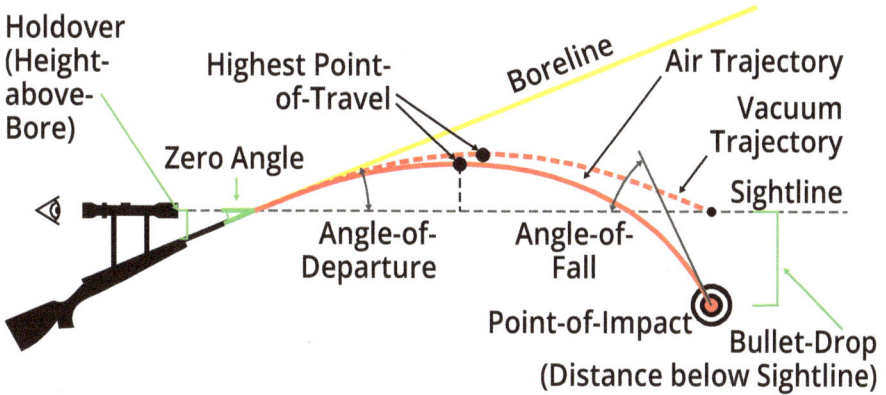

Holdover (Height-above-Bore)

Highest Point-of-Travel

Zero Angle

Boreline

Air Trajectory

Vacuum Trajectory

Sightline

Angle-of-Departure

Angle-of-Fall

Point-of-Impact

Bullet-Drop (Distance below Sightline)

Image 111: Bullets do not follow a straight line, even though they may appear to follow one at short distances. In a vacuum, gravity would make them follow a parabola (i.e., a curve that mirrors about the apex). However in air, they slow down and have a **steeper angle-of-fall than angle-of-departure**. What bullets never do, is rise above the line-of-departure, which is identical to the boreline.

Trajectory Vocabulary

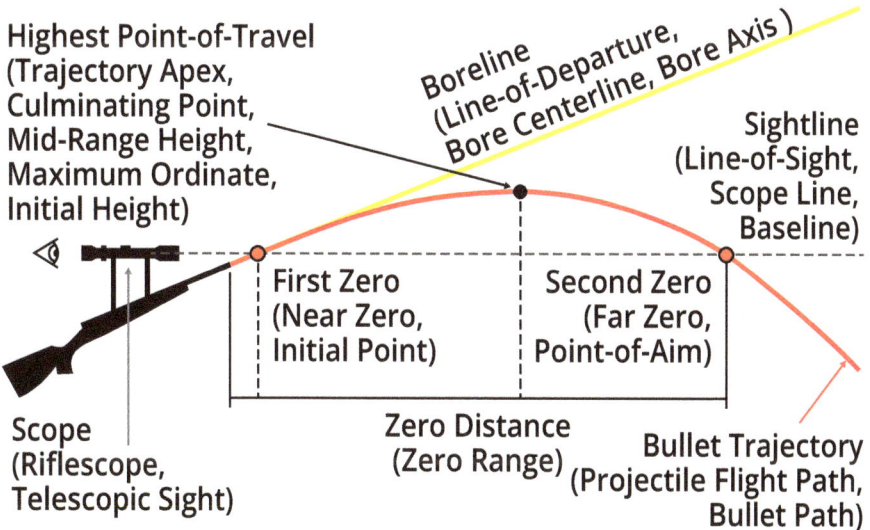

Highest Point-of-Travel (Trajectory Apex, Culminating Point, Mid-Range Height, Maximum Ordinate, Initial Height)

Boreline (Line-of-Departure, Bore Centerline, Bore Axis)

Sightline (Line-of-Sight, Scope Line, Baseline)

First Zero (Near Zero, Initial Point)

Second Zero (Far Zero, Point-of-Aim)

Scope (Riflescope, Telescopic Sight)

Zero Distance (Zero Range)

Bullet Trajectory (Projectile Flight Path, Bullet Path)

Image 112: Because firearm culture has always been decentralized, many concepts have **multiple terms or phrases that mean the same thing**. For example, the sightline can also be called the sightline, scope line, or baseline. There is no good solution but to be familiar with all the terms.

a marksman fires. Therefore, bullets impact at different elevations on targets at different distances. **Scopes help marksmen predict a bullet's elevation on a target based on the distance to the target.** (See Image 116, Pg. 134.)

Acceleration is different from movement. Bullets do initially move upwards because gun barrels are angled upward to ensure that the bullet's path-of-travel (i.e., trajectory) intersects with the scope's sightline. (See Image 111, Pg. 129.) However, after bullets exit the muzzle, they universally accelerate downward, even if they are initially moving upwards. (The lift force from aerodynamics may provide some initial upward force, but it is negligible compared to gravity.) This means that bullets never move higher than the imaginary line that points out of a barrel (i.e., the "boreline").

In a vacuum, gravity would cause the bullet to travel in a curved (parabolic) path. However, since bullets are not fired in a vacuum, they are affected by air resistance. The air resistance is the result of bullets having to push air out of the way to move forward, and it is formally called "**aerodynamic force.**" (See Image 114, Pg. 131.)

Aerodynamic force is a consequence of Newton's law of equal and opposite reactions; when the bullet pushes on the air, the air pushes back on the bullet. That is, aerodynamic force only exists when an object (e.g., a bullet) moves through a fluid (e.g., the air). While gravity is a consistent force, the aerodynamic force is variable. For example, air density can vary, resulting in thicker or thinner air, leading to higher or lower air resistance. (See Air Density, Pg. 189.)

The component of aerodynamic force that slows a bullet down is called the "**drag force.**" (Contrary to its name, drag does not pull objects. It is actually the air in front of an object that pushes objects backwards.) In contrast, the component that pushes a bullet off its path-of-travel is called the "**lift force.**"

Rifle bullets are designed to minimize the effects of aerodynamic force by slicing through the air with a pointed tip and a thin, long body. Therefore, bullets only experience relatively little drag as they move forward. However, any loss in speed is permanent and cumulative. Therefore, drag effectively accelerates bullets backward, decreasing velocity, even as they move forward. In other words, **bullets slow down as they travel forward.**

As a bullet slows, gravity has more time to pull it downward over a given distance. For example, a bullet may take twice as much time to travel from 900 to 1,000 m or yd as the time it would take to travel from 200 to 300 m or yd. Because of this, gravity has twice as much time to act, causing the bullet to accelerate downwards more over each unit of distance as it travels farther

Interaction of Bullets and Air

Image 113: This bullet was photographed mid-flight to show the effect it has on the air. The curved lines at the front are the bullet's **sonic boom**. The sonic boom does not directly touch the bullet because air sticks to the bullet (a.k.a., the "**air jacket**") as it flies and acts as a barrier. The chaotic air directly behind the bullet is the **turbulence** created from having a flat back. Both of these effects are results of aerodynamic force (i.e., air resistance) acting on the bullet.

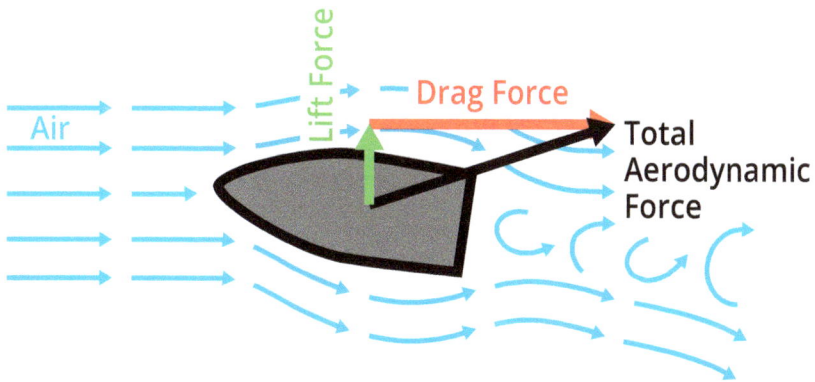

Image 114: When a bullet moves through the air, the air hits the front of the bullet. The force of the air pushing the bullet is called the "**aerodynamic**" force. The aerodynamic force can be broken down into two components: **lift and drag**. Lift force is the amount of force pushing the bullet perpendicular to its direction of movement, while drag is the amount of force pushing the bullet backwards (i.e., parallel to the direction of its movement).

---

from the marksman. The net effect is that a bullet's trajectory (i.e., its path-of-travel) is not a parabola, but a higher-order curve. (See Image 111, Pg. 129.)

All that being said, it is crucial to understand that bullets always fall at the same speed. A useful thought experiment to illustrate this is to imagine firing a bullet horizontally while simultaneously dropping another bullet from a hand. Both bullets would hit the ground at virtually the same time (the only difference being any insignificant lift forces.)

The trajectory of a bullet can only be altered in three ways. First, a marksman can reduce the amount of time gravity has to act on the bullet by making the bullet move faster. This can be achieved by increasing muzzle velocity with, for example, a higher-powder load or a longer bore.

The second method is reducing drag, or in other words, minimizing the force of air resistance. This can be accomplished by altering the physical shape and surface of the bullet, allowing it to travel more efficiently through the air.

Finally, the mass of the bullet can be changed, which has an interesting effect on its trajectory. Bullets with more mass (a.k.a., heavier bullets) have more inertia, and so they traverse a bore and exit the muzzle at slower speeds. However, because more mass travels slower, they have more time to absorb energy from exploding gases in the bore and so retain more energy to resist air resistance. This means that, everything else being equal, bullets with more mass travel slower at the start but faster at the end of their trajectory compared to bullets with less mass. In other words, bullets with more mass have more curved trajectories at closer distances and flatter trajectories at farther distances. (I.e., their trajectories curve down or decelerate more slowly.)

13.c Standard Reticles

A reticle is a **pattern of lines, dots, or markings** that can be seen within an optical sighting device (e.g., scope, binoculars, rangefinder). The simplest and most recognizable reticle design is crosshairs. It consists of one horizontal line and one vertical line that intersect at the center. (See Image 115, Pg. 133.)

The reticle sits in front of the lenses and is used as a reference point(s) when aiming. In other words, reticles are used to align the marksman's point-of-aim (a point on the reticle) with the bullet's intended point-of-impact (the point where the bullet impacts). To ensure this alignment happens, marksmen use the zeroing process (See Zeroing, Pg. 79.)

The center of the reticle only aligns to the point-of-impact at a specific distance (i.e., the zero distance). (See Picking a Zero Distance, Pg. 77.) At other distances to the target, the bullet is higher or lower due to the pull of gravity. (See Image 116, Pg. 134.) Therefore, at different distances to the target, the marksman must aim using a different point on the reticle to achieve the same point-of-impact. To enable precise shooting at the same point-of-impact at different distances, a reticle with multiple markings is necessary.

Each marking on a reticle used for aiming is referred to as a "**hashmark**." Hashmarks on standard reticles are always spaced at set intervals on the reticle and form a repeating pattern. The distance between hashmarks is referred to as a "**subtension**." Each subtension measures an angular distance, such as 1 mil or 1 MOA. (See Angular Distance, Pg. 237.)

Standard Reticles

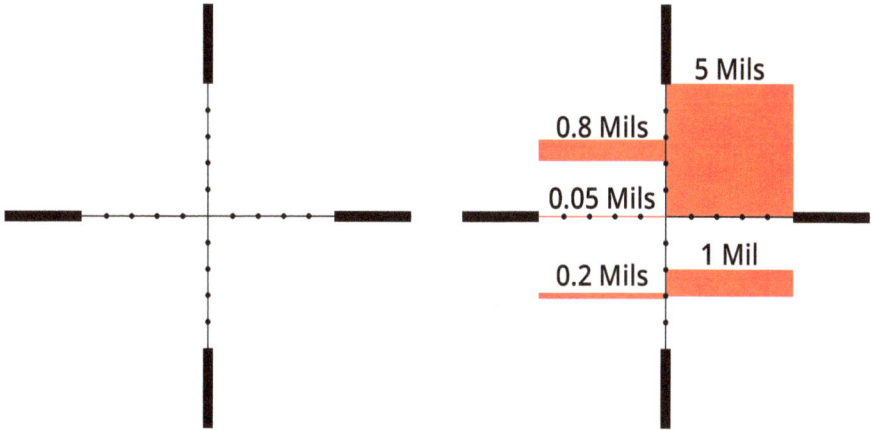

Image 115: A mil-dot reticle contains one vertical and one horizontal crosshair, with a set of dots atop them. The center of the dots are always 1 mil apart. Depending on the manufacturer, the diameter of the dots may be 0.2 or 0.25 mils. In this example, the dots are 0.2 mils in diameter, and therefore the space between the dots is 0.8 mils. The width of the crosshairs themselves can also be used as a very fine measure, although the crosshair width is often not given.

All reticles can use either MOA or mil for their angular measurements. Either standard works for a single marksman, although the gun market is shifting towards the mil standard. What is important is that both the reticle and turret use either the mil standard or the MOA standard. Otherwise, turret clicks cannot shift the reticle by whole units (for example, a click of 1 MOA would shift a reticle by approximately 0.29 mils).

All hashmarked reticles have additional hashmarks below the center point. This allows a marksman to choose their point-of-aim from multiple, vertical reference points (i.e., the hashmarks) in their sight picture. This is useful because where the point-of-impact appears in the sight picture depends on the distance that a bullet travels. As bullets travel, gravity pulls them down to the Earth and so a bullet fired farther has a lower impact in a scope's sight picture (past its second zero point). (See Image 116, Pg. 134.)

Therefore, predicting **a lower impact requires using a lower point-of-aim (i.e., a lower reference point or hashmark)**. This divergence between where the zeroed center of the reticle points (i.e., the zero point) and the actual point-of-impact is called "bullet-drop" (i.e., the linear distance on the target that a bullet drops from the zeroed point-of-aim).

Standard Reticle

Distance between hashmarks (i.e., the subtensions) is **constant from the center**.

Each hashmark on the reticle aligns to a point on the bullet's trajectory.

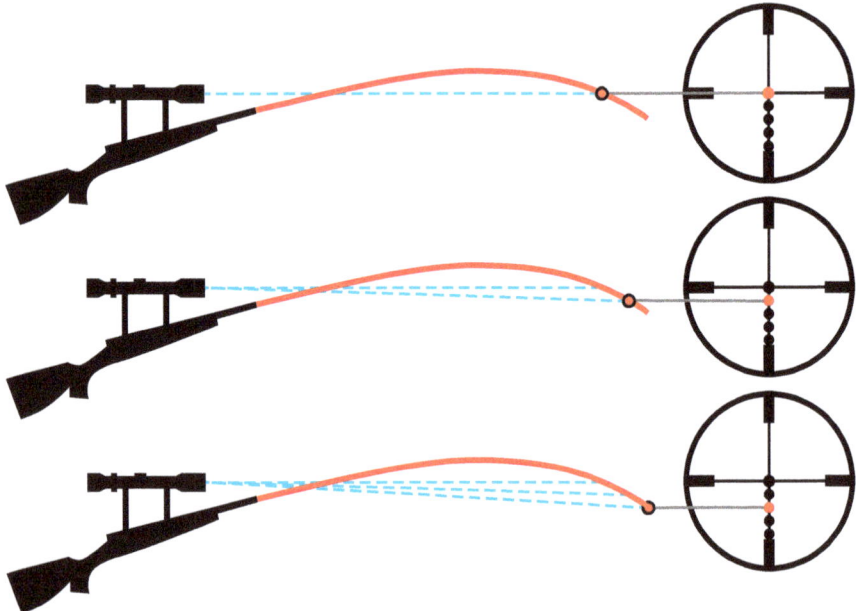

0 Mils
1 Mils
2 Mils
3 Mils
4 Mils

Image 116: A standard reticle has hashmarks that are spaced at a constant angular distance from the center. Lower hashmarks point to lower points on the bullet's trajectory, which are farther from the marksman. To find the distance (or the point on the trajectory) to which a hashmark points, a marksman measures that distance beforehand to create a conversion chart (See Image 120, Pg. 139.), or they predict the distance with a ballistic calculator (See Ballistic Calculators, Pg. 221.).

Standard reticles, including the mil-dot reticle, measure angular distance and so require the extra step of converting the linear distance of bullet-drop to a mil or MOA measurement. (See Zeroing, Pg. 79.) Marksmen

Grid Reticles

Image 117: Grid reticles have additional hashmarks in the bottom half to enable simultaneous holding for elevation and windage.

Image 118: Wind-dot reticles have hashmarks that form an upsidedown parabola-like shape emanating from the center of the reticle.

often use calculators or data cards to minimize mistakes. (See Conversion Charts, Pg. 222.)

One of the most basic and classic systems of reticle hashmarks can be seen on the mil-dot reticle. (See Image 115, Pg. 133.) It consists of a series of evenly spaced dots overlaid along both crosshairs. The vertical dots account for bullet-drop, and the horizontal dots account for crosswind, which pushes bullets sideways. (See Crosswind Deflection, Pg. 162.) Both series of dots can also be used for milling. (See Using a Standard Reticle (Milling), Pg. 113.) The distance between the centers of two adjacent dots is 1 mil, hence the name "mil-dot." Depending on the manufacturer, each dot is either 0.25 mils in diameter and spaced 0.75 mils apart or 0.2 mils in diameter and spaced 0.8 mils apart. These distances can be subtracted and added for more specific measurements. Manufacturers also produce tactical mil-dot reticles that use precise hashmarks instead of dots.

Proponents of the mil-dot reticle value its clean and less cluttered sight picture. However, since mil-dot reticles only have one horizontal line and one vertical line, marksmen cannot precisely adjust for wind (horizontal movement) and bullet-drop (vertical movement) at the same time. Although more complex reticles may seem difficult to use at first, once mastered, they become fast, accurate, and reliable.

The most common reticles that are more complex than mil-dot reticles use a grid design, with hashmarks that allow for simultaneous adjustments for both wind and bullet-drop. Many gridded reticles consist of a pure grid, with dots and lines conforming exactly to a horizontal and vertical grid. (See Image 117, Pg. 135.)

Wind-dot reticles have additional windage markings, often dots, which do not measure angular distance per se, but rather the effect of the wind on a bullet. The effect of a constant crosswind (e.g., 10 km/h or mi/h) increases with distance. The farther a bullet travels to a target, the more wind is able to push it to the left or right. Therefore, the windage markings on lower rows are spaced farther apart to account for the fact that bullets traveling farther are also more affected by the same wind.

That being said, as bullets travel farther, they drop faster and so have less time to be affected by wind per unit distance fallen. In other words, the relationship between bullet-drop and a constant wind effect is not linear; for example, a bullet that drops 1 mil may be pushed by wind 1 mil to the right, but if the same bullet drops 3 mils, the same wind may push the bullet only 2.5 mils to the right. Therefore, wind dots form an upsidedown parabola-like shape emanating from the center of the reticle. (See Image 118, Pg. 135.)

Because both grid reticles and wind-dot reticles have more hashmarks at the bottom than at the center, they are sometimes called "christmas-tree reticles" because the pyramid shape with dots resembles a pine tree with ornaments.

13.d First and Second Focal Plane Reticles

A scope contains many lenses in a row to redirect light. Two of the lenses are focal-plane lenses. These lenses are locations within the scope where the image of the object or target is brought into focus, and they are also where a reticle (crosshairs or aiming point) is typically placed. (See Image 16, Pg. 35.) If the reticle is located on the focal-plane lens closer to the target (i.e., the first-focal-plane lens), the reticle changes size as the magnification is adjusted. Conversely, if the image is placed on the focal-plane lens closer to the marksman (i.e., the second-focal-plane lens), the reticle does not change size with magnification. (See Image 119, Pg. 137.) This is because the magnification assembly is between the first and second focal planes.

Reticle Focal Plane

First Focal Plane Reticle/Scope

1x Magnification   12x Magnification   24x Magnification

Second Focal Plane Reticle/Scope

1x Magnification   12x Magnification   24x Magnification

Image 119: Reticles can respond in one of two ways to magnification. If the reticle is placed in the front of the scope (i.e., "first focal plane" (FFP)), then it changes size when the sight picture is magnified. (See Image 16, Pg. 35.) The hashmark measurements in first-focal-plane scopes always stay constant, so for example, a 1-mil subtension is always 1 mil at any magnification. In contrast, when the reticle is placed in the rear of the scope (i.e., "second focal plane" (SFP)) the reticle is always the same size no matter the magnification. For a second focal plane scope, subtensions are only accurate at maximum magnification; however, this is rarely a problem since most marksmen only use an elevation hold at the maximum magnification (or half-max where subtension numbers are doubled) on a second-focal-plane scope.

13.e Elevation Hold and Dial

In the context of scopes, "elevation" refers to vertical angular distance in the scope. In contrast, "bullet-drop" refers to the vertical linear distance that a bullet falls on the target from the zero point (usually where the center of the reticle points). In other words, **elevation is how a marksman compensates for bullet-drop**. For example, consider a rifle that is zeroed at 100 m (109 yd). If at 500 meters (547 yd) a bullet impacts 1 m (1.1 yd) lower on a target and 2 mils lower in the sight picture, then the marksman would have to adjust up the scope's reticle (i.e., increase elevation) by 2 mils to realign the point-of-aim with the point-of-impact.

"**Hold**" refers to using a hashmark on the reticle that is below the zero point as a point-of-aim. That is, the marksman physically holds up the reticle. For example, if a bullet drops by 5 mils at a certain distance, the marksman can hold their reticle 5 mils higher. That 5 mils would be the "elevation hold" for that distance.

To determine the correct elevation hold for a particular distance, modern marksmen rely on ballistic calculators. (See Ballistic Calculators, Pg. 221.) These devices, once set up according to the manufacturer's instructions, require marksmen to input into the calculator: the target distance, and various other information such as temperature, altitude, and type of equipment. Then the calculator outputs an elevation hold for the marksman to use. Many ballistic calculators are available online for free. Others come in standalone devices.

Using ballistic calculators is a vast improvement over the traditional method, which was trial-and-error. To determine accurate holds, a traditional marksman would have set up targets at various distances, shoot at them, and measure the bullet-drop at each distance. The marksman would then record each drop in a log book for future reference, called the "Data-On-Previous-Engagements" (DOPE) book. (See Recording, Pg. 217.) DOPE books are not limited to bullet-drop data, and may contain other valuable information on every relevant variable (e.g., air temperature and wind effects).

After collecting all their data, the marksman would plot each bullet-drop measurement against the corresponding shooting distance. Such a graph can help to understand how bullet-drop changes with different target distances. So for example, if a bullet had previously dropped 5 mils at 500 m (547 yd), a marksman could reasonably assume that future bullets would exhibit a similar drop. And if a marksman needed to shoot at 450 m (492 yd) but only had bullet-drop information for 400 m (437 yd) and 500 m (547 yd), they could calculate the average between these two distances.

Instead of holding the reticle up, another option is **dialing** the elevation turret and adjusting it to realign the zeroed point with the desired point-of-impact. For example, if a bullet drops by 5 mils, the marksman can rotate the elevation turret to raise the reticle within the scope by 5 mils ("elevation dial").

yards	drop	clicks	Wind 10
100	1.6	+6	0.2
200	0	0	2.5
250	-2.8	4	4.0
300	-7.1	9	5.9
350	-13.0	14	8.2
400	-20.6	20	10.8
450	-29.9	25	14.0
500	-41.2	31	17.5
550	-54.7	38	21.6
243 105 Amax			

yards	drop	clicks	Wind 10
600	-70.3	45	26.2
650	-88.5	52	31.3
700	-109.2	60	37.0
750	-132.9	68	43.3
800	-159.6	76	50.2
850	-189.7	85	57.7
900	-223.4	95	66.0
950	-261.1	105	75.0
1000	-303.0	116	84.7
243 105 AMax			

Image 120: This marksman has attached to his rifle a table that compares target distance to the marksman (yards) against 1) bullet-drop (inches), 2) dial rotation to compensate for the drop (clicks), and the wind deflection from a 10 mi/h crosswind (inches). The scope is a 1/4 MOA scope, so each click corresponds to a rotation of the reticle by 1/4 MOA. At 1000 yd, 1 MOA is 10.47 in, so at 1000 yd, each click corresponds to 2.62 in. Because windage was not also converted to clicks, it can be presumed that this marksman dials for distance but holds for wind.

This approach eliminates the need to rely on vertical hashmarks, meaning a marksman can always aim with the center of their reticle. Some marksmen choose this method because basic reticles only offer windage adjustments on the baseline (e.g., the mil-dot reticle). (See Image 120, Pg. 139.)

However, when shooting at distances where both wind and bullet-drop affect a bullet's trajectory, it is advisable to upgrade to a modern reticle with a grid pattern of horizontal and vertical hashmarks. (See Image 117, Pg. 135.) This is because dialing for elevation: takes more time than holding; diverts a marksman's attention from the sight picture more; and a marksman may forget to reset the elevation turret when a new target appears.

That being said, there are optional scope features for both holders and dialers. Holders prefer scopes that feature lockable turrets. Dialers prefer scopes with zero stops and tactile rotation indicators to allow for precise dial adjustments without looking at the turrets.

If a marksman does choose to dial their turrets for each shot, it is crucial that they consistently start and end with a zero elevation and windage setting to develop reliable habits. Additionally, for rapid firing, marksmen must learn to adjust the turrets while only glancing at turrets briefly to verify their accuracy or even adjusting them without looking at them at all.

13.f Precisely Adjusting the Sight Picture

At long distances, even tiny movements of the scope can have outsized effects on the accuracy of a rifle; therefore, precise adjustment is paramount. To precisely move a rifle requires very fine and controlled movements by the marksman. Movements must also be done quickly so a marksman can both avoid fatigue and also hit time-limited targets.

An advanced technique for precisely adjusting elevation hold involves placing a sandsock under the rifle's buttstock. (See Image 17, Pg. 37.) A sandsock is a small, durable bag filled with sand, plastic beads, or other granular material. Then the marksman uses their support hand to hold the sandsock in position. (See Image 121, Pg. 141.)

By squeezing the sandsock, it elongates and pushes the rifle's buttstock up, causing the entire rifle (including the reticle) to rotate. As the buttstock goes up, the muzzle goes down. This causes the reticle to point lower. On the other hand, relaxing the squeeze lowers the sandsock and rotates the rifle in the opposite direction; the buttstock goes down and consequently, the muzzle goes up. This causes the reticle to point higher.

Because the support hand is no longer supporting the front of the rifle but is positioned at the marksman's chest, this technique requires excellent front support. This often means being in the prone position and with the assistance of a bipod. The support hand must then be able to stabilize both the sandbag and the rifle together, and the marksman must be careful not to unintentionally move the rifle when operating the bolt of the weapon.

Using a sandsock is helpful but not strictly necessary. A skilled marksman can squeeze their fist below the stock, effectively achieving the same sandsock effect. However, when only using their hand, it is important for marksmen to avoid accidentally squeezing their support hand when pulling the trigger, as that would disrupt the point-of-aim at a critical moment.

13.g Bullet-Drop Compensator Reticles

Similar to the mil-dot reticle, the bullet-drop compensator (BDC) reticle features a horizontal and vertical line. However, the BDC reticle differs in that the hashmarks are placed to correspond to points in the bullet's trajectory

Using Sandsock to Adjust Sight Picture

Image 121: In the center is a neutral-position sandsock under the buttstock of the rifle. (See Image 17, Pg. 37.) **Squeezing the sandsock** (top) rotates the front of the rifle down. Relaxing the sandsock (bottom) rotates the front of the rifle up.

that are evenly spaced distances from the marksman, rather than correspond to fixed angular intervals like the mil-dot does. (See Image 122, Pg. 142.)

For example, a 100-m (109-yd) BDC reticle has a hashmark for each 100 m distance to the target. If the center point-of-aim is zeroed at 100 m, the first hashmark below center predicts the impact at 200 m (219 yd), the second hashmark below center at 300 m (328 yd), and so on. Some BDC reticles have labels on the reticle denoting the distance to the target for each hashmark.

Subtensions (the spacing of the hashmarks) on the BDC reticle increase exponentially down the vertical crosshair to account for the increasing drop that bullets experience as they travel further from the marksman.

BDC reticles are efficient because they work directly with rangefinders, using linear distance as input. For example, suppose a rangefinder indicates a target is at 300 m (328 yd). In this case, a marksman

Bullet-Drop Compensator Reticle

Vertical distance between hashmarks (i.e., the subtensions) inceases from center.

100 m (109 yd) ①
200 m (219 yd) ②
300 m (328 yd) ③
400 m (437 yd) ④

100 m

200 m

300 m

Image 122: A bullet-drop compensator (BDC) reticle has hashmarks that correspond to set intervals of distance from the marksman to the target. The hashmarks in this example (labeled 1, 2, 3, and 4) align to a point on the bullet's trajectory that is a specific distance (here 100 m) farther from the marksman than the previous one.

Bullet-drop compensators bypass the conversion of distance to mils, MOA, or clicks. (See Image 120, Pg. 139.) They are typically only **used for 500 m or yd and below** because they only work for a specific setup, and quickly lose accuracy in different environments. For example, while the second hashmark may correspond to 300 m for one setup, it could correspond to 343 m for another setup.

can immediately aim with the second hashmark (assuming the center point is zeroed at 100 m (109 yd) and assuming each additional 100 m of distance represented by one hashmark downward). This contrasts with the standard angular-distance reticles which require the extra step of converting linear distance to the target into a mil or MOA measurement on the reticle.

There are three issues with BDC reticles. First, the bottom half of the reticle cannot be used for milling to estimate the distance to a target. (See Using a Standard Reticle (Milling), Pg. 113.) To solve this, many BDC reticles have standard hashmarks in their upper half; or marksmen can use the horizontal hashmarks on the reticle.

Second, the location of each BDC hashmark depends on the specific rifle and ammunition being used. That is, faster ammunition would hit higher on a target than a BDC would predict, and slower ammunition would hit lower. For example, if a marksman switches to more powerful, faster ammunition, each dot may represent impacts at 103, 211, and 336 m (respectively 113, 231, and 367 yd), instead of a perfect progression of 100, 200, and 300 m (respectively 109, 219, and 328 yd).

Finally, if the scope is a second-focal point scope, the BDC hashmarks only correctly measure at a single magnification setting, typically the highest. (See Image 119, Pg. 137.)

Since the BDC hashmarks may denote somewhat arbitrary distances and cannot account for individual weapon system variations, their intended purpose of simplifying the translation of distance to a hashmark is not fully achieved. This error compounds with distance, **making BDC reticles rarely used beyond 500 m** (547 yd).

13.h Danger Distance

Essentially, the danger distance is the base of a rectangle at a specific distance from the marksman. The height of the rectangle is delimited by an upper and lower offset from the point-of-aim chosen by the marksman. The bottom of the rectangle is the subset of the bullet's trajectory in which a bullet would hit a target within that defined height. (See Image 123, Pg. 145.)

For example, consider a target area (i.e., killzone) that is 10 cm (3.9 in) tall. The bullet may fall 10 cm (3.9 in) from 50 to 200 m (55 to 219 yd). That means that the target can be anywhere within that range and if the marksman aims at the center of the target area, the bullet would impact somewhere within the target area (although not necessarily the center of the target).

Put simply, a bullet always hits a target smaller than the killzone within the killzone's corresponding danger distance because the bullet never exceeds the killzone's bounds along the danger distance. If, for example, a bullet flies faster and flatter, its danger distance is longer because it can travel even farther without exceeding the bounds of the killzone. A marksman can achieve a flatter bullet trajectory through factors such as more powder, a longer barrel, or a bullet with a higher ballistic coefficient.

Calculating a danger distance is a useful technique to hit targets without calculating an elevation hold. Although uncommon among marksmen with modern scopes, knowing a danger distance is vital when using a scopes that lacks hashmarks on its reticle. Knowing the danger distance is also useful for marksmen who 1) know their target (e.g., a deer), but 2) do not know that target's exact distance, and 3) they do not want to fiddle with a rangefinder.

For example, a North American whitetail deer may have 20 cm (7.9 in) of vital organs below the vertical middle of their torso, behind their shoulder. Therefore, hunters use this 20 cm (7.9 in) killzone to determine a danger distance. For example, a marksman might zero their rifle to have the far-zero at 50 m (55 yd), meaning that after 50 m (55 yd), the bullet always drops lower. Then, the marksman might test the rifle and discover that the bullet drops to 20 cm (7.9 in) at 250 m (273 yd). Therefore, if the marksman aims behind the shoulder in the middle of the deer's ribcage anywhere within the danger distance of 50 to 250 m (55 to 273 yd), they are guaranteed to hit the vital organs and thereby secure a humane kill.

For a given target area (i.e., killzone), danger distance is usually derived from the distance to the target. However in reverse, the distance to the target can also be derived from the danger distance. This occurs when the distance to the target is unknown, and the marksman must determine their allowable margin of error when estimating that distance. For example, if a target is actually 500 m (547 yd) away and 25 cm (9.8 in) tall, and the bullet drops 25 cm (9.8 in) from 475 to 525 m (520 to 574 yd), then the marksman's estimate of the target's distance can be anywhere within that 475 to 525 m (520 to 574 yd) range and they would still hit the target if they aim at its center. In summary, knowing the danger distance is 50 m (55 yd) at 500 m (547 yd) for a 25 cm (9.8 in) target lets the marksman know that they have a 5% margin-of-error in their distance-to-target estimate at 500 m (547 yd) for a 25 cm (9.8 in) target. (See Image 107, Pg. 124.)

A specific case of danger distance is the point-blank distance, which is the danger distance starting from the muzzle for a given target height. Since a bullet momentarily travels above the sightline of a scope during its trajectory, the top of the killzone for the point-blank distance must be higher than the point-of-aim. For example, the maximum trajectory of a bullet may be two

Danger Distance and MPBR

Image 123: A danger distance for any given trajectory is the distance in which a bullet **is present within a vertical range (i.e., the killzone)**. Together, the danger distance and the killzone form a rectangle, with the max height being one corner and the minimum height being the opposite corner. The maximum point blank range (MPBR) is **a special case of the danger distance** where the danger distance is defined as starting at the muzzle, instead of at an arbitrary point in the trajectory. The minimum height is still defined at whatever point the marksman wants to cover their desired target area.

cm (0.79 in) above the sightline. However, after reaching the maximum trajectory, the highest point never changes regardless of how much farther the bullet travels because it always begins to fall after that point. Point-blank distance is not particularly useful for long-range marksmen because, by definition, they do not shoot at point-blank range.

14. Eliminating Rifle Cant

"Cant" refers to an inclination or tilt in an object that deviates it from the "vertical" or "horizontal." For this book, "vertical" is the direction of gravity, and "horizontal" is perpendicular to gravity. Therefore for example, a canted scope is one that is not above the barrel with respect to gravity. It can also help to define a rifle according to the three anatomical planes (See Image 124, Pg. 146.):

Anatomical Planes

**Median Plane
(Sagittal Plane)
Pitch Rotation**

**Transverse Plane
(Axial Plane)
Yaw Rotation**

**Frontal Plane
(Coronal Plane)
Roll Rotation**

Image 124: There are **three anatomical planes**, each of which corresponds to a dimension of rotation. A rifle is oriented perfectly vertical in the median plane. It is rotated in the median plane to shoot at closer or nearer targets. A rifle is rotated in the transverse plane to aim to the left or right. **A rifle is never rotated in the frontal plane. To do so introduces cant.**

Rifle System Alignment

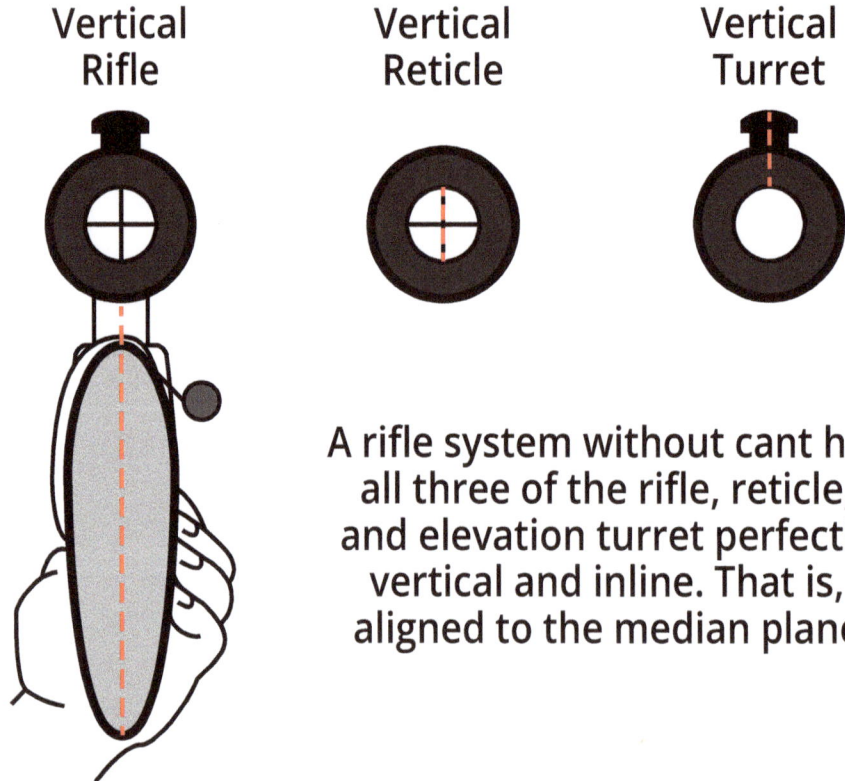

Vertical Rifle

Vertical Reticle

Vertical Turret

A rifle system without cant has all three of the rifle, reticle, and elevation turret perfectly vertical and inline. That is, aligned to the median plane

Image 125: A perfectly aligned rifle is vertical and centered on the **median plane**.

Median (Sagittal Plane) – This plane divides the rifle into left and right sections. It helps describe the tilting of a rifle up or down (i.e., pitch rotation). Pitch rotation is how a marksman changes how far, or at what elevation, a rifle shoots.

Transverse (Horizontal Plane) – This plane divides the rifle into upper and lower sections. It helps describe side-to-side movements, such as swinging the rifle left or right (i.e., yaw rotation). Yaw rotation is how marksmen aim left and right, either at new targets or to account for wind.

Front (Coronal Plane) – This plane divides the rifle into front and back sections. It helps describe rotational movements, such as spinning the rifle along its long axis (i.e., roll rotation). Marksmen only roll-rotate their rifles when they have no other choice for whatever reason (e.g., a restricted loophole). Otherwise, any roll rotation is what is known as "**cant**."

Cant is bad because of how bullets exit a barrel. Barrels always point towards the sightline, which emanates straight out of the scope. When a rifle has no cant, bullets exit the barrel going slightly up to meet the sightline; but when cant is present, bullets must go slightly to the left or right to meet the sightline. A bullet then keeps that sideways moment when falling as well. **That is, cant forces a bullet to travel sideways a bit.**

Therefore, shooting with a cant is difficult for two reasons. First, rifles are zeroed without cant, and so introducing it deviates a bullet from its zero. That is, it is far easier to keep a rifle vertically upright as opposed to rotated by an arbitrary amount. Second, the sideways deviation from cant increases linearly with the distance the bullet travels (i.e., bullets shot farther deviate more). Zero cant is far easier to deal with because zero always "scales" to zero.

14.a Formulas for Cant Offset

Marksmen must attempt to eliminate cant, so calculating the amount of cant offset is almost never useful. That being said, the offset between where a bullet would impact from a rifle in a proper orientation versus a canted orientation is presented here for completeness and is determined by the following:

Variables:

Velocity towards the scope (*v towards scope*) – The component of a bullet's muzzle velocity that directs the bullet towards the scope (as opposed to forward). For a non-canted rifle, this would be the vertical component of the bullet's muzzle velocity.

Cant offset occurs because bullets are not fired straight; they are fired slightly upwards so that the trajectory of the bullet can meet the sightline. When a rifle is canted, some of this normally vertical velocity is converted into sideways velocity.

Bullet travel time *(travel time)* – The amount of time the bullet takes to hit the target in seconds. Because velocity is distance per unit of time, knowing the velocity and time allows a marksman to calculate distance, or in this case, the cant offset.

Horizontal Offset *(Ox)* – The horizontal change in distance from a bullet impact from a vertical rifle compared to that of a canted rifle.

Vertical Offset *(Oy)* – The vertical change in distance from a bullet impact from a vertical rifle compared to that of a canted rifle.

θ – The cant angle, where 0 is perfectly vertical.

Sin and Cos – Abbreviations for "sine" and "cosine" respectively, trigonometric functions that translate a degree to a percent. (See Image 178, Pg. 210.)

Formulas:
- *Horizontal Offset (Ox)* = *(v towards scope)* × *(travel time)* × *sin(θ)*
- *Vertical Offset (Oy)* = *(v towards scope)* × *(travel time)* × *(cos(θ) - 1)*

Examples:
If θ = 0°, then:
- The rifle is vertically upright.
- Both $\sin(\theta) = 0$ and $(\cos(\theta) - 1) = 0$, so there are no offsets.

If θ = 90° or 270°, then:
- The rifle is sideways.
- $\sin(\theta) = 1$, so all of the energy that pushes the bullet up towards the scope now pushes the bullet sideways towards the scope.
- $(\cos(\theta) - 1) = -1$, so the bullet never rises because it is fired perpendicular to gravity. Because the presence of an upwards velocity is set to "0" (i.e., $\cos(90°) = 0$), the "-1" indicates the bullet would fall below that baseline.

If θ = 180°, then:
- The rifle is upsidedown.
- $\sin(\theta) = 0$, so there is no vertical offset.
- $(\cos(\theta) - 1) = -2$, so the bullet not only lacks the energy of being pushed up, but is actively fired at a downwards angle towards the scope.

Conclusions:

▸ Because *(v towards scope)* and *(travel time)* are both constant for a given muzzle velocity and distance to target, for any particular setup the horizontal and vertical offsets are only determined by the cant angle. Because the cant angle is input into sine and cosine formulas, **canting a rifle makes bullets follow a circle** (i.e., the cant offset circle), where the zero point is the apex of that circle. (See Image 127, Pg. 151.)

▸ At the zero distance, "*(v towards scope)* × *(travel time)*" is calibrated to be exactly equal to the bullet-drop at that distance. Therefore, the radii of the boreline circle (i.e., where the boreline points as a rifle is rotated around the sightline) and the canting error circle are both exactly equal to the vertical travel of the bullet. A side effect is that, at the zero distance, cant is unaffected by whether a scope is mounted high or low on a rifle.

▸ Generally, beyond the zero distance, a low-mounted scope gives less canting error, while before the zero distance, a high-mounted scope gives less canting error.

▸ *(v towards scope)* can be deduced from other variables: *(v towards scope)* = *(Ox)* / *(travel time)* where a rifle shot at 90° (sideways), so that $\sin(\theta) = \sin(90°) = 1$. That is, fire a zeroed rifle at 90° at its zero distance. Then measure the *(Ox)* on the target and measure *(travel time)* using a chronometer to measure the muzzle velocity. Divide *(Ox)* by *(travel time)* to find *(v towards scope)*.

▸ Close to 0, sine changes much more quickly than cosine. For example, at 1°, $\sin(1°) = 1.75\%$ which is 116 times $(1 - \cos(1°)) = 0.015\%$. At 5°, $\sin(5°) = 8.72\%$ which is 51 times $(1 - \cos(5°)) = 0.172\%$. This means that the vertical component of cant is almost always negligible for most marksmen. In contrast, the horizontal component can be a significant cause of a bullet's deviation to the left or right.

For example, if a bullet is fired with a muzzle velocity of 1000 m per second, it would take 0.1 s to reach a target 100 m away. Assuming the scope is mounted 5 cm above the bore, and the rifle is zeroed to reach the sightline and fall again on its way to the 100 m zero point, the bullet travels towards the scope line at a rate of 10 cm per 0.1 s, or 1 m/s at the muzzle. If the bullet slows down as it travels to a target at 1000 m away, it may have a travel time of 2 s. Using the horizontal offset formula, 1 m/s × 2 s × 1.75% = 3.5 cm or 1.4 in of horizontal offset for 1°of cant.

Effect of Cant Side View

Image 126: This diagram shows a rifle that is firing at the zero distance, where the vertical-hold point-of-aim impacts at the far-zero. (See Image 112, Pg. 129.) Canting a rifle makes bullets impact lower and to the side. The vertical difference between the vertical-hold point-of-impact and the canted-hold point-of-impact is the "**elevation error**." The horizontal difference between the vertical-hold point-of-impact and the canted-hold point-of-impact is the "**windage error**."

14.b Holding Cant

Holding cant is the simplest type of cant. This occurs when a marksman rolls their weapon simply because they are handling it wrong. Holding cant is not due to a mismeasurement, misalignment, nor a miscalibration; rather the marksman is simply rotating their rifle.

Rotation can happen, for example, with new marksmen who may naturally hold their rifle with a cant. This often happens because they have weak or untrained neck muscles, causing them to lean the rifle toward their cheek to achieve a comfortable cheek weld. On the other hand, excessive pressure on the cheek weld can cause a marksman to push the rifle away from them, thereby canting it in the opposite direction.

During teaching and zeroing, one way to identify these patterns is with the assistance of an observer who can correct improper posture. Observing shot groups can also indicate whether the rifle is canted. When the marksman cants the rifle towards their head (for right-handed marksmen), shots tend to be to the left, while for left-handed marksmen, shots tend to be to the right.

Effect of Cant Rear View

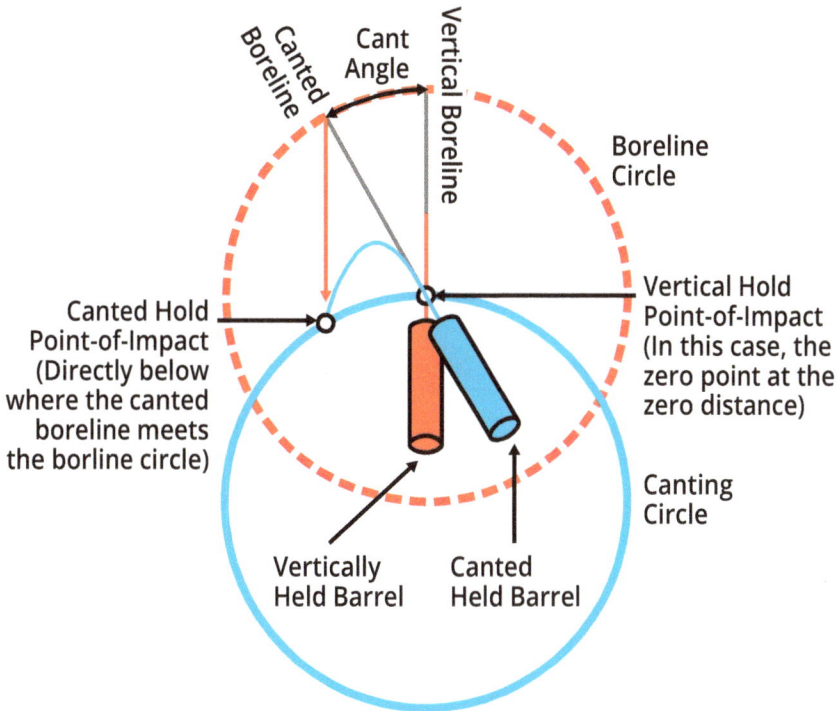

Canted Boreline

Cant Angle

Vertical Boreline

Boreline Circle

Canted Hold Point-of-Impact (Directly below where the canted boreline meets the borline circle)

Vertical Hold Point-of-Impact (In this case, the zero point at the zero distance)

Canting Circle

Vertically Held Barrel

Canted Held Barrel

Image 127: Small amounts of cant **have a noticeable windage error** but only a nominal elevation error. (See Image 126, Pg. 150.) This is because canting a rifle moves the point-of-impact around the canting circle, and moving away from the apex of the circle does not have much vertical change. The canting circle is based on the boreline circle. The boreline circle is the circle on the target that the boreline (a straight imaginary line coming out of the center of the bore) points at when the rifle is rotated about the sightline. The canted-hold point-of-impact on the canting circle is directly below where the canted boreline points on the boreline circle.

Rifle canting can be prevented before it occurs by **checking the orientation of the rifle before shooting**. This is done with a bubble level attached to the rifle. (See Image 128, Pg. 153.) When the bubble is between the two interior lines, the rifle is level. When the rifle is canted to the left, the bubble moves outside the right line and vice versa. Using a level is particularly useful when shooting from uneven terrain that can disrupt a marksman's natural sense of balance by a few degrees.

However, a bubble level is not a precise tool for determining orientation, and with sufficient practice, marksmen may even become more accurate than a bubble level. Additionally, while marksmen can purchase highly accurate leveling attachments, no external level can be observed while also looking through the scope. As a result, some modern scopes now incorporate an internal level into the sight picture.

It is worth noting that cant is different from an offset-mounted scope. That is, some cross-dominant marksmen (e.g., left-eye-dominant, right-hand-dominant marksmen) intentionally offset their scope from the midline of their rifle to make their setup more comfortable. Despite the offset, these special setups still have the bullet exit the muzzle with zero side-to-side velocity when properly aligned. They do this by zeroing the rifle with the point-of-aim to the left or right of the point-of-impact at a distance equal to the scope's horizontal offset from the bore.

14.c Reticle Cant

Many marksmen level their rifle by aligning the reticle (i.e., the lines seen projected onto the sight picture) (See Standard Reticles, Pg. 132.) to the horizon. However, this only works if the reticle is properly aligned so that its lines are exactly parallel or perpendicular to the force of gravity. Otherwise, rolling a rifle to level out the reticle may actually introduce cant to the rifle.

While **aligning a reticle to gravity is part of the scope mounting process**, it can be done whenever a marksman suspects their reticle is crooked. Realigning a reticle requires remounting the scope. To do so, a marksman can place a bubble level or electronic level on their rifle. Then, they place the reticle in the scope mount, but do not tighten it yet.

Then, the marksman creates a plumb line, which is a bright piece of string with a weight at the end (i.e., the plumb). (See Image 129, Pg. 153.) The plumb pulls the string into an exactly vertical position. The plumb line is positioned against a white wall or paper 10 m or yd away from the marksman. Then the marksman looks through their scope and rotates the scope in its mount until the vertical crosshair is aligned to the plumb line, at which point the reticle is level. When both the rifle and the reticle are simultaneously level, the marksman secures the scope to the rifle.

14.d Mechanical Adjustment Cant

Scopes have two separate components that are intended to be "plumb" (i.e., aligned to the direction of gravity). First, the reticle (i.e., the lines seen projected onto the sight picture) (See Standard Reticles, Pg. 132.) is built to have lines that are either exactly vertical or exactly horizontal with respect to gravity. Second, the adjustment turrets move the reticle within the

Tools to Eliminate Cant

Image 128: This is a scope mount with a built-in **bubble level**. The level faces the marksman so they can observe it and eliminate holding cant just before looking into the sight picture. It can also be used during scope mounting to eliminate reticle cant.

Image 129: A **plumb line** is a bright string with a weight (i.e., the plumb) attached to the end that pulls the string straight down. "Plumb" comes from the Latin word for "lead" and while these plumbs are specialized, any heavy weight works equally well.

scope housing exactly vertically or horizontally with respect to gravity. (See Adjusting Elevation and Windage Settings with Turrets, Pg. 36.) Therefore, since both components must be plumb, **a perfect scope would have the reticle aligned with its adjustment mechanisms**.

However, almost all scopes have some degree of misalignment between their reticle and their adjustment mechanisms. That is, if the reticle is aligned, then all adjustments pull the reticle slightly diagonally; and if the adjustments are aligned then the reticle would be crooked. While most manufacturers claim a limit on this misalignment, ranging from 0.5% to 5% depending on the company, scopes with perfect alignment are rare.

Also, misalignment error is random, so while some scopes of the same make and model are manufactured perfectly, others may have significant error. Therefore, to account for this error, marksmen must test the actual scope they intend to use.

Even after being properly calibrated, most scopes retain some degree of mechanical adjustment cant. This is because, unlike holding cant or reticle cant, **mechanical adjustment error is a feature of the equipment's construction itself and cannot be fixed**. A marksman must choose between leveling their reticle and leveling their adjustment mechanisms. And a plumb reticle is preferable to plumb adjustment mechanisms because reticles are used more often and are easier to plumb than adjustment mechanisms.

The best way to measure the diagonal mechanical adjustment error in a rifle system is to perform a **tall target test**. To conduct a tall target test, a marksman must first ensure that their reticle is plumb and that their rifle

Mechanical Adjustment Cant

Reticle and Elevation Turret Aligned

Level Turret

Level Reticle

Level Rifle

Reticle and Elevation Turret Not Aligned

Level Turret

Unlevel Reticle

Level Rifle

Misalignment angle

Step 4: Shoot 1 mil (or 4 MOA) higher.

Step 5: Find the center of each group (here, blue).

Step 3: Shoot 1 mil (or 4 MOA) higher.

Step 6: Connect the centers with a line

Step 2: Shoot 1 mil (or 4 MOA) higher.

Step 7: Find the angle of the line against vertical to get the misalignment angle.

Step 1: Shoot at zero point.

Image 130: When turrets are level, reticles are also supposed to be level (top). However, **due to manufacturing errors**, many scopes are produced with reticles and turrets that are misaligned (bottom). If a marksman vertically aligns the turrets of a misaligned scope, then holding for elevation introduces windage error. If a marksman vertically aligns the reticle then dialing for elevation introduces windage error.

Image 131: A marksman can conduct a **tall-target test** to determine the misalignment angle between a reticle and an elevation turret. The marksman shoots groups at increasing elevation (either dialing or holding) and then connects the group's centers. If the line that connects the group's centers is plumb (here, the dotted line), then the reticle and turrets are aligned. If there is an angle between the line and plumb, then they are misaligned.

is zeroed. In fact, the tall target test is a good addition to the end of the zeroing process for any scope new to the marksman. (See Ensuring Accuracy (Grouping and Zeroing), Pg. 74.)

First, the marksman sets up a 1.5-m-or-yd-tall target at a distance of 100 m or yd (or a target of the same relative height at the rifle's zero distance). On the target, they use a plumb line to draw a perfectly vertical line from the top to the bottom of the target. Because both the reticle and the target have a plumbed line, the marksman can ensure no scope cant by aligning the reticle to the target's line.

Thereafter, the bottom of the line becomes the point-of-aim for all subsequent shots. The marksman then fires a five-shot group at the bottom of the line. (See Grouping, Pg. 74.) After this initial group, the elevation turret is adjusted upward by 3 mils (or 10 MOA), and another five-shot group is fired while aiming at the bottom of the line. This process is repeated two more times, resulting in a total of four groups. The center-averages of these groups are then connected to form a line. If this line is angled in relation to the vertical line, the angle represents the difference between the reticle and turret adjustments.

Entirely separately from reticle cant, the results of the tall target test can also tell the marksman what one click on their scope mechanism actually represents. For example, if a scope advertises that each click is $1/10$ of a mil, then rotating the elevation turret 30 times would elevate the reticle by 3 mils. At 100 meters, that is a vertical difference of 30 cm. However, if the actual average distance between the center-averages of the four groups is 27 cm, then a marksman knows that each click only actually rotates a turret by 10% less than advertised by the manufacturer, or about $1/11$ of a mil.

Significant discrepancies between what a click is advertised to rotate versus what it actually rotates can cause marksmen to over or under adjust. And even if a marksman becomes accustomed to their own equipment, inaccurate click adjustments can cause issues when they use other equipment or when equipment is shared. In fact, whenever marksmen complain about variations in ammunition performance across different rifles, the culprit may actually be the performance of the scopes (i.e., the tools used to measure the performance of the ammunition).

Expert Contents

15. Wind   157

Wind Value   159
Crosswind Deflection   162
Vertical-Wind Deflection   163
Wind Gradient   164
Multiple Winds over Distance   167
Changing Wind over Time   171
Windage Holds and Dials   173
Converting a Wind to a Windage Hold (Summary)   177

16. Wind Estimation   178

Flags and Vanes   179
Vegetation   182
Mirage   184

17. Climate   187

Temperature   188
Air Density   189
Simple Mirage   191
Inferior and Superior Mirage   194
Precipitation   194

18. Inherent Imprecision   197

Lateral Throwoff   198
Probabilistic Shooting (Mil and MOA-Accuracy)   201
Estimation (Range and Wind) Uncertainty   206

19. Shooting Uphill or Downhill (Inclination)   207

Perpendicular and Parallel Components of Gravity   208
Change in Air Density with Change in Altitude   212

Experts
(300 to 600 Meters or Yards)

An expert is someone who has made all the mistakes which can be made in a very narrow field.

—*Niels Bohr, Danish, Nobel-Prize-winning physicist*

Defining skill level with distance is arbitrary. However, the 300-m-or-yd mark often serves as a practical cutoff for human vision, and the beginning of considering environmental conditions. That is, even seasoned hunters and military marksmen must rely on advanced optics and calculations to maintain accuracy beyond 300 m or yd. These calculations take into account variables such as wind, probability, and inclination. And the required skills for these calculations include reading and understanding: tables and charts, environmental conditions, and basic statistics and trigonometry.

15. Wind

Wind is the fluid movement of air. It affects the air resistance that bullets experience as they travel because air resistance is directly correlated to the differential speed between a bullet and the air surrounding it. If air moves against the bullet, resistance increases; and if air moves with the bullet, resistance decreases.

At close distances, wind has a minimal impact on bullets and is often discounted as a factor that can affect accuracy. This is because wind acts on bullets over time, and at close distances a bullet simply does not spend enough time in the air for wind to significantly change its trajectory.

However, as bullets travel longer distances, the influence of wind becomes more significant as wind has more time to act on the bullet. In fact, **the impact of wind grows exponentially with distance**, as not only does the bullet have farther to travel, but it also slows down during flight. Therefore, accounting for wind becomes critical for accurate shooting at longer distances. Specifically, concepts such as wind value, crosswind deflection, vertical wind deflection, wind gradient, and changing winds are crucial in determining a projectile's trajectory. To combat these wind effects and ensure consistent accuracy, marksmen must employ windage holds and windage charts.

Effect of Wind on Precision

	±2 km/h (±1 mi/h)	±5 km/h (±3 mi/h)	±8 km/h (±5 mi/h)

500 m or yd

1000 m or yd

Image 132: Each red dot represents a bullet impact. The impacts are more dispersed at 1000 m or yd than at 500 m or yd because **wind has more time to deflect the bullets**. In contrast, the target size appears to shrink as it gets farther. These two contrasting effects cause the skill of estimating wind to quickly go from unimportant to very important as targets get farther from the marksman.

Accuracy of Wind Estimates versus Probabiltiy of Hitting a Target

(Using a very accurate rifle at a 25.4 cm (10 in) circlular target at 640 m (700 yd) away, excluding estimates outside two standard deviations.)

Probability of Hitting the Target

Wind speed estimate accuracy km/h (mi/h)

(E.g., ±8.0 means the true wind is within 8 km/h of the estimated wind)

Image 133: This graph shows that **becoming more accurate at estimating wind speed can quickly make a marksman more accurate**. This curve is a "sigmoid," which means that it starts slow, ramps up quickly, and then plateaus. In other words, marksmen who are very bad and very good at wind calls may not see too much immediate benefit from more accurate estimates; but marksmen who are in the middle skill area see fast improvement. (Of course, an advanced marksman at one distance can be just an intermediate marksman at a farther distance.)

Finding Wind Direction

Image 134: Marksmen with electronic wind meters (a.k.a., "anemometers") rotate their meter 360 degrees. The **direction with the highest reading** is the direction of the wind.

Image 135: The classic way to find the direction of the wind is to hold, drop, or watch something light, such as fabric or leaves, in the air and watch which way the wind blows it.

Image 136: This man measures the wind's speed with a wind meter, and the wind's direction with a compass.

Image 137: If a marksman is staying in one place for a while, they can set up a portable tripod vane.

15.a Wind Value

Wind can come from any direction in three-dimensional space: left-right (crosswind), forward-back (headwinds and tailwinds), and up-down (vertical wind). However, headwinds and tailwinds have little effect on bullets, and vertical winds are uncommon. Therefore, the most important "wind" for a marksman to understand is not the actual wind, but the component of the wind that blows perpendicular to the bullet's path. This left-right component is known as the "equivalent crosswind," and the equivalent crosswind's percent of the overall wind is the "wind value" (measured in percent).

The calculation of the wind value involves two steps. First, the marksman determines which direction the wind is blowing from. (See Finding Wind Direction, Pg. 159.) This is not a precise measure because wind constantly

Wind Value Chart

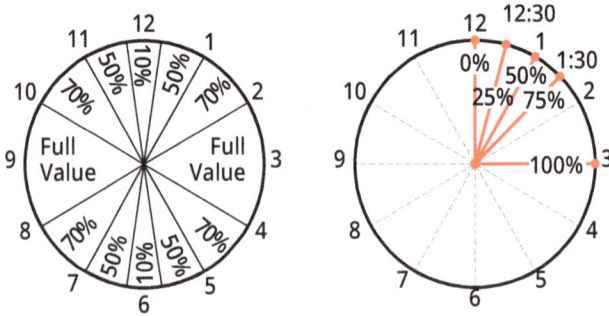

Image 138: Wind value is based on where the wind comes from. The most common method of finding a wind value is to use the clock system, where the marksman is pointing at 12 o'clock. A parallel wind (12 and 6 o'clock) has no value, while a crosswind (9 and 3 o'clock) has a full value. On the left is the simplest system a marksman can use, with only 4 values: 100%, 70%, 50%, and 10%. This becomes less precise as the wind comes closer to 12 or 6 o'clock, and the percentages change quickly. On the right are the wind values for specific clock locations.

Image 139: This wind value table uses the same values as the table to the right, but in a clock format. To use this chart, first, find the actual wind speed in either yellow slice. Then rotate around the circle to the direction the actual wind is coming from. The resulting cell contains the equivalent crosswind in km/h (mi/h).

Actual Wind to Equivalent Crosswind

Wind Direction		Wind Speed in Kilometers (Miles) per Hour									
Degrees	O'clock	5 (3)	10 (6)	15 (9)	20 (12)	25 (16)	30 (19)	35 (22)	40 (25)	45 (28)	50 (31)
0°	12:00, 6:00	0 (0)	0 (0)	0 (0)	0 (0)	0 (0)	0 (0)	0 (0)	0 (0)	0 (0)	0 (0)
10°		1 (1)	2 (1)	3 (2)	3 (2)	4 (3)	5 (3)	6 (4)	7 (4)	8 (5)	9 (5)
15°	11:30, 12:30, 5:30, 6:30	1 (1)	3 (2)	4 (2)	5 (3)	6 (4)	8 (5)	9 (6)	10 (6)	12 (7)	13 (8)
20°		2 (1)	3 (2)	5 (3)	7 (4)	9 (5)	10 (6)	12 (7)	14 (9)	15 (10)	17 (11)
30°	11:00, 1:00, 5:00, 7:00	3 (2)	5 (3)	8 (5)	10 (6)	13 (8)	15 (9)	18 (11)	20 (12)	23 (14)	25 (16)
40°		3 (2)	6 (4)	10 (6)	13 (8)	16 (10)	19 (12)	22 (14)	26 (16)	29 (18)	32 (20)
45°	10:30, 1:30, 4:30, 7:30	4 (2)	7 (4)	11 (7)	14 (9)	18 (11)	21 (13)	25 (15)	28 (18)	32 (20)	35 (22)
50°		4 (2)	8 (5)	11 (7)	15 (10)	19 (12)	23 (14)	27 (17)	31 (19)	34 (21)	38 (24)
60°	10:00, 2:00, 4:00, 8:00	4 (3)	9 (5)	13 (8)	17 (11)	22 (13)	26 (16)	30 (19)	35 (22)	39 (24)	43 (27)
70°		5 (3)	9 (6)	14 (9)	19 (12)	23 (15)	28 (18)	33 (20)	38 (23)	42 (26)	47 (29)
75°	2:30, 3:30, 8:30, 9:30	5 (3)	10 (6)	14 (9)	19 (12)	24 (15)	29 (18)	34 (21)	39 (24)	43 (27)	48 (30)
80°		5 (3)	10 (6)	15 (9)	20 (12)	25 (15)	30 (18)	34 (21)	39 (24)	44 (28)	49 (31)
90°	3:00, 9:00	5 (3)	10 (6)	15 (9)	20 (12)	25 (16)	30 (19)	35 (22)	40 (25)	45 (28)	50 (31)

Image 140: This table takes the actual horizontal wind speed (X-axis) and the wind's direction (Y-axis), and outputs the **equivalent crosswind (cell contents)**. The value is found by multiplying the wind speed by the sine of the degree (See Image 178, Pg. 210.), when the direction the marksman faces is 0 degrees.

changes direction. At a distance, marksmen can observe vegetation or flags to see in which direction they are blowing. (See Wind Estimation, Pg. 178.)

Because wind measurements are imprecise, they typically are put into an "o'clock system." (See Image 138, Pg. 160.) Imagine the marksman is standing in the middle of a clock, with the target straight ahead at 12 o'clock. If the wind is blowing directly from the front, it's coming from 12 o'clock. If it's coming from directly behind, it's at 6 o'clock. Winds from the right side are at 3 o'clock, and winds from the left are at 9 o'clock.

The second step is to input the wind direction into a lookup table or chart to find the wind-value percentage. The wind value is then multiplied by the full wind speed to obtain the equivalent crosswind speed. (See Image 139, Pg. 160.) For example, if the marksman estimates a 15 km/h (9.3 mi/h) wind blowing at a 90° angle to the bullet, the wind value is 100%, resulting in an equivalent crosswind speed of 15 km/h (9.3 mi/h). However, if the wind is blowing at a 45° angle to the trajectory, the wind value is only 71%, leading to an equivalent crosswind speed of approximately 11 km/h (15 km/h × 71% ≈ 11 km/h) or 6.8 mi/h (9.3 mi/h × 75% ≈ 6.8 mi/h).

15.b Crosswind Deflection

To deflect is to cause something to change its direction. That is, "deflection" is a technical way of saying "push." So "crosswind deflection" is how a left-right wind pushes a bullet sideways off its original path-of-travel. A crosswind is named based on the direction it comes from. For example, a "left wind" blows from left to right. To account for a crosswind, a marksman uses a windage hold or dial. (See Windage Holds and Dials, Pg. 173.)

Crosswind deflection is more important than tailwind and headwind deflection because bullets are designed to be aerodynamic (i.e., ignore air resistance) in the direction they travel. But their design does not prioritize preventing a crosswind from pushing them sideways. Crosswinds are more important than vertical-winds (updrafts and downdrafts) simply because crosswinds are far more common, although their effect is the same.

Wind-deflected bullets behave in a way that might seem counterintuitive: they actually point into the wind, not away from it. This happens because the front of the bullet, having a smaller surface area, experiences less wind force than the thicker rear section. As a result, the wind pushes the rear more than the front, causing the bullet to rotate and align itself toward the direction of the wind. (See Image 141, Pg. 163.) Also, the same gyroscopic effect from spinning that causes bullets to point forward causes them to point into all incoming air resistance, including wind. Pointing into incoming air resistance is referred to as "weathervaning," as it is similar to how weathervanes align with the wind.

The amount of rotation a bullet experiences is directly proportional to the ratio of the crosswind to the bullet's speed. For example, a crosswind of 20 km/h (12 mi/h) is approximately 1% of the wind that a bullet traveling at 2,000 km/h (1,240 mi/h) encounters. This means that the bullet may rotate about 1% or roughly 1° into the wind. The rotation happens rapidly as the bullet travels through the air, and the new orientation persists as long as the wind persists. Therefore if the wind stops, a bullet's tip realigns itself with the direction-of-travel (i.e., the direction of incoming air resistance).

Crosswind Deflection

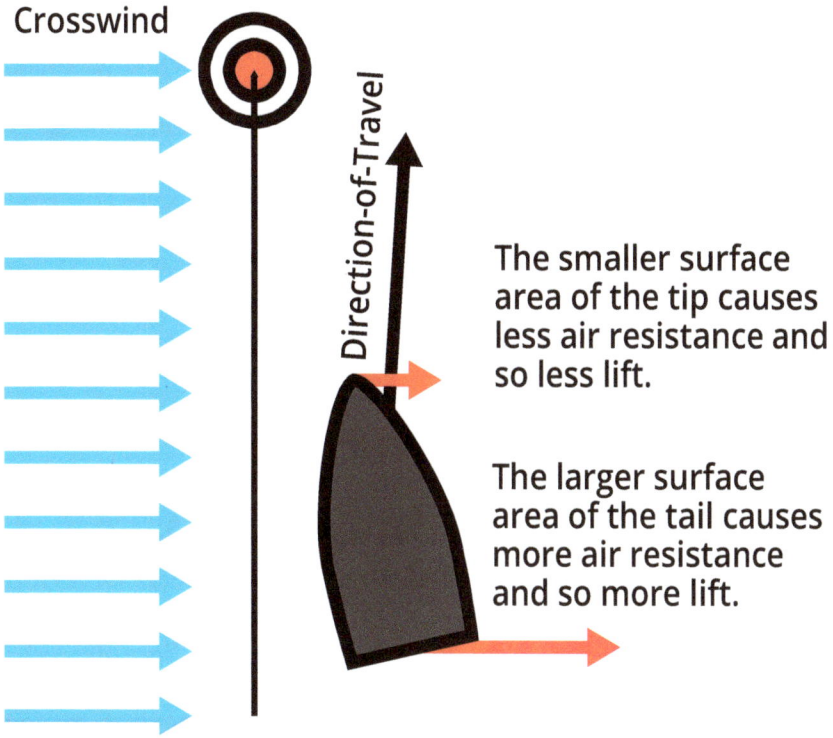

Crosswind

Direction-of-Travel

The smaller surface area of the tip causes less air resistance and so less lift.

The larger surface area of the tail causes more air resistance and so more lift.

Image 141: Crosswinds push bullets sideways, deflecting them off of their original course onto a new trajectory. In this diagram, the left-to-right crosswind has deflected the bullet sideways, so its direction-of-travel is now shifted to the right. However, at the same time, the bullet rotates to face the crosswind because its tip is less affected by the crosswind compared to its tail. Although this diagram is intended to be from a top-down perspective, this diagram would also apply equally to vertical winds if it were a side view.

15.c Vertical-Wind Deflection

Vertical-wind deflection caused by updrafts and downdrafts operates identically to crosswind deflection but at a 90° angle. Most long-range shooting locations have natural features that cause updrafts and downdrafts which the wind must pass over. However, these obstacles often generate an upward wind on one side and a downward wind on the other side, effectively neutralizing each other. Therefore, to observe a vertical-wind effect, a bullet must only travel over a portion of an obstacle or have a significant distance between the updraft and downdraft.

Therefore, the most common situation where significant vertical wind occurs is when targets are positioned on large cliffs. When wind hits a cliff, it has nowhere to go but up. When a wind falls over a cliff, it must go down. (See Image 149, Pg. 169.) This is most often a consideration for inclined shooting. (See Shooting Uphill or Downhill (Inclination), Pg. 207.) Overall however, vertical wind is always connected to nearby terrain features. (See Image 142, Pg. 165.) So, the most effective way to detect vertical wind is to remember that air behaves like a fluid, and therefore consider how a fluid may interact with the surrounding terrain.

Vertical wind deflects bullets up and down, and so marksmen must compensate for it with an elevation hold or dial (See Elevation Hold and Dial, Pg. 137.) and not a windage hold or dial (See Windage Holds and Dials, Pg. 173.) which is specifically used for crosswind deflection.

15.d Wind Gradient

At higher altitudes, the wind blows faster due to the wind's reduced friction with the ground. This increase in wind speed at higher elevations causes bullets to **experience a stronger wind** than would be implied by the wind speed at ground level alone.

Even at a height of 5 m (5.5 yd) above the ground, the wind can be significantly faster than at ground level, and this speed increases every 5 m (5.5 yd) above that. The severity of the gradient depends on how frictional the ground is. For example, trees greatly impede ground wind from blowing, and so they cause a more severe wind gradient than grass does. In fact, a forest may experience a wind speed increase from 0 to 50 km/h (31 mi/h) from below to above its canopy. In contrast, a desert or frozen lake may have almost no wind gradient (i.e., wind speed at ground level would be the same as the wind speed at a higher altitude). Consequently, when determining wind speed at a particular location on a range, the marksman must take into account the bullet's elevation along its trajectory and account for the wind speeds at those elevations in addition to the ground-level wind speed.

A bullet's precise trajectory is not necessary to account for a wind gradient, and instead **the wind gradient can be estimated** by finding the apex of the bullet's travel. This happens because wind direction and speed are constantly changing, so any precise measurements quickly become outdated and turn into rough estimates anyway.

To estimate the apex, a marksman can use the following process. First, they picture their bullet traveling in a straight line upward for the first half of the distance to the target, and then in a straight line downward for the second

Image 142: A Belgian special operations sniper takes aim at targets across a valley. Hochfilzen Training Area, Austria, 11 Sep 2018. When shooting over a valley or depression, marksmen must consider **wind gradients**. Marksmen near cliffs, especially those who shoot parallel to a cliff, must also be aware of the possibility of **updrafts** deflecting their bullets up and **downdrafts** deflecting them down.

half. For example, if the target is positioned 1,000 m (1,090 yd) away from the marksman, halfway to the target would be at 500 m (547 yd). Then, the marksman determines the apex of the bullet's travel. For example, assume a marksman compensates for bullet-drop by aiming the center of their reticle 10 mils high (i.e., an elevation hold of 10 mils), which translates to a height of 5 m (5.5 yd) at 500 m (547 yd). (1 mil is defined as 1 m of height at a distance of 1000 m. So, at 500 meters away, 10 mils would have a height of 5 m.) This calculation determines the height of the apex for a triangular shape, but since bullets follow a curved trajectory, the apex can be estimated as half of the height of the theoretical triangle. Therefore, in this example, the bullet's maximum height may be approximately 2.5 m.

Accounting for wind gradients is typically necessary only for very long shots. However, as the distance to the target increases, the height of the bullet's apex rises rapidly. This is because the upward angle at which marksmen must aim **grows exponentially** with the distance to the target. In the example provided, if the distance is doubled, the bullet's apex rises to three times its original height.

That being said, wind gradient remains relevant in a different capacity when shooting at distances of 500 m (547 yd) and beyond. Marksmen must take into consideration **faster winds that occur over valleys or depressions** and remember that the maximum speed in these cases does not apply to the

Wind Gradient

Wind Speed   **Wind Deflection in cm (~in)**

Image 143: The graph on the left shows what a wind gradient is: as altitude increases (Y-axis), wind speed (X-axis) also increases. The diagram on the right shows how powerful a wind gradient can be for high-altitude shots. The shorter curve shows how much a bullet fired in a crosswind of constant power would be deflected. The longer curve shows what happens when the crosswind speed is low (close to the ground) and high (far from the ground) (i.e., a wind gradient). When a bullet is fired in a wind gradient, the wind speed quickly surpasses the speed of the constant crosswind, thereby deflecting a bullet much farther overall.

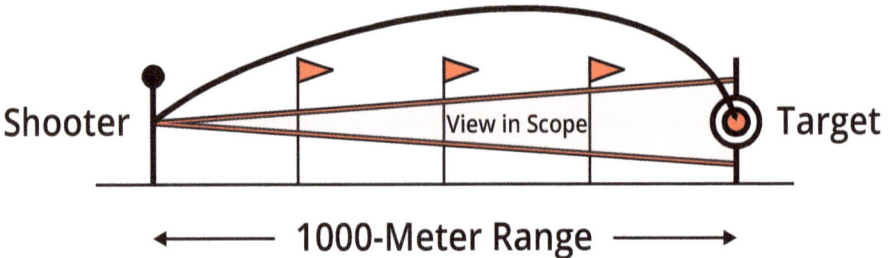

Image 144: **Wind gradients also affect the ability of marksmen to read the wind** because the indicators that marksmen use to determine wind speed, such as mirage or flags, can be below the trajectory of the bullet.

entire trajectory of the bullet and must be considered as a separate wind. (See Multiple Winds over Distance, Pg. 167.)

Even if a marksman can determine a bullet's apex, precisely determining a wind gradient is difficult because the typical ways to estimate wind speed do not work in the sky. (See Image 144, Pg. 166.) For example, vegetation slows the wind, so cues from vegetation do not accurately reflect the wind speed above it. Additionally, simple mirage disappears at 10 m (11 yd) above the ground. (See Mirage, Pg. 184.) To compensate for this, most marksmen simply add a few km/h or mi/h to the speed of the prevailing wind for every 5 m (5.5 yd) above the ground that the bullet travels.

That being said, one indicator that can inform a marksman about the wind speed in the sky is **the presence of gusts** on the ground. A "gust" is a sudden, strong rush of wind that overrides the prevailing wind for a short period of time. When a gust occurs, the marksman can infer that the gust is moving at the same speed as a higher wind.

DOPE books can also be helpful; marksmen may be able to determine the wind gradient of an area by looking at their previous engagements and figuring out how each variable affected their shooting. If every variable except wind is accounted for, the marksman can reasonably infer information about the wind in an area, and thereby deduce what wind gradient must be present.

15.e Multiple Winds over Distance

Because wind is fluid, it is never consistently uniform across a range. Typically, the wind is consistent enough that a single wind measurement at any point is sufficient for generalization. However, this is not always the case, and marksmen may encounter significant variations at different areas along a bullet's path, especially if it travels far. (See Image 145, Pg. 168.)

In fact, **when shooting over different kinds of terrain, marksmen can assume that there are multiple winds, one for each terrain type**. To calculate the overall wind effect, a marksman can divide the bullet's trajectory into equal sections, assign each section a wind speed and a wind value, and then average the equivalent crosswinds for each section. (See Wind Value, Pg. 159.) Marksmen typically only divide the bullet's trajectory into two sections (near and far) or three sections (near, middle, and far) to minimize the number of measures they take.

After calculating the equivalent crosswind (EC) for each section (See Wind Value, Pg. 159.), these sections are combined using a **weighted average**. A weighted average is a method of computing an average where some data points contribute more than others. In this case, weights are necessary because wind affects a bullet's trajectory more when the wind is closer to the marksman than when it's closer to the target. This is because winds change the bullet's trajectory, and the earlier this happens, the more time the bullet has to deviate from its original trajectory. Therefore, winds closer to the marksman are given more weight in the averaging process. (See Image 146, Pg. 168.)

Multiple Winds over Distance

Image 145: **Wind follows terrain**, and so marksmen must take multiple wind readings when shooting at longer distances over changing terrain.

Image 146: **Winds nearer to the marksman have a greater effect** because they deflect bullets sooner (left). For the same effect as a near wind, a far wind must deflect the bullet at a steeper angle (right).

Total influence of wind attributed to sectional thirds (Near, middle, and far from shooter.)

- Far third of target distance
- Middle third of target distance
- Near third of target distance

Image 147: The wind near a marksman has more effect on a bullet's trajectory than wind far from a marksman. This diagram also shows that the far third has more influence as total distance to the target increases.

Influence of wind speed across a bullet's trajectory

Image 148: This shows the same idea as the graph to the left in a different form. Wind deflects bullets less as bullets get closer to their target. However, this reduction is not linear; it follows a convex curve.

Eddy Currents

Windward-Side Eddies

Leeward-Side Eddies

Image 149: **An eddy is a circular movement of air** that deviates from the main flow. On the windward side, where the wind directly hits the barrier, eddies form less frequently, especially when the barrier is smooth. Wind may instead move smoothly, but vertically up the barrier. However, small-scale eddies can occur if the wind is turbulent or the surface of the barrier is rough. **On the leeward side (the side with wind blowing away from it), eddies are much more prominent**. The length of the eddies formed in the wake of the barrier can extend five to fifteen times the barrier's height.

For example, assuming three sections of equal distance, the marksman can allocate the first third of the bullet's travel distance to have a weight of one-half of the weighted average equivalent crosswind (AEC), the second third to have a weight of one-third of the weighted AEC, and the last third to have a weight of one-sixth of the weighted AEC. That is:

► $\frac{1}{2}$(near-third EC) + $\frac{1}{3}$(middle-third EC) + $\frac{1}{6}$(far-third EC) = AEC.

A formula for only two sections may look like this:

► $\frac{2}{3}$(near-half EC) + $\frac{1}{3}$(far-half EC) = AEC.

These formulas use simple, approximate fractions to make the math easy. The true fractions would change based on the distance a bullet travels and the rifle equipment used. (See Image 147, Pg. 168.)

Marksmen can also use sections of different distances with equal weight (i.e., a simple average, or mean). For example, a marksman may consider calculating the mean-average equivalent-crosswind by simple-averaging the effects of wind in the first quarter of the distance, the second quarter of the distance, and the second half of the distance with equal weight. That is:

► $\frac{1}{3}$(first-quarter EC) + $\frac{1}{3}$(second-quarter EC) + $\frac{1}{3}$(second-half EC) = AEC.

This method gets its wind readings closer to the marksman, and may be preferable if a marksman can get better wind estimates at closer locations.

A third way to section a bullet's trajectory is to avoid fractions altogether, and instead section the trajectory using natural dividers such as valleys or open areas. Then the marksman assigns a weight to each section based on their personal experience. This approach works well when a bullet travels through terrain that is unusual or has an unusual wind for some reason. However, it also requires a lot of practice shooting over different terrain, as the connection between an environmental feature and a certain kind of wind can be both subtle and dynamic. There cannot be a general rule either, as the terrain must inherently be very unusual or non-standard for this method to be superior to the aforementioned fractional-sectioning methods.

For example, valleys are notorious for channeling wind to high speeds that do not match the surrounding area. In contrast, an area surrounded by vegetation may have significantly slower winds than surrounding areas. Solid obstacles such as large walls or cliff faces can even create **eddy currents** (air moving in chaotic circles) that reverse or confuse the wind direction. (See Image 149, Pg. 169.) While single, odd trees or reversed wind flags can be ignored, a few factors together may indicate a significant reverse wind.

There are two final options to shoot accurately through multiple winds while avoiding calculations altogether. First, a marksman can simply **wait for there to be only one wind present**. This can occur if either all the winds merge into a single wind, or the wind dies entirely. (See Changing Wind over Time, Pg. 171.) Alternatively, if all winds are relatively constant and a marksman can afford to miss, they can fire a test shot and make adjustments for following shots based on the first shot's impact.

Despite the science that says otherwise, some marksmen have been taught that farther winds affect bullets more. The logic behind this is that bullets slow down as they travel, so farther winds have more time to act on a bullet per unit of distance traveled compared to closer winds. While bullets do slow down, they do not slow down enough for the effect to be stronger than the earlier deflection caused by closer winds. For example, wind in the final third of a trajectory would need to work on a bullet traveling three times slower than a wind in the first third to have an equivalent effect, At one third the speed, most bullets would become subsonic and start to tumble. (It is possible to check if a projectile tumbled by observing its impact on the target since tumbling in the air typically leads to an oblong hole on the target (a.k.a., a "keyhole.").

Another reason marksmen may believe that farther winds are stronger is that they often measure wind in enclosed spaces with no wind. This is common because people naturally seek non-windy areas to be more comfortable. With that said, **it is recommended that marksmen consider the effects of wind outside their immediate, comfortable area.**

Another flawed method of dealing with multiple winds used by some marksmen is to simply use the highest recorded wind speed. While this approach may work to some extent, it is important to recognize that a strong wind is likely not present along the entire shooting range.

15.f Changing Wind over Time

Wind is constantly changing its direction, speed, or both over time, so the wind that a marksman expects is not necessarily the wind that the bullet experiences after being shot from the rifle. This mismatch mostly occurs because of the delay between when a marksman observes the wind and when they shoot. However, **there is also a fundamental delay from the moment a marksman pulls the trigger to when the bullet travels through the air**.

Typically, changing wind is not a problem as it remains stable enough from second to second. However, there are instances when the wind changes too rapidly for any estimate to be timely, such as when gusts are common. In those cases, it is best to **measure and shoot when the wind is consistent**, rather than when it is completely calm. This is especially true where wind only experiences brief and random pauses (i.e., most places). If wind is never consistent, marksmen can take multiple measurements to calculate an average.

If the wind is changing too rapidly to measure quickly, a marksman can rely on a visual cue instead. For example, if the marksman notices that their shots consistently hit the target when the wind causes the tips of two branches to come into contact, they can time their shots and only fire when the branches touch. The only downside of using a visual cue is that they are difficult to find in the first place.

The patterns by which wind changes can be found, but are not immediately obvious. To measure a wind pattern, marksmen can create a diagram using equivalent crosswind velocity on the Y-axis and time on the X-axis. (See Image 150, Pg. 172.) Every minute or less, the marksman records the wind value. After ten minutes and ten measurements, a pattern may emerge.

With this information, marksmen can make reasonably accurate predictions about when a preferred wind is likely to occur. However, more importantly, such a diagram also illustrates the limits of the equivalent crosswind speed, or in other words what winds do not occur. This is useful because it allows marksmen to avoid shooting when a gust is likely to occur. Knowing the limits of wind speed is also useful when using a "hold-target" windage hold. (See Image 154, Pg. 177.) The most common equivalent wind

Tracking Wind Speed over Time

Equivalent Crosswind Speed over Time

Image 150: Wind in an area during a period is bounded by high and low speed limits. Also, **wind may have multiple conditions and switch between them**. This example shows two conditions. A marksman can chart wind patterns in a specific area over different times and seasons to identify local wind trends. By understanding how long wind conditions typically remain stable, the marksman can determine the optimal window to take a shot before the wind changes.

Windrose Plot for [DFW] DALLAS/FT WORTH

Image 151: Wind roses are circular histograms that show winds' origins during a period of time. This one for Dallas, Texas in October 2022 (left) and 2023 (right) shows that during October the wind primarily comes from the North and South.

speed is the wind's "condition." A wind can have multiple conditions. (See Image 151, Pg. 172.).

(For the statistically oriented, the equivalent crosswind speed can be modeled with a normal distribution or a multimodal normal distribution. A "condition" is the mean or mode and not technically sticky, and a "limit" is where winds become very improbable according to the standard deviation.)

15.g Windage Holds and Dials

Once the equivalent crosswind is determined, the marksman can convert it into a windage hold, which indicates which horizontal hashmark on the reticle the marksman must use as their point-of-aim. To counteract the wind's force, the marksman **must always shoot into the wind**. So, if a marksman encounters a left-to-right crosswind, they aim to the left by using a windage hashmark on the right side of the reticle.

The fact that wind from the left requires using the right side of the reticle, and vice versa, can cause confusion and errors when an assistant or spotter communicates wind information to a marksman, especially under stressful conditions. To clear up all of this confusion, it can be helpful to always add "point rifle" or "wind from" at the beginning of any communication. For example, "wind from the left, 2 mils," meaning: align a target with the 2-mil hashmark on the right side of the reticle.

Reticles can compensate for wind effects because they measure angular distance, and wind deflects bullets at an angle. (See Image 141, Pg. 163.) That is, wind deflects bullets from their original trajectory onto a new trajectory, so the windage hold is equal to the angle of that deflection. For example, a 20 km/h (12 mi/h) crosswind is 1% of the airflow that a bullet traveling at 2,000 km/h (1,240 mi/h) experiences. Therefore, the bullet may rotate approximately 1% or about 1° into the wind. In sum, countering the wind's effect can be achieved by firing the bullet at the exact opposite angular distance to cancel out the wind's angular deviation.

The amount of deflection is directly correlated to the speed of the crosswind. For example, if in an 8 km/h (5 mi/h) crosswind, a bullet would be deflected by 10 cm (2.5 in) at a distance of 100 m, a 16 km/h (10 mi/h) crosswind would deflect the bullet by 20 cm (5.1 in) at 100 m. That is, a wind that is double the power always deflects twice as much.

Rifles can be called by the amount of crosswind velocity required to deflect them by 1 mil. For example, a setup that is deflected by 1 mil by an 8 km/h (5 mi/h) crosswind would be an "8-km/h gun" or a "5-mi/h gun."

Crosswind to Windage in Mils (MOA)

Equivalent Crosswind in km/h (mi/h)	Distance to Target in m (yd)									
	100 (109)	200 (219)	300 (328)	400 (437)	500 (547)	600 (656)	700 (766)	800 (875)	900 (984)	1000 (1094)
2 (1)	0.0 (0.1)	0.1 (0.2)	0.1 (0.2)	0.1 (0.3)	0.1 (0.4)	0.1 (0.5)	0.2 (0.6)	0.2 (0.8)	0.3 (0.9)	0.3 (1.0)
5 (3)	0.1 (0.2)	0.1 (0.4)	0.2 (0.6)	0.2 (0.8)	0.3 (1.1)	0.4 (1.3)	0.5 (1.6)	0.6 (1.9)	0.7 (2.3)	0.8 (2.6)
10 (6)	0.1 (0.4)	0.2 (0.8)	0.3 (1.2)	0.5 (1.6)	0.6 (2.1)	0.8 (2.6)	0.9 (3.2)	1.1 (3.8)	1.3 (4.5)	1.5 (5.2)
15 (10)	0.1 (0.5)	0.3 (1.1)	0.5 (1.8)	0.7 (2.5)	0.9 (3.2)	1.2 (4.0)	1.4 (4.8)	1.7 (5.7)	2.0 (6.8)	2.3 (7.9)
20 (13)	0.2 (0.7)	0.4 (1.5)	0.7 (2.4)	1.0 (3.3)	1.2 (4.2)	1.5 (5.3)	1.9 (6.4)	2.2 (7.6)	2.6 (9.0)	3.1 (10.5)
25 (16)	0.3 (0.9)	0.6 (1.9)	0.9 (3.0)	1.2 (4.1)	1.5 (5.3)	1.9 (6.6)	2.3 (8.0)	2.8 (9.6)	3.3 (11.3)	3.8 (13.1)
30 (19)	0.3 (1.1)	0.7 (2.3)	1.0 (3.5)	1.4 (4.9)	1.9 (6.4)	2.3 (7.9)	2.8 (9.6)	3.3 (11.5)	3.9 (13.5)	4.6 (15.7)
35 (22)	0.4 (1.3)	0.8 (2.7)	1.2 (4.1)	1.7 (5.7)	2.2 (7.4)	2.7 (9.2)	3.3 (11.2)	3.9 (13.4)	4.6 (15.8)	5.3 (18.3)
40 (25)	0.4 (1.4)	0.9 (3.0)	1.4 (4.7)	1.9 (6.6)	2.5 (8.5)	3.1 (10.6)	3.7 (12.8)	4.4 (15.3)	5.2 (18.0)	6.1 (21.0)
45 (28)	0.5 (1.6)	1.0 (3.4)	1.5 (5.3)	2.2 (7.4)	2.8 (9.5)	3.5 (11.9)	4.2 (14.4)	5.0 (17.2)	5.9 (20.3)	6.9 (23.6)
50 (31)	0.5 (1.8)	1.1 (3.8)	1.7 (5.9)	2.4 (8.2)	3.1 (10.6)	3.8 (13.2)	4.7 (16.0)	5.6 (19.1)	6.5 (22.5)	7.6 (26.2)

Image 152: This table takes the distance to the target (X-axis) and the equivalent crosswind (Y-axis), and **outputs the windage hold or dial (cell contents) in mils and MOA.** The values here are specifically for a .308 match cartridge (155-grain bullet, 43.0 grains powder). Note that the additional windage hold (e.g., for a 50 km/h (31 mi/h) wind) is almost twice as much between 900 and 1000 m (7.6 - 6.5 = **1.1**) as between 100 and 200 m (1.1 - 0.5 = **0.6**). Marksmen can also convert the hold into clicks on their scope if they are dialing for wind.

This table is only an example because equipment characteristics, such as barrel length, affect a rifle's muzzle velocity, which in turn influences the windage adjustments. In fact, elevation and windage charts require constant adjustments anyway because the correction elevation and windage holds change as a barrel experiences wear over its lifetime. Both windage and elevation tables are also **only valid for one set of environmental conditions** (e.g., temperature, altitude, etc.); however, it only takes minor changes to make a table made for one setup in one environment valid in another environment.

Ready-made charts such as this can often be found online for specific rifles and ammunition. This chart is just an example.

Actual Wind to Windage at 500 m

Wind Direction		Actual Wind Speed in km/h (mi/h)									
Degrees	O'clock	5 (3)	10 (6)	15 (9)	20 (12)	25 (16)	30 (19)	35 (22)	40 (25)	45 (28)	50 (31)
0°	12:00, 6:00	0 (0)	0 (0)	0 (0)	0 (0)	0 (0)	0 (0)	0 (0)	0 (0)	0 (0)	0 (0)
10°		0.1 (0.4)	0.1 (0.4)	0.1 (0.4)	0.1 (0.4)	0.3 (1.1)	0.3 (1.1)	0.3 (1.1)	0.3 (1.1)	0.6 (2.1)	0.6 (2.1)
15°	11:30, 12:30, 5:30, 6:30	0.1 (0.4)	0.1 (0.4)	0.1 (0.4)	0.3 (1.1)	0.3 (1.1)	0.6 (2.1)	0.6 (2.1)	0.6 (2.1)	0.6 (2.1)	0.9 (3.2)
20°		0.1 (0.4)	0.1 (0.4)	0.3 (1.1)	0.3 (1.1)	0.6 (2.1)	0.6 (2.1)	0.9 (3.2)	0.9 (3.2)	0.9 (3.2)	1.2 (4.2)
30°	11:00, 1:00, 5:00, 7:00	0.1 (0.4)	0.3 (1.1)	0.6 (2.1)	0.6 (2.1)	0.9 (3.2)	0.9 (3.2)	1.2 (4.2)	1.2 (4.2)	1.5 (5.3)	1.5 (5.3)
40°		0.1 (0.4)	0.3 (1.1)	0.6 (2.1)	0.9 (3.2)	0.9 (3.2)	1.2 (4.2)	1.2 (4.2)	1.5 (5.3)	1.9 (6.4)	1.9 (6.4)
45°	10:30, 1:30, 4:30, 7:30	0.1 (0.4)	0.3 (1.1)	0.6 (2.1)	0.9 (3.2)	1.2 (4.2)	1.2 (4.2)	1.5 (5.3)	1.9 (6.4)	1.9 (6.4)	2.2 (7.4)
50°		0.1 (0.4)	0.6 (2.1)	0.6 (2.1)	0.9 (3.2)	1.2 (4.2)	1.2 (4.2)	1.5 (5.3)	1.9 (6.4)	2.2 (7.4)	2.5 (8.5)
60°	10:00, 2:00, 4:00, 8:00	0.3 (1.1)	0.6 (2.1)	0.9 (3.2)	1.2 (4.2)	1.2 (4.2)	1.5 (5.3)	1.9 (6.4)	2.2 (7.4)	2.5 (8.5)	2.8 (9.5)
70°		0.3 (1.1)	0.6 (2.1)	0.9 (3.2)	1.2 (4.2)	1.5 (5.3)	1.5 (5.3)	2.2 (7.4)	2.5 (8.5)	2.5 (8.5)	2.8 (9.5)
75°	2:30, 3:30, 8:30, 9:30	0.3 (1.1)	0.6 (2.1)	0.9 (3.2)	1.2 (4.2)	1.5 (5.3)	1.9 (6.4)	2.2 (7.4)	2.5 (8.5)	2.8 (9.5)	3.1 (10.6)
80°	.	0.3 (1.1)	0.6 (2.1)	0.9 (3.2)	1.2 (4.2)	1.5 (5.3)	1.9 (6.4)	2.2 (7.4)	2.5 (8.5)	2.8 (9.5)	3.1 (10.6)
90°	3:00, 9:00	0.3 (1.1)	0.6 (2.1)	0.9 (3.2)	1.2 (4.2)	1.5 (5.3)	1.9 (6.4)	2.2 (7.4)	2.5 (8.5)	2.8 (9.5)	3.1 (10.6)

Image 153: This table takes the actual horizontal wind speed (X-axis) and the wind's direction (Y-axis), and outputs the wind hold (windage) for a 500 m target. **This table is not valid at any other distance.** Also, because the Crosswind to Windage table used increments of 5 km/h, the conversion to mil (MOA) is rounded to the nearest 5 km/h.

This kind of table is useful for marksmen who only shoot at one distance. It can be made by taking a wind value conversion table (See Image 140, Pg. 161.), and converting the cell contents (equivalent crosswind) into a corresponding windage by using a single column of a windage table (See Image 152, Pg. 174.). In fact, marksmen can drastically **simplify the table even further** by eliminating columns and rows for uncommon winds that they do not expect to see where they are shooting. If a marksman is dialing their windage, they can go one step further and convert windage hold into the number of windage dial clicks.

Ready-made charts such as this can often be found online for specific rifles and ammunition. This chart is just an example.

Doubling the distance to the target, however, more than doubles the deflection. This is because bullets slow down as they travel, so a crosswind has more than twice as much time to act on a bullet at twice the distance. (That being said, when bullet speed is relatively stable, for example before 300 m or yd, the angle of wind deflection can be treated as if it were linear because wind speed is an estimate anyway.)

A wind meter and a ballistic calculator are very helpful to determine the necessary windage hold to counteract a crosswind. (See Meters, Pg. 216.) (See Ballistic Calculators, Pg. 221.) However, the traditional method is simple trial and error. That is, a marksman fires into a well-measured crosswind at targets of various distances and measures how far sideways the bullet was deflected. The marksman can then construct a table where the X axis is the distance to the target, the Y axis is the wind speed, and each cell has the angle of deflection (i.e., the windage hold). Although these tables take considerable time to create for an individual marksman and their specific equipment, generic wind tables are readily available online. (See Image 152, Pg. 174.)

Wind can gust, die down, or even switch directions. To account for these swift changes in wind, **windage is not typically adjusted by dialing turrets**. Instead, marksmen use the faster approach of holding the reticle to the side and using an off-center hashmark as their point-of-aim.

However, there are a few exceptions to not dialing for wind. First, in the case of a second-focal plane scope, the reticle hashmarks only represent whole-number angular measurements at specific magnifications. (See Image 119, Pg. 137.) If the marksman wants to use an arbitrary magnification setting, they cannot utilize the reticle to hold for wind. Nevertheless, this scenario is not particularly common, as marksmen often only magnify to a zoom level where the clicks on the turret align with the reticle hashmarks.

Another exception to not dialing for wind includes when marksmen take multiple shots at a target in relatively consistent wind conditions; then they can adjust the windage dial so that the center of the reticle aligns with the center of the target when the wind blows at its most common speed. That is, they can dial to the wind's primary condition. (See Image 150, Pg. 172.) Furthermore, marksmen who engage in shooting at extreme distances can employ both dialing and holding techniques to compensate for shooting into extremely strong winds, as relying solely on one method could be insufficient.

There are also techniques to compensate for wind without using reticle hashmarks nor dialing. The simplest technique is to "**hold target**" left or right. (See Image 154, Pg. 177.) This technique points the center of the

"Hold Target" Windage Hold

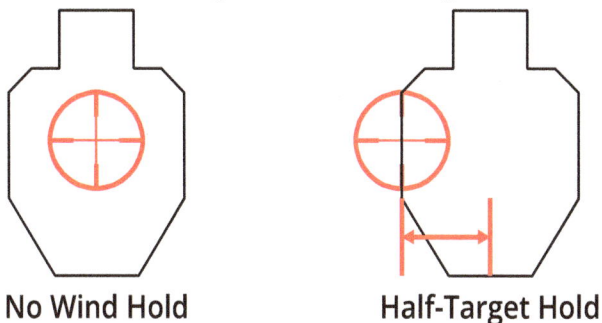

No Wind Hold   Half-Target Hold

Image 154: If the direction of the wind is known, a marksman can use a half-target hold, where the marksman aims into the wind at the maximum edge of the target. Thereby, **the wind has twice as much distance to blow the bullet** before the wind deflects the bullet completely off the target.

reticle at the maximum left edge of the target if the wind blows from left to right, and the maximum right edge if the wind blows from right to left. The logic behind this is that a constant crosswind only pushes a bullet in one direction, and so it can only push the bullet from the edge towards the center. Therefore, if the wind weakens, a bullet hits the edge of the target, and if the wind remains strong, it pushes the bullet toward the opposite side of the target. Either way, the bullet hits the target.

After the first shot, this technique can also be used to refine a marksman's aim. For example, if a bullet aimed at one edge is blown across the target and hits the opposite edge, the marksman can adjust their aim by half of a target's width off the edge opposite of the impact to hit the center of the target.

Another technique that works irrespective of the wind is **for the marksman to change their position**. If there is a 90° crosswind, the marksman can rotate their shooting position 90° around the target to turn the crosswind into a headwind or tailwind. Alternatively, the marksman can try to get closer to the target. At a closer distance, bullets are less affected by many external factors including wind.

15.h Converting a Wind to a Windage Hold (Summary)

Converting wind to a windage hold is a complicated process. In fact, tournament competitors often arrive hours before a competition to assess the wind conditions on the range. Prepared marksmen may even study maps and pictures to determine the typical wind patterns, and understand any potential atypical patterns and obstacles that could alter the prevailing wind's direction and strength. Bowl-shaped areas, in particular, can create non-standard wind patterns, requiring extra preparation. Therefore to help simplify the process, the above section on determining a windage hold is summarized here:

1) **Measure the wind at a specific location.** This can come from an electronic tool or through an estimate. (See Wind Estimation, Pg. 178.)
   a) If the wind changes over time, wait for the best wind. (See Changing Wind over Time, Pg. 171.)
   b) If shooting near any cliffs or mountains, consider the effects of a vertical wind component. (See Vertical-Wind Deflection, Pg. 163.)
   c) If shooting long or over a valley, consider the wind gradient and add to the measured or estimated wind speed. (See Wind Gradient, Pg. 164.)
   d) If there are multiple winds, divide the range into sections and measure the wind in each section. (See Multiple Winds over Distance, Pg. 167.)

2) **Convert the actual wind into the equivalent crosswind** using a wind-value table. (See Wind Value, Pg. 159.) (See Image 140, Pg. 161.)
   a) If shooting at a long range, combine the equivalent crosswinds at different locations into an average equivalent crosswind for the bullet's entire trajectory. (See Multiple Winds over Distance, Pg. 167.)

3) **Convert the equivalent crosswind into a windage hold** using a windage-hold table. (See Windage Holds and Dials, Pg. 173.) (See Image 152, Pg. 174.)
   a) If shooting at one distance, combine a wind-value table and a windage table into a wind-to-windage table. (See Image 153, Pg. 175.)

16. Wind Estimation

Predicting wind is the most challenging aspect of shooting because wind is invisible and is constantly changing. And while electronic wind meters are extremely accurate, they can only measure the wind speed from the marksman's position (and they need batteries). Therefore, when shooting on a range (besides high-tech ranges with remote electronic wind meters), wind speed is not measured, but rather it is estimated. Marksmen can estimate wind by observing certain indications in the air, such as flags, surrounding vegetation, or mirages. When estimating, marksmen must **be bold** with their estimates, as wind speed is most commonly underestimated.

To improve accuracy, multiple people, such as a marksman-spotter team, can each **make independent estimates and then average them.** A group's average tends to be more accurate than a single individual's guess.

Marksmen can also enhance their wind estimation skills through training. For example, they can make estimates and compare them to readings from a wind meter. Some marksmen enjoy estimating wind as a pastime, for example, while walking their dog. With practice, a marksman can improve the accuracy of their wind speed estimates based on any cues they see. They can also discover more wind cues, enabling them to consider more information.

That said, the practice of estimating wind using environmental cues may soon become obsolete due to technological advancements, much like how laser rangefinders replaced manual distance estimation. For example, ZX Lidars' product, the ZXTM, can measure wind speed and direction with a high level of precision, up to a distance of 550 m (602 yd). Today, its total adoption is only hindered by its weight of 90 kg (198 lb), but its future generations will undoubtedly be smaller and much more portable.

16.a Flags and Vanes

Flags and vanes (devices that rotate to face the wind) are useful tools for estimating wind conditions at a distance from a marksman, as they can show both wind speed and direction.

Streamer flags (i.e., normal flags) are simple strips of cloth or plastic that flutter in the wind. Flags point opposite the wind's direction, with a raised flag indicating strong wind and a hanging flag suggesting light or no wind. They are often brightly colored for visibility. Streamer flags can indicate wind direction, but they are not reliable for showing wind speed.

Wind cones (a.k.a., windsocks or wind sleeves) are also made of fabric and flutter in the wind; but in contrast to regular flags, they are cone-shaped fabric tubes that catch the wind and point in the direction it is blowing. They are often designed with sections that indicate the wind speed by inflating at specific speeds. (See Image 155, Pg. 180.)

Vanes (a.k.a., wind vanes or weathervanes) are mounted on a pivot, and they consist of a tail and a pointer. The tail catches the wind, while the pointer indicates the wind direction. Pinwheel vanes are a kind of vane that have a small pinwheel or propeller attached to their vane. The direction of the pinwheel indicates wind direction, and the speed at which it spins provides an indication of wind speed. (See Image 157, Pg. 181.)

While streamer and vane flags are useful, marksmen often do not know precisely how one reacts to wind. Even two flags or vanes of the same type

Wind Flags and Wind Cones

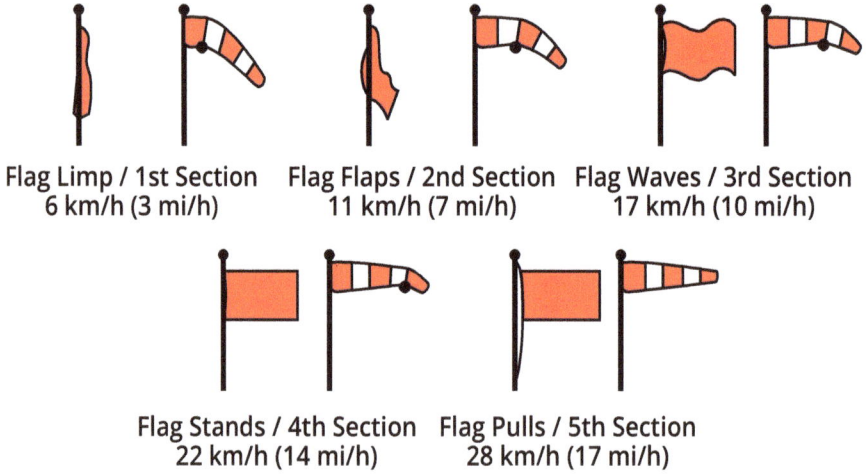

Flag Limp / 1st Section
6 km/h (3 mi/h)

Flag Flaps / 2nd Section
11 km/h (7 mi/h)

Flag Waves / 3rd Section
17 km/h (10 mi/h)

Flag Stands / 4th Section
22 km/h (14 mi/h)

Flag Pulls / 5th Section
28 km/h (17 mi/h)

Image 155: A wind flag can either be a regular, streamer flag (left) or a special tool called a "wind cone" (right). Wind cones have hollow, stripped sections that represent 3 knots each. So, the second section extends at 6 knots, the third at 9, etc. (A knot is an imperial unit of distance equal to 1.14 miles or 1.85 kilometers.)

Direction flag (parallel to sightline).

Target

Velocity flag (perpendicular to sightline).

Direction flag

Flag Straight Away

Wind Direction Changing

Flags from Shooter's Point-of-view

Image 156: A marksman can use **multiple wind flags** to better read the wind. This is best done by placing two flags, one at 45 degrees forward left and one at 45 degrees forward right. Then, no matter which direction the wind blows from, one flag is more parallel to the marksman, and one is more perpendicular. The parallel flag is easier to read wind direction from, and is the direction flag (left). The perpendicular flag is easier to read wind speed from, and is the speed flag (first diagram above).

Wind Vanes

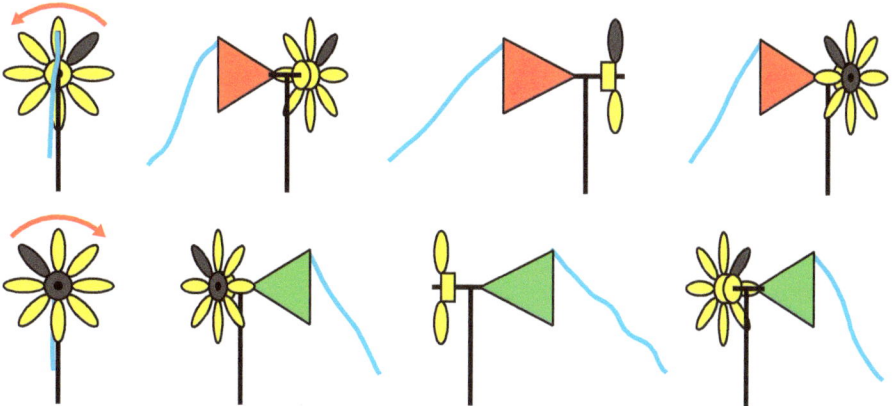

Image 157: A wind vane is a stick mounted on a swivel that points in the direction of the wind. Some wind vanes may have attached propellers that can also tell the wind speed. This is made easier by coloring one blade a different color so that the speed of rotation is more obvious. This example also has a few other indicators, such as a tail painted with two colors to more easily know which direction the vane is slanted towards, and also a ribbon tail that flows in the wind. Wind vanes are **cheap and easy to transport**, so a marksman can stick many into the ground.

may perform differently depending on how they are mounted, their age, and the environmental conditions in which they operate. Therefore, it is important that marksmen have experience using their particular devices And novice marksmen may have to rely on backup measurements from other devices or techniques.

Marksmen can enhance the utility of flags by using multiple flags simultaneously. For example, a marksman can place a flag 45° forward left and 45° forward right of their shooting location. (The two flags form a 90° angle with the marksman.) (See Image 156, Pg. 180.) Since two identical flags respond identically to the same wind conditions, the marksman can effectively gather information from two different viewpoints (one for each flag). The flag that points towards or away from the marksman can provide more precise information on wind direction, while the one perpendicular to the marksman gives a more accurate representation of wind speed.

Multiple flags can also be useful when placed on either side (left and right) of a shooting range. Then regardless of whether a wind blows left-to-right or right-to-left, at least one flag is always upwind, showing information about the wind entering the shooting range. In contrast, downwind flags are not as useful because they only describe winds that have already left the shooting range.

Of course, marksmen are not limited to a specific number of flags or vanes and can use many. Additional flags help gather information in challenging areas, such as terrain gaps with gusts. However, it is important to note that each reading from a flag takes time, during which wind conditions can change.

16.b Vegetation

In instances where flags are not available, the surrounding vegetation can be used to estimate wind. The effects of wind on vegetation vary depending on the intensity of the wind. While the impact may not be as distinct or as precise as with specifically designed wind flags, some general guidelines from the Beaufort wind scale can be useful (See Image 158, Pg. 183.):

- 0 to 5 km/h (0 to 3 mi/h) : Wind is barely noticeable.
- 5 to 10 km/h (3 to 5 mi/h): Leaves slightly sway.
- 10 to 15 km/h (5 to 8 mi/h): Leaves are in continuous motion.
- 15 to 20 km/h (8 to 12 mi/h): Stray leaves become airborne.
- 20 to 25 km/h (12 to 15 mi/h): Thin tree trunks slightly bend.
- 25 to 30 km/h (15 to 20 mi/h): Thin tree trunks bend significantly.
- 30 to 40 km/h (20 to 25 mi/h): Thick tree trunks slightly bend.

Although it is clear that these are estimations, the accuracy of these guidelines or any guidelines is actually worse than it may appear. This is because when a guide mentions "thin tree trunks sway," each word can have a completely different meaning depending on the tree, the location, and the marksman. This is because different trees and leaves have varying weights and structures, and react to wind accordingly. So, an expert marksman who becomes skilled at reading wind in North America may not be as accurate in other parts of the world or even in different states with different vegetation.

That being said, any estimate is better than no estimate. And marksmen can become extremely accurate if they practice correlating the wind speed with the wind's effect on local vegetation in a specific location. Some expert marksmen keep notebooks filled with information on how local vegetation reacts to wind for later reference, and can be specific enough to reference the reaction of individual trees. For example, two branches may touch on an oak tree in their backyard when the wind is 21 km/h (13 mi/h).

Another wind estimation method is known as the **divide-by method**. Here, the marksman either drops a leaf, observes a falling leaf, or looks at a string or flag along the path to the target. The marksman estimates the angle at which the object is falling or hanging and then divides that angle by 2.5

Vegetation to Wind Speed Chart

0 km/h
0 mi/h
Smoke Rise Up

2-5 km/h
1-3 mi/h
Smoke Drifts

6-11km/h
4-7 mi/h
Leaves Rustle
Flags Stir

13-19 km/h
8-12 mi/h
Leaves and
Twigs Move

21-29 km/h
13-18 mi/h
Branches Move
Flags Flap

31-39 km/h
19-24 mi/h
Small Trees
Sway
Flags Ripple

40-50 km/h
25-31 mi/h
Large Branches
Move,
Flags Beat

51-61 km/h
32-38 mi/h
Whole Trees
Move
Flags Extend

63-74 km/h
39-46 mi/h
Twigs Break
Walking
is Difficult

76-87 km/h
47-54 mi/h
Signs, Antennas
Blown Down

88-117 km/h
55-73 mi/h
Trees Unrooted
Considerable
Damage
to Buildings

119 km/h
or above
74 mi/h
or above
Countryside
Devastated

Image 158: Charts that convert environmental cues to wind speed are called "Beaufort scales." **Beaufort scales are not exact** because vegetation of different locations responds differently to different winds. They are also not subjective, as what one marksman may consider a "small" tree or a "large" branch may differ from another marksman's opinion. Therefore, for maximum accuracy, marksmen make their own scales for the specific locations they shoot on for later reference.

if using km/h, and by 4 if using mi/h. The resulting number is the estimated wind speed. For example, a leaf that falls at a 45° angle. 45 divided by 2.5 equals 18. So, the estimated wind speed is 18 km/h (11.2 mi/h).

The location from which the wind is estimated makes a significant difference for whether it can be generalized to the whole range. Gaps between hard structures, such as valleys or tree groves, are noticeably windier than the surrounding areas, even if they seem calm. However, if wind from a gap is

Image 159: This is a **shooting berm**, a raised strip of dirt behind shooting targets. It serves as a backstop to stop bullets from going further. The numbers indicate a shooting lane. The puff of dust under the "35" sign starts on the left (thinner) and is being pushed right (thicker) by the wind. While **impacted dirt-dust** is a consistent wind indicator, it only occurs behind the target, where the wind does not necessarily match the wind on the rest of the shooting range.

parallel to the marksman, it is a headwind or tailwind. Those have little impact on bullets and can be disregarded. Wind can even become erratic when it spills over the tops of trees and swirls back (known as an eddy current). This swirling or eddying can deceive the marksman because an eddy wind moves branches without indicating a specific direction. (See Image 149, Pg. 169.)

Marksmen can also calibrate their wind calls by observing the **dust** created when a bullet hits a dry berm. (See Image 153, Pg. 175.) In competition, an astute marksman waits for others to shoot into the backstop and reads the dust generated by their shots to calibrate their own estimates.

16.c Mirage

Mirage is an optical distortion whereupon objects appear wavy due to fluctuating air temperatures. This happens when different areas of air have varying temperatures and densities. Since light bends differently through air of different densities, pockets of hot and cold air create the wavy visual effect effect known as a "mirage."

Mirages appear as vertical lines because hot air rises straight up, creating columns between cooler, denser air. As the sun heats the ground, a layer of hot air forms just above the surface, beneath the cooler air. The intense heat at ground level causes the air to move quickly and shimmer, producing wavy lines that resemble moving water or oil.

Certain conditions are more prone to creating a mirage than others. For example, sunnier days in open areas cause the ground to heat up faster, and higher humidity prolongs the cold air's coolness. Since mirage depends on

Simple Mirage

Image 160: Firefighters, assigned to the Ohio National Guard's 180th Fighter Wing, extinguish controlled fires. Swanton, Ohio in Swanton, OH, 19 May 2020. The firefighter on the left is **obscured by mirage,** while the firefighter on the right is in the clear. Mirage is difficult to capture in a still image, which is why these images are of intense flames. However, even hot rocks can create moving mirage.

Image 161: This image shows how mirage forms **vertical columns**. Note the pole centered in the image, and how the mirage forms vertical streaks on it.

relative temperature rather than absolute temperature, it exists universally. Even snowy mountains experience mirage.

The fact that mirage forms alternating, vertical columns is very important. When wind blows mirage, it blows the top farther than the bottom making the columns bend over at an angle. (See Wind Gradient, Pg. 164.) Faster winds blow the columns of mirage over at a steeper angle. Therefore, a skilled marksman can infer wind speed by reading the angle of the mirage. (See Image 162, Pg. 186.) However, this cannot be done beyond approximately

Mirage in Wind

0 Wind:
Boiling

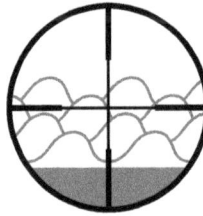

1-5 km/h (1-3 mi/h)
Leaning

6-10 km/h (4-6 mi/h)
Flowing Gently

11-15 km/h (7-9 mi/h)
Flowing Rapidly

16-20 km/h (10-12) mi/h
Slick

Image 162: Mirage moves with the wind: the stronger the wind, the steeper and faster the mirage appears to flow. Although mirage itself behaves consistently at any given wind speed, **every marksman interprets mirage differently**. Therefore, to interpret mirage and correlate it with wind speed, a marksman must practice observing real mirage in the field and determine how it appears to them.

---

15 km/h (9.3 mi/h) because wind faster than that merges the pockets of hot and cold air, thereby removing the mirage effect.

Here are some rough estimates of wind speed based on mirage:

- 0° (boiling): 0-1 km/h   (0-1 mi/h)
- 30° mirage (leaning):   2-5 km/h   (1-3 mi/h)
- 45° mirage (flowing):   6-10 km/h   (4-6 mi/h)
- 90° mirage (rapid):   11-15 km/h   (7-9 mi/h)
- Slick or scrambled mirage: 16-20 km/h   (10-12 mi/h)
- No mirage in ideal conditions: 21 + km/h   (13 + mi/h)

The pattern of mirage can vary in both intensity and direction. Weak mirage appears as faint lines visible only through a scope, while strong mirage with clearly separated columns can be so severe that even the bullet holes on a target become difficult to see. Occasionally, when various temperature layers are present, multiple mirages may blend together. The strength of a mirage indicates nothing about the wind; wind speed can only be inferred from the mirage's angle.

Compared to flags, mirage is a more valuable indicator because it occurs **directly within the air that a bullet travels through**. Marksmen are best served by looking at a mirage that is either close to the marksman or at the midpoint of a bullet's trajectory because closer winds affect a bullet's trajectory more than farther winds. (See Image 147, Pg. 168.) Mirage at the target is only useful if it represents wind across the bullet's entire trajectory. Additionally, mirage responds instantaneously to sudden gusts because mirage is made of the air itself.

Mirage is easier to see where two contrasting surfaces meet, such as a bright white piece of paper against a dark background. To observe a strong mirage in nature, a marksman can seek a sunlit rock in front of a contrasting background such as a shaded treeline. Usually, mirage is observed at a low magnification setting since zooming too far loses the macro pattern of vertical lines when viewing a weaker mirage. However, if intense lines are present, marksmen can zoom in on a contrasting edge to observe the mirage's angle.

Even when a mirage is correctly observed, there is an incredibly rare occurrence known as a false mirage where the mirage appears to flow into the wind instead of away from it. This phenomenon happens when the wind encounters a solid object, such as a group of trees, causing the wind to spin in a descending arc against the object's surface (i.e., an eddy). (See Image 149, Pg. 169.) When this arc reverses direction, it essentially creates a reverse wind. If a mirage is observed in this reverse wind, it would indicate a wind direction contrary to the prevailing wind in the surrounding area. While eddies themselves are not uncommon, it is unusual for marksmen to be firing very close to large, solid objects that might be capable of creating eddies.

17. Climate

Climate refers to the characteristics of the air, including air density (which depends on pressure, temperature, and humidity) and precipitation. When bullets travel through the air, these characteristics influence the bullets' trajectories. Also, changes in air density change how light refracts, and can thereby interfere with a marksman's sight picture.

Given the complexity of the interactions between a climate and a bullet's trajectory, one may ask how marksmen managed to compensate for changes in climate before the age of calculators and computers. The short answer is that they didn't; marksmen would always re-zero their rifles and scopes whenever they encountered a new climate. (See Ensuring Accuracy (Grouping and Zeroing), Pg. 74.) They also didn't shoot as far, and so didn't need to account for as many variables in the first place.

17.a Temperature

Scientifically, temperature is a physical quantity which describes how quickly molecules are moving in a material. In more understandable terms, hotter objects contain more heat energy.

This increase in temperature has three effects that are relevant to shooting. First, hotter air is less dense than colder air. This is because the molecules are moving faster with more energy, and so bounce off each other with more force. This fact has implications for air density. (See Air Density, Pg. 189.)

The second effect is an increase in the speed of sound. Temperature is by far the most important factor that determines the speed of sound. The speed of sound is proportional to the square root of the absolute temperature, giving an increase of about 0.6 m/s per 1°C (1.1 ft/s per 1°F) at normal, outdoor temperatures. Bullets are designed to be the most stable when they travel at supersonic speeds, and that stability can quickly deteriorate once a bullet travels through transonic speeds into subsonic speeds. Therefore in hotter air, bullets become unstable at higher velocities (i.e., sooner in the trajectory).

The third major effect is how temperature affects gunpowder. Gunpowder is a chemical compound that burns into a gas after reaching a certain temperature. The chemical reaction of gunpowder creates a set amount of energy. Therefore, colder gunpowder explodes into a colder gas, and hotter gunpowder into a hotter gas. As mentioned, hotter gas is less dense than colder gas at the same pressure; therefore, hotter gas exerts more pressure at the same density. Given that the bore confines the gas to a set density, hotter gas then exerts more pressure on bullets in the bore than colder gas does. More pressure means higher muzzle velocity; so in sum, there is a direct relationship between the temperature of gunpowder and a bullet's muzzle velocity.

The relationship between gunpowder temperature and muzzle velocity isn't exactly linear, but it is close enough that it can be treated as such. For most rifles, raising the powder temperature by 1°C increases the muzzle velocity by about 0.5 to 1 m/s. Similarly, increasing the powder temperature by 1°F increases the muzzle velocity by about 1 to 2 ft/s.

This temperature effect usually isn't a major concern, especially before 300 m or yd. However, it can become significant in certain situations. For example, if an extreme-long-range marksman zeros their rifle on a cold winter morning but later uses it on a hot summer afternoon, that difference in temperature can impact performance. A rifle might rise about 0.2 mils (0.7 MOA) more at 300 m or yd, 1 mil (3.4 MOA) more at 1000 m or yd, and exponentially more thereafter if the muzzle velocity may increase by 3% due to a temperature difference of 40°C (72°F).

Again, all of these calculations are approximations because the chain of events (i.e., an increase in temperature, to greater chamber pressure, to a higher muzzle velocity, to decreased bullet-drop) is highly dependent on the specific rifle and ammunition in question. Also, bullet-drop is related to the time a bullet spends in the air before hitting the target, and that variable's relation to muzzle velocity is not straightforward. Therefore, exact calculations require the aid of ammunition-temperature versus muzzle-velocity calculators that can account for specific bullets and gunpowders. The best estimates may also require a bit of trial and error to dial-in estimates to specific equipment.

Of course, the simple answer is to **always shoot ammunition at the temperature it was zeroed with.** This can be done by shooting in a similar climate, or by warming (or cooling) ammunition before shooting it.

But marksmen must also not allow ammunition to heat up too much during a shooting session. Ammunition must be stored out of direct sunlight, not just as a safety measure, but also to keep it more accurate. Additionally, rifles must be able to cool down enough, and marksmen fire quickly enough so that rifle cartridges cannot absorb extra heat from the rifle. In fact, a primary purpose of brass casings is to absorb heat when fired, allowing that heat to be expelled as the casings are ejected, thereby ensuring that subsequent rounds remain cooler.

17.b Air Density

Bullets do not follow a perfect parabolic arc due to gravity because they encounter air resistance along the way. As a bullet travels, it hits air molecules and must push them aside. To push away the air molecules, the bullet must transfer some of its forward momentum to the air, which prevents the bullet from retaining that momentum to move itself forward. In denser air, there are more molecules to push through, increasing air resistance. Therefore, higher air density causes bullets to move slower and drop sooner.

To measure air density, an observer must consider the three climatic factors influencing air density: **air pressure, temperature, and humidity.** (Notably, altitude is only relevant insofar as it changes these three.) Before computers, these values were challenging to translate into a change in windage and elevation holds. However, nowadays, marksmen can read this information from the internet, and then input it into a ballistic calculator

Air Pressure

Air pressure falls significantly as altitude increases.

Air pressure does not change significantly with different locations of the same altittude.

Lower pressure (i.e., fewer air molecules) at higher altitudes means less resistance.

Higher pressure (i.e., more air molecules) at lower altitudes means more resistance.

Image 163: Altitude is the strongest determining factor of air pressure. At the altitudes almost all bullets are fired from, for every 1 km (0.6 mi) of elevation gained, pressure drops by approximately 10%.

and easily obtain the ballistic output. (See Ballistic Calculators, Pg. 221.) If weather predictions are reliable, marksmen can even use forecasts to precalculate air density at different times of day. However, because factors influencing air density change throughout the day, marksmen get the most accurate results by periodically reentering climate data into their calculators.

For whatever reason, a marksman may not want to use the internet, and instead prefer to get analog readings for air pressure, temperature, and humidity respectively from a barometer, a thermometer, and a hygrometer. However, they would still need to input the data into a ballistic calculator, or else perform very arduous trial and error processes as well as calculations to learn how the climate affects their equipment.

When reading the air pressure off the internet, observers must get the station pressure and not the barometric pressure. **Station pressure** is the actual air pressure at a specific location and elevation, reflecting the force exerted by the atmosphere due to the weight of the air column directly above that point.

In contrast, **barometric pressure**, also known as sea-level pressure or adjusted pressure, is the station pressure corrected to a standard reference level, usually sea level. Barometric pressure does not change much based on location. The highest recorded barometric pressure was 108 kilopascals (kPa) (16 pounds per square inch (psi)) in Siberia, while the lowest was 87 kPa

(13 psi) during a typhoon in the Pacific Ocean, representing only about 20% difference in barometric pressure.

Station pressure can change a lot based on location. Not only does barometric pressure influence station pressure, but altitude also affects station pressure. (See Image 163, Pg. 190.) For every 1 km (0.6 mi) of elevation gained, pressure drops by approximately 10 kPa (1.5 psi). (Although the real relationship is curved, this approximated linear relationship holds well enough for everywhere except the high-altitude Andes and Tibetan Plateaus.)

Although not as significant as air pressure, **temperature** also affects air density. As hotter air expands, it becomes less dense at the same pressure if it can expand. However, the reason that temperature does not affect local air pressure as much as it could, is because it cannot freely expand. For air in an area to expand, it must push the rest of the air on earth out of its way in. Regardless, the effect of temperature on air density is just as difficult to determine as the ambient temperature's effect on gunpowder. Marksmen who want extreme accuracy must use quality ballistic calculators.

Finally, air density is also affected by **humidity**, which is the percent of air that is comprised of water in its gaseous form. More broadly speaking, air is a composition of gases, and each gas has a different pressure and density relationship. At the same station pressure, water vapor is less dense than air. Therefore, as the air becomes more humid, it becomes less dense even if the air pressure remains the same. That being said, the effect of humidity is too small for anything but a calculator to consider; going from 0 to 100% humidity may cause a bullet to impact 0.1 mils (0.3 MOA) higher at 1,000 m or yd.

17.c Simple Mirage

Mirage is an optical phenomenon that distorts air and causes objects to appear distorted by wavy or often described as "dancing" or "vibrating" lines. It was briefly explained as it related to wind (See Mirage, Pg. 184.), but mirage is also relevant to shooting in other ways. To explain the concept of mirage again, it occurs when different parts of the atmosphere have varying temperatures and, therefore, different densities. The way light travels through the air is dependent on its density, so alternating pockets of hot and cold air with different refractive indexes result in the wavy vertical lines as hot air rises between columns of denser, colder air (the scientific name is "Snell's law").

In the context of shooting and using a scope, **mirage can make the target appear distorted or blurry**, making it challenging for the marksman to interpret the scene and thereby aim accurately. To determine the actual

Image 164: A Soldier testing an XM250. This machine gun is fed by a belt of ammunition, so it can fire hundreds of rounds per minute. Such rapid firing creates **mirage on the barrel** (seen here above the suppressor), which can distort a marksman's sight picture through the scope. U.S. Army Cold Regions Test Center, Delta Junction, AK, 20 Feb 2024.

location of an object obscured by mirage, it is best to select a specific part of the target. As the image dances in the moving air, there are moments when the mirage momentarily disappears, causing the entire target to snap back into a relatively clear image. By focusing on the chosen section, the marksman can gauge the severity of the mirage shift by observing how much their selected target section snaps back. Because the shift is relatively small, marksmen usually measure the snapback by using their reticle.

Additionally, better optics with lenses that contain fewer defects or better technology may provide a clearer view of both the target and the mirage, making it easier for the marksman to interpret what they see. However, because mirage is a physical property of the air, no lens can completely eliminate it. Lower-end scopes may appear to have more mirage simply because their initial clarity is lower than that of higher-end scopes.

With that said, a common technique for interpreting mirage through a scope is to reduce the aperture of the objective lens. (See Image 165, Pg. 193.) This can be done with a specialized, cheap plastic cap that has a hole in the middle, or simply putting opaque tape over the sides of the objective bell. The size of the aperture directly affects the depth-of-field. (See Image 41, Pg. 62.) With a deeper depth-of-field, the image sharpness for the marksman is enhanced over a greater distance.

However, one limitation of this technique is that the human eye already limits the objective lens's aperture. In daylight, the human eye's pupil is only about 2 mm in diameter. Reducing the objective lens aperture of a scope won't affect the depth-of-field until the exit pupil of the scope becomes smaller than

Reduced Aperture

Image 165: A German Army sniper with Jager Battalion 292, looks through his scope. Rukla, Lithuania, 11 June 2015. He has partially covered his objective bell. Covering the aperture increases depth-of-field.

Image 166: An infantryman assigned to the 1st Armored Division uses his M110 semi-automatic sniper system. Fort Bliss, TX, 13 Aug 2019. Covering the aperture prevents the lens from reflecting light (i.e., "glint").

2 mm. (See Image 28, Pg. 49.) Another limitation is that when the exit pupil is smaller than the human pupil, the image is less bright to the eye.

A second technique is to reduce the scope's magnification. Again, the effectiveness varies among marksmen and nullifies the effect of reducing the aperture. But the point of this technique is to reduce the relative size of mirage shift to allow the brain to more easily see and interpret the scene as a whole.

An expert marksman knows, however, that the most effective way to mitigate mirage is to **avoid it entirely**. For example, a marksman can avoid direct midday sunlight or mornings when dew cools the ground, both of which are events that commonly create mirage. Alternatively if waiting is not an option, a marksman can look to the sky for moving clouds and wait for one to provide shade, as this can significantly reduce any effects of mirage.

Mirage is not only induced by the sun hitting the ground, but also by the rifle itself. A heated barrel can create a large enough temperature differential with surrounding air that mirage appears to float away from it. (See Image 164, Pg. 192.) Although this mirage is very limited in area, it occurs directly in front of the scope, obscuring a marksman's entire sight picture. In fact, this problem is so common that it has its own name: "scope mirage."

To fix scope mirage, marksmen can cover the hot barrel or suppressor. There is specialized equipment available to purchase; however, the traditional fix is to put a piece of cardboard, such as a used ammo box, over the hot suppressor. Ignition of the cardboard is rarely a concern because scope mirage is a clear indicator to a marksman that their barrel is getting too hot, and that they must slow down their rate-of-fire to make the mirage go away.

17.d Inferior and Superior Mirage

Apart from dancing air, mirage may also cause macro effects that shift an entire section of the marksman's sight picture. Generally, these macro mirage-shifts distort the image of an object towards the hotter air. If there is wind present, the mirage may even shift the image sideways towards hotter air.

An "**inferior mirage**" occurs when hot air is closer to the ground and displaces the image of a distant object downwards. This type of mirage is commonly observed in places where the ground is significantly warmer than the air above it, such as roads or deserts. (See Image 167, Pg. 195.) When shooting at the center of a downwardly distorted image, shots land low on the target object.

In contrast, a "**superior mirage**" is when the image of the object appears above its actual position. This type of mirage typically occurs in colder regions or areas that experience evaporative cooling in the morning, causing the air near the ground (or water) to be cooler than the air above it. As light passes from the cooler air into the warmer air, it is refracted downwards, making the object appear higher. (See Image 168, Pg. 195.) When shooting at the center of a higher image, shots land high on the target object.

This information is provided not for the sake of precise calculations, but rather to explain why a rifle that is zeroed in the morning might drift in the afternoon. The equipment itself is not shifting, but **the sight picture is drifting** instead. Applying this information to practical shooting is challenging because predicting mirage is as difficult as predicting slight variations in air temperature throughout the bullet's flight path. However, as a general guideline, marksmen can remember that heavy mirage may cause a vertical displacement of up to a maximum of 0.5 mils (2 MOA) up or down, although almost all vertical displacements are actually much lower than that.

17.e Precipitation

Unlike air density, precipitation has **minimal impact** on a bullet's trajectory. While water concentrated at the muzzle could cause the bullet to deviate, rain is spread out over a distance, and a few raindrops in a light rainstorm are unlikely to have a significant impact on the bullet's trajectory. In fact, due to the bullet's supersonic air jacket (a layer of air that surrounds

Inferior and Superior Mirage

Image 167: Inferior mirage shows a reflection of the real image below the real image.

Image 168: Superior mirage shows a fake, right-side-up image of the real image above the real image.

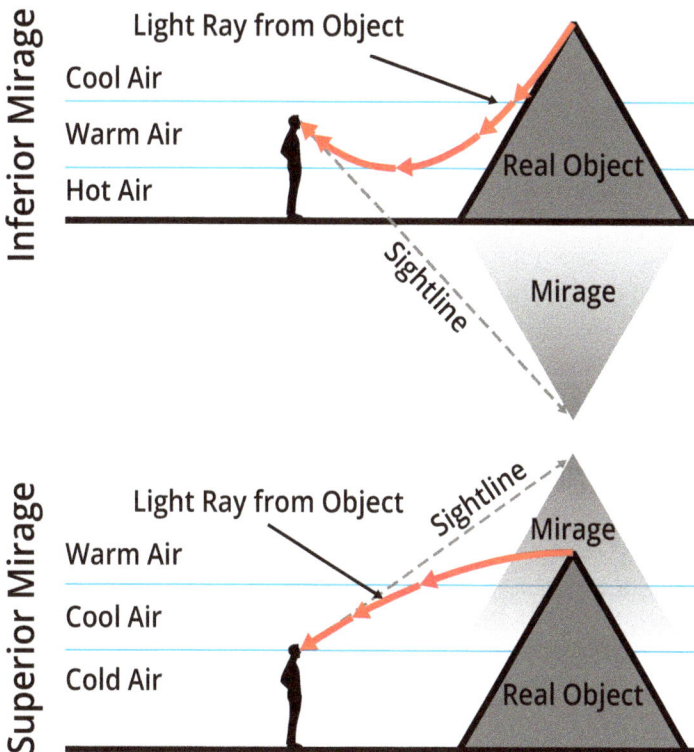

Inferior Mirage

Light Ray from Object

Cool Air

Warm Air

Hot Air

Real Object

Sightline

Mirage

Superior Mirage

Light Ray from Object

Sightline

Mirage

Warm Air

Cool Air

Cold Air

Real Object

Image 169: Differences in air temperature and therefore air density cause light to bend. However, the brain assumes all light travels in a straight line (i.e., the sightline). Therefore, bent-light images, called "mirages," appear to be in a different place than the object actually is. For inferior mirage, the object appears flipped and located below the real object's actual location. For superior mirage, the object appears right-side-up, but above the real object's actual location. The point of these diagrams is not to say that marksmen may mistake the mirages for the real thing, but rather that **mirage is capable of shifting an entire sight picture**.

the bullet (See Image 113, Pg. 131.)), precipitation may never even come into contact with the bullet's surface in the first place.

To put this into perspective, consider a candidate for the world record for the heaviest rainfall in an hour (i.e., rainfall rate): 36 cm/h (14 in/h). Even in such extreme conditions, this equates to only 0.01 cm/s (0.004 in/s) of water falling, and a normal heavy rain would be less than 1/10 of that.

In sum, **rain feels denser than it actually is**. This can be proven with calculations. Normal terminal velocity of rainfall is 7 m/s, but this example rounds up to 10 m/s for easier calculations. Combining the rainfall rate with the rainfall velocity gives the rainfall density in the air:

(Rainfall Rate)   = 36 cm/h = 36 cm ÷ 3,600 s = 1/100 cm/s
(Rain Velocity)   = 10 m/s = 1,000 cm/s
(Rain Density)   = *(Rainfall Rate)* ÷ *(Rain Velocity)*
(Rain Density)   = (1/100 cm/s) ÷ (1000 cm/s)
(Rain Density)   = 1 ÷ 100,000 = 0.001%

Separately, the volume of a bullet's trajectory can be determined, which represents the space that a bullet occupies during its flight. The volume can be calculated by multiplying an area *(Bullet Cross-Section)* by a length *(Bullet Trajectory)*. In this case, the cross-section is simplified to 1 cm² for ease of calculation, although the cross-section of most bullets is significantly smaller. Also, this example uses a trajectory of 1,000 meters for easier calculations.

(Volume of Trajectory) = *(Bullet Cross-Section)* × *(Length of Trajectory)*
(Bullet Cross-Section) = 1 cm²
(Length of Trajectory) = 1,000 m = 100,000 cm
(Volume of Trajectory) = 1 cm² × 100,000 cm
(Volume of Trajectory) = 100,000 cm³

By multiplying the two results, *(Rain Density)* and *(Volume of Trajectory)*, the volume of water in the flight path can be determined.

(Volume of Water) = *(Rain Density)* × *(Volume of Trajectory)*
(Volume of Water) = (1/100,000) × 100,000 cm³
(Volume of Water) = 1 cm³

The primary concern in considering precipitation's effect on bullets, is **rain getting inside the barrel**. That is why protecting the muzzle is important. To prevent this, some marksmen also use tape to cover the rifle's barrel, but tape must be reapplied after each shot. The more common (and practical) solution is to shoot under covering. Also, when shooting outdoors in the rain or when

Image 170: An Albanian sniper and his spotter prepare for a shot. Zagan Training Area, Poland, 20 May 2016. The sniper **keeps his scope covered** until he is prepared to take a shot, and relies on his spotter to search for targets. This photo is very clear despite intense rainfall, showing that rain is always sparse in the air.

rain is expected, it is advisable to bring waterproof clothing and some kind of cover to keep the ammunition dry.

If lightning is expected, it is best to consult with the range officer to determine if it is safe to continue shooting, as lightning strikes have killed many people. Mist and fog are only relevant insofar as they obscure the sight picture and affect humidity. Shooting in limited visibility conditions is always dangerous and not recommended.

18. Inherent Imprecision

Precision refers to how closely multiple shots hit relative to each other. A marksman can increase their precision by, for example, steadying their rifle or using timed breathing correctly. With that said, there are some sources of imprecision that **cannot be eliminated**. For example, one cartridge may have gunpowder positioned slightly differently than in another cartridge, causing exploding gas to push on a bullet differently.

The imprecision that is inherent to the equipment being used can be reduced by switching to different equipment; for example, buying more consistently manufactured ammunition. Another example would be buying bullets with lower air resistance. This is because air, being a moving fluid, has different, constantly changing densities from one cubic meter to the next, and the only way to reduce the influence of air resistance is to use high-quality bullets that experience less aerodynamic forces. (See Image 114, Pg. 131.)

Even then, imprecision cannot be eliminated; all equipment has some amount of inherent imprecision. In other words, part of the deviation of every bullet from the point-of-aim is effectively random. Therefore, accounting for precision is an important part of shooting, and it is the reason why using groups of shots is better than using single shots to zero a weapon. (See Grouping, Pg. 74.) At the expert level, marksmen assign a probability to each shot hitting their target with even perfect technique. This concept of thinking probabilistically instead of in certain terms is called "**probabilistic shooting**."

Probabilistic thinking is often applied to estimates, which are notorious for being imprecise. Laser rangefinders have largely mitigated the imprecision of estimates of a target's distance from a marksman. However, estimation still remains an important part of determining wind speed, even in today's technologically modern shooting era.

18.a Lateral Throwoff

Lateral throwoff refers to a bullet's unpredictable sideways deflection from the marksman's intended trajectory as the bullet exits the firearm. This deflection occurs because the bullet experiences uneven force, either from exploding gas applying uneven force to its rear, or lift applying uneven force to its front.

The phenomenon is a "throwoff" because it is sudden. In the bore, the bullet follows the bore; however, when the bullet leaves the bore, it is suddenly no longer stabilized by the bore's tight fit, and quickly deflects laterally.

Not all of the sideways deflection is random; however, any consistent deflection is accounted for during the zeroing process. Therefore by definition, lateral throwoff is the change to a bullet's trajectory that is impossible to predict or account for.

Lateral throwoff has multiple causes. The first cause discussed here is the **static imbalance** caused by a difference between a bullet's center-of-form and its center-of-mass. The center-of-form is the point that is the furthest on average from every point on the bullet's surface; and the center-of-form axis is the line drawn through the center-of-form for each two-dimensional cross-section of a bullet. When a bullet spins inside the barrel, it rotates around this center-of-form axis because the shape of the bullet is confined within the bore.

On the other hand, the center-of-mass is the point where the bullet's mass is evenly distributed in all directions. For any circle (e.g., the cross-section of a bullet) made of a single material, the center-of-form is the center-of-mass.

Lateral Throwoff

Balanced Bullet

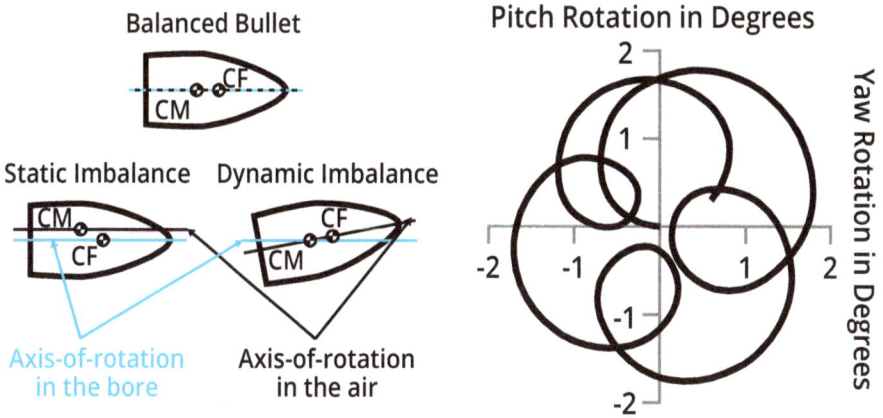

Pitch Rotation in Degrees

Static Imbalance   **Dynamic Imbalance**

Axis-of-rotation
in the bore

**Axis-of-rotation
in the air**

Image 171: A perfectly balanced bullet rotates from the tip to the tail on the same axis in the bore and in the air after it leaves the muzzle. However, **perfect bullets don't exist** and always have some (perhaps imperceivable) static and dynamic imbalance. Static imbalance occurs when the center-of-mass (CM) through the bullet is on a different axis from the center-of-form (CF) through the bullet. This occurs when multi-material bullets do not have their different materials equally rotationally distributed. A statically imbalanced bullet deflects due to inertia as the bullet changes from one axis of rotation to another. Dynamic imbalance occurs when a bullet is slanted in the bore, so that the center-of-form is not aligned with the boreline. A dynamically imbalanced bullet catches air as it exits the muzzle, causing lift force to deflect the bullet. The example graph on the right shows where an example bullet-tip pointed as it left the muzzle for the first 15 yd (13.7 m) due to either static or dynamic instability.

However, in bullets made of multiple materials, such as those with a copper jacket and a lead core, the center-of-mass might not align with the center-of-form. (An easy-to-visualize example of this is a bottle that is halfway-filled with water, laid on its size, and frozen. Then, the remaining half is filled with liquid water. Because ice is less dense than water, the bottle's center-of-mass is within the water, but the center-of-form is in the middle of the bottle.)

If the center-of-mass does not fall on the center-of-form axis, and the bore contains the bullet and forces it to rotate around the center-of-form axis, then that means that the center-of-mass travels through a bore in a helical or spiral path. However, when the bore no longer confines the bullet (i.e., when the bullet exits the muzzle), the bullet is free to spin around its center-of-mass. Therefore, the center-of-form must then travel in a helical path.

When the center-of-mass spins in a helix, it **retains angular momentum** in the direction it is spinning. (See Momentum, Pg. 246.) Therefore, when the center-of-mass is allowed to travel straight, it retains that angular momentum and is slightly diverted into the direction it was spinning in. This diversion is referred to as lateral throwoff.

In fact, this cause of lateral throwoff is the limiting factor on bullet spin. The more a bullet spins, the greater the angular momentum of its center-of-mass, and the greater the lateral throwoff. As such, bullets are not spun as fast as the materials would otherwise allow, as greater rotation would result in increased lateral throwoff. In other words, long-range marksmen prefer to use the slowest possible twist rate that stabilizes a bullet in flight.

As previously mentioned, a likely cause of a misaligned center-of-mass is manufacturing defects in the bullet, resulting in one side of the bullet jacket being thinner than the other side. Since the marksman does not know the orientation of the center-of-mass (cartridges are radially symmetrical and can be loaded into rifles with any rotation about their axis), the final vector of momentum is also unknown, making the lateral throwoff effectively random. In fact, excellent mass balance to reduce lateral throwoff is a major selling point for high-grade ammunition and shorter bullets (consistent jacket thickness is easier to maintain on shorter bullets).

Besides imperfect mass distribution in the construction of a bullet, a second reason for lateral throwoff is the **dynamic instability** caused by a mass misalignment from exiting the bore with a spin axis that is not aligned to the axis of the bullet's rotational symmetry (in simple terms, the bullet tip doesn't point forward). (See Image 171, Pg. 199.)

This can be caused by a bullet entering the bore at an angle instead of straight. An angled entry could in turn be caused by any number of reasons: bullets that are not aligned properly in their case; a chamber or throat that is not concentric with or sized comparatively to the bore; a configuration that leads bullets to catch the rifling and continue down the bore in that slightly angled orientation; etc.

With an angled entry into the bore, the tip of the bullet would rotate spirally, causing the center-of-mass to also misalign with the center of the bore. Equipment with low part tolerances is especially prone to angled bullets. Military and hunting rifles, for example, require some level of clearance between the rifling and the bullet to accommodate the inevitable ingress of dirt and grime into the action and bore of the rifle. On the other hand, while marksmen can buy high-tolerance equipment to reduce lateral throwoff, getting that equipment dirty may greatly impede performance in other ways.

Besides mass imbalance, **uneven gas escape** as the bullet exits the bore is another cause of lateral throwoff. If gas escapes from one side of the bore faster than from the other side, the bullet experiences uneven pressure and therefore undergoes a slight turn away from the stronger-pushing gas. This is due to a few reasons. First, gas is a fluid and as such is not always predictable. Second, uneven lateral gas escape can physically push the rifle sideways, causing unpredictable secondary effects. Finally, escaping gas can exacerbate the aforementioned mass-imbalance effects.

Uneven gas escape can, in theory, be caused by the crown (i.e., the final part of the bore to touch the bullet) not being concentric around the bullet. The concentricity of the crown is the variation in the distance between the crown and the tail of a bullet as it exits the rifle (ideally both are perfect circles, but in practice both have microscopic or macroscopic defects). However, in practice, groupings made with rifles with heavily damaged crowns are indistinguishable from those made with fresh crowns at 100 m or yd. It is actually the boreline alone that shifts. That is, error is introduced, but it is the same error for every shot and is therefore predictable and accounted for with a new zero. A crown that is damaged in the middle of a shooting session may only appear to lose precision as the boreline shifts from one location to another, causing marksmen to not trust their original zero.

18.b Probabilistic Shooting (Mil and MOA-Accuracy)

When bullets are fired, they invariably encounter random forces that cause their trajectory to deviate from the marksman's point-of-aim. The resulting imprecision is not just difficult to avoid, but it is impossible to avoid. Therefore, expert marksmen must learn to account for imprecision caused by random forces, and hit their targets anyway.

The first step is to reframe shooting as the act of firing groups instead of single shots. (See Grouping, Pg. 74.) This is because, whereas a single bullet may follow a relatively linear trajectory, a group of bullets, when fired together, forms a pattern of dispersion that widens as they travel farther from the muzzle. This pattern is referred to as a "**conical trajectory**" or "**cone-of-fire**" meaning it resembles the shape of a cone for which the point is at the muzzle and the base spreads out as distance increases. (See Image 172, Pg. 203.)

When the bullets in the cone-of-fire meet the target, they randomly impact around a central point (i.e., the point-of-aim for an accurate marksman). The bullet impacts create an elliptical, or circle pattern, with more bullets at the center of the ellipse and fewer around the edges. (See Image 173, Pg. 203.) This pattern in statistics is known as a "bivariate normal distribution." The definitions of each term are:

Bivariate – refers to involving two variables; in this case, the horizontal and vertical distances from the point-of-aim in the center.

Mean – is the average value of a set of data points; in this case, the mean is the center of the impacts (i.e., the point-of-aim for an accurate marksman).

Dispersion – is how a set of points is distributed or spread over a wide area. In shooting, high dispersion is low precision, and vice versa. Higher dispersion means a larger ellipse.

Standard deviation – measures the dispersion of data points around the mean. A smaller standard deviation means the impacts are clustered around the mean, showing higher precision. A larger standard deviation indicates that the impacts are more spread out, showing lower precision.

Normal distribution – describes how data points (here, bullet impacts) are distributed for a given mean and standard deviation so that most are near the center (i.e., the mean), and fewer are farther away.

The goal of probabilistic shooting is to ensure that the normal distribution of a marksman's grouping fits within their target area well enough, allowing only an acceptable number of bullets to miss. While it's theoretically impossible to achieve an error rate of 0%, it can be reduced to extremely low levels, such as 0.0001% of shots missing the target.

To decrease their error rate, marksmen can either reduce standard deviation or enlarge the target area. Standard deviation can be decreased and precision increased through methods such as more practice and better equipment. The target area can be made bigger by bringing it closer.

For example, big-game hunters often target the heart of an animal because hitting it instantly secures the kill. Assuming that the heart is a circle with a radius of 10 cm (3.9 in), using traditional thinking, bullets follow the marksman's point-of-aim in a straight line to be able to hit the deer's heart 100% of the time regardless of distance. However, in reality, dispersion means that bullets do not follow a single, predicted trajectory; and to hit the deer's heart 100% of the time, the deer must be close enough to the marksman so that the marksman's cone-of-fire overlays the target area.

There are instances where a marksman may struggle to narrow their cone-of-fire enough to reduce their impact area to a size smaller than the target area. In such cases, some bullets hit the target while others miss. **Shooting multiple times to ensure that at least one bullet hits the target**

Cone-of-Fire

Individual Bullet Trajectory

Cone-of-Fire of Multiple Trajectories

Individual Trajectories

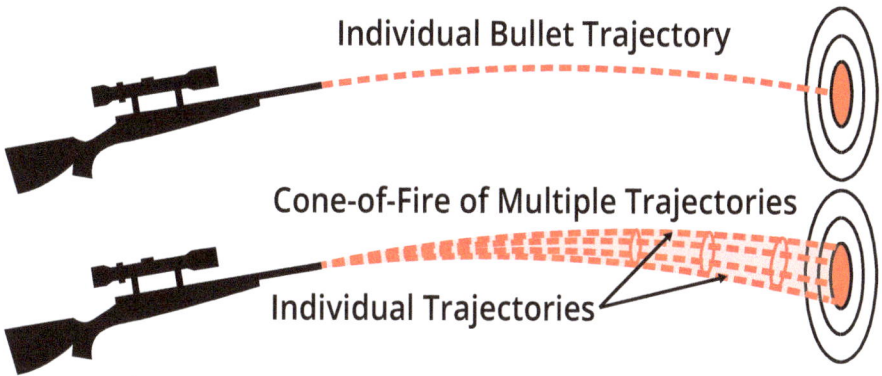

Image 172: Every bullet fired from a rifle has its own trajectory. However, they all originate from the rifle's muzzle. Therefore, all the trajectories fit within a (bent) cone shape called the "cone-of-fire." A major goal of marksmen is to make the cone **as narrow as possible**.

Normal Distributions of Shot Dispersion

Probability in X-Axis

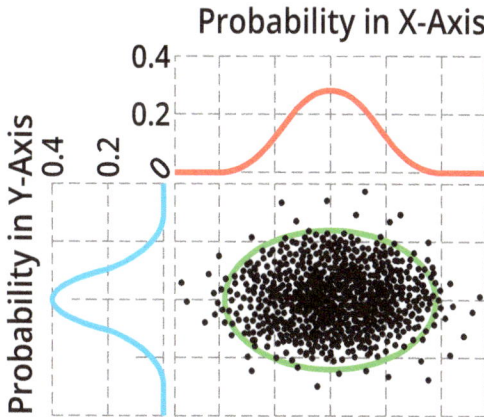

Image 173: The black dots represent example shots on a target. The center ellipse encircles 95% of the general pattern of the shots. The top graph shows the probability that a shot impacts at a location under the graph. It is elevated in the center, which represents that **shots are more likely to impact at the center** of the X-axis. Similarly, the left graph shows that shots are more likely to impact at the center of the Y-axis than on the edges. The top graph is a different shape (wider) than the left graph (narrower) because most effects work independently to affect either windage (e.g., wind) or elevation (e.g., estimating range), but not both. Both the top and left curves are called "**normal distributions**," and entire disciplines of statistics are dedicated to them.

is a viable strategy. Machine gunners are more familiar with this concept, as their weapon systems are designed to compensate for extreme dispersion with an extreme rate-of-fire.

However, before a marksman can effectively work with their rifle's dispersion pattern, they first need to identify what that pattern is. In practice, this can be done during the process of zeroing and grouping (See Ensuring Accuracy (Grouping and Zeroing), Pg. 74.) Each group of shots on the target represents the marksman's best effort to be precise under ideal conditions, making it a good indicator of the bullets' minimum dispersion. By measuring the angular distance between the bullet holes in a group (i.e., the diameter (See Grouping, Pg. 74.)) the marksman can obtain a simplified measure of the diameter of their cone-of-fire (e.g., 0.5 mil or 2 MOA). (The measure is simplified because a normal distribution only has a diameter that corresponds to a percent, so for example 99% of bullets may be within 0.5 mils.) However, getting the diameter for a few groupings is very useful because a marksman can determine whether that diameter fits within the area of their target.

This diameter of a rifle's cone-of-fire is often advertised by rifle sellers and manufacturers. They advertise with terms such as "mil-accurate" or "MOA-accurate" which refer to the precision or dispersion of the cone-of-fire. For example, if a firearm is fired from a vice and the impact area has a dispersion of 0.3 mils, then the rifle is considered a "0.3 mil-accurate" rifle. As explained previously, using a number to describe width without stating what percent of shots that width encompasses is problematic. However, comparing a single-number mil-accuracy or MOA-accuracy between two rifles is far easier than comparing their normal distributions.

Different shooting disciplines have different standards on what error rates or dispersion patterns are acceptable. For example, benchrest marksmen, compared to hunters, have much smaller target areas at the same distance. Therefore, they are much more stringent about reducing potential sources of dispersion. To be as precise as possible, they shoot from rigid setups.

The general task of reducing dispersion starts with identifying the various sources of imprecision. Of course, **the greatest source of imprecision is the marksman** themself, which is why benchrest marksmen place their rifles in vices and other similar devices to remove human interference. Similarly, good marksmen practice a lot to reduce their error rate as much as possible.

However, reducing error from external factors is more complicated, and often depends on the distance one is shooting at. At close ranges, equipment becomes a significant source of dispersion. On the other hand, wind exerts

Mil-Accuracy and MOA-Accuracy

Dispersion Patterns and Hit Percentage for Four Levels of Rifle Precision

1 MOA	0.5 MOA	0.3 MOA	0.1 MOA
(~0.29 Mils)	(~0.14 Mils)	(~0.09 Mils)	(~0.03 Mils)
70% Hit	78% Hit	80% Hit	80.7% Hit

Image 174: These shot patterns represent four rifles with different precisions. The hit percentage is the number of shots that landed on the gray, circular target. Even though the 1 MOA rifle is 10 times as precise as the 0.1 MOA rifle, the former was only ~10% more precise on the target. This is because of other sources of inherent imprecision. For example, these dispersion patterns included the same wind uncertainty, which is why all of the widths are roughly the same. The point is that often, **equipment is not the largest source of inherent imprecision**.

Hit Percentage versus Distance to Target for Three Levels of Rifle Precision

— High Precision: 0.5 MOA (~0.14 Mils)
— Medium Precision: 1.0 MOA (~0.29 Mils)
— Low Precision: 1.5 MOA (~0.43 Mils)

Image 175: This graph reinforces the diagram above: at longer ranges and outdoors, modern equipment is usually not the largest source of imprecision. Even at the largest vertical difference, doubling the rifle precision did not double the hit percentage. This is because other factors, such as wind estimates, distance uncertainty, and inconsistent gunpowder power all cause imprecision as well.

a continuous, albeit small, effect on a bullet, becoming a more significant source of dispersion over longer trajectories.

Using a DOPE book, marksmen can effectively record and analyze information from each shooting session to identify the factors causing variation in their dispersion. Data points recorded for a given distance might include: the mil-accuracy of their rifle; the standard deviation of muzzle velocity of their ammunition; and weather data on wind speed variability and the likelihood of gusts. This recording process requires a significant amount of shooting, as each new variable introduced in multivariate testing requires collecting an increased number of data samples.

After all the data is collected, it can be input into statistical computer programs for analysis. However, even with a computer, statistical identification of the source of standard deviation (i.e., the source of imprecision) can be a complicated and challenging subject. This complex subject is best explored in and left to a comprehensive book on statistics.

18.c Estimation (Range and Wind) Uncertainty

Modern equipment has severely improved the precision of the data with which marksmen determine their trajectories. Laser rangefinders give exact distances to the target within less than 0.1% accuracy up to 1000 m or yd, and electronic wind meters also give extremely precise wind information (albeit only regarding the wind next to the meter).

Therefore, it is worth explaining how more precise inputs lead to much more precise bullet trajectories. This explanation not only shows the limitations of human estimates, but also shows why even novice marksmen can improve their precision by using specialized equipment.

Range uncertainty (expressed as a percentage) is the accuracy of an estimate of the distance from a marksman to their target. For example, a high-quality laser rangefinder can boast an accuracy of less than 0.1% error over distances up to 1000 m or yd, or 0.1%. In contrast, a marksman is considered a good estimator if their guess is within 10% of the actual distance.

Range uncertainty affects vertical placement (bullet-drop) on the target, and the elevation hold that marksmen use to compensate for that drop. Therefore, to ensure a hit despite an inaccurate range estimate, the elevation hold must compensate for any bullet-drop that would occur within the range of inaccuracy. For example, if a marksman estimates with 10% accuracy, and they estimate a range of 500 m or yd, they must use an elevation hold that would compensate for a bullet both at 450 m or yd and at 550 m or yd (i.e., 500 m or yd minus and plus 10%). (See Image 107, Pg. 124.)

The problem of range uncertainty is associated with its traditional solution, danger distance. (See Danger Distance, Pg. 143.) To restate that section, a bullet always hits a target of a given height within the corresponding distance because the bullet never exceeds the bounds of that height along that distance. That property can be reversed, so that if the height is known, a marksman can determine the danger distance, which is the allowable ranging error.

A much larger problem for marksmen is **wind uncertainty** because wind changes in real time and there exists no portable device that can give accurate wind readings from a distance. (Distance to target can change too, but aside from vehicles, it is extremely rare for a target to be traveling fast enough to cause range uncertainty).

Accounting for wind uncertainty uses a similar concept to danger distance. First, a marksman determines the angular width of their target using the hashmarks on their reticle. For example, assuming a target is 1 m or yd wide at 500 m or yd from the marksman, the target is 2 mils (6.9 MOA) wide. Therefore, if the marksman aims in the center, they can be 1 mil (3.4 MOA) to the left or the right and still hit the target.

To convert that 1 mil leeway into how accurate a wind estimate must be, the marksman can use the wind-to-wind-hold-conversion in reverse. (See Converting a Wind to a Windage Hold (Summary), Pg. 177.) For example, a wind hold of 1 mil at 500 m or yd may correspond to an equivalent crosswind of 15 km/h (10 mi/h). (See Image 140, Pg. 161.) Assume that the wind value is 100% because the wind direction is 90°to the marksman. Therefore in this case, the marksman's estimate of wind speed would have to be plus or minus 15 km/h (10 mi/h) of the true wind speed to still hit the target.

However, obviously narrower targets at farther distances require much more accuracy. A 1/2 m or yd target (half the width) at 1000 m or yd (double the distance) would require an estimate more than 4 times as accurate because **target-width scales linearly** with allowable wind estimation error; but **distance-to-target scales exponentially**.

19. Shooting Uphill or Downhill (Inclination)

Shooting uphill or downhill is common in hilly or mountainous areas. Changes in elevation affect how gravity affects the bullet's trajectory and how the marksman must adjust their point-of-aim vertically, but it does not change the effect of wind or how the marksman must adjust their point-of-aim horizontally.

To understand the effects of shooting uphill or downhill, marksmen must be aware of altitude-related climate effects, but also the "inclination." **Inclination is the angle** between any straight line and an X-axis. For shooting purposes, the straight line is the sightline from the marksman to the target, and the X-axis always lies on the **imaginary plane that is perpendicular to the direction of gravity**. With the inclination, a marksman determines the inclined elevation hold.

(Unrelated to the rest of the section, it is important to note that keeping a good posture during angled shooting is important. Specifically, marksmen who aim up must ensure that they do not rotate their scope towards their eye so much that recoil would force their scope backwards into their eye.)

19.a Perpendicular and Parallel Components of Gravity

When the bullet flies horizontally, 100% of gravity pulls the bullet away from its trajectory. And if a bullet is shot straight up in the air, 0% of gravity affects the trajectory, as the bullet just flies directly up and down (in a vacuum). Between horizontally and vertically, when a bullet travels at an inclination, somewhere between 100% and 0% of **gravity acts perpendicular to the bullet's trajectory**.

Converting gravity into its perpendicular components requires the use of trigonometry, specifically taking the cosine (cos) of the inclination. (See Image 178, Pg. 210.) This equation can be computed precisely, although it is usually approximated:

- cos(0° to 15°)   ≈ 1.0   = 100% of gravity is perpendicular
- cos(15° to 30°)   ≈ 0.9   = 90% of gravity is perpendicular
- cos(30° to 40°)   ≈ 0.8   = 80% of gravity is perpendicular
- cos(45°)   ≈ 0.7   = 70% of gravity is perpendicular
- cos(90°)   = 0   = 0% of gravity is perpendicular

Applying gravity percentages to bullets is complex, so marksmen use a method known as the "Rifleman's Rule." This rule simplifies the calculation by considering only the distance that a bullet travels horizontally (i.e., perpendicular to gravity) when calculating an elevation hold, ignoring any changes in altitude. (See Image 180, Pg. 211.)

Inclined Shooting

Image 176: A Finnish Soldier participates in the Advanced High Angle Sniper Course. Lizum, Austria, 27 Apr 2015. Most inclined shots are between a cliff and flat ground, meaning **sightlines are usually less steep than a hillside's incline**.

Image 177: The Mildot Master is an **analog calculator** that calculates bullet-drop for a given distance in both the imperial and metric systems. It can also give a correct elevation hold for any inclination. This makes it useful for milling with a reticle.

For example, if a bullet is fired at a 45° angle toward a target 400 meters (438 yards) away, the rule calculates the horizontal distance that the bullet travels by using the cosine of the inclination. The cosine of 45° is about 0.7, so multiplying this by the total distance (400 meters) gives 280 meters (306 yards) of horizontal travel. Therefore, to compensate for bullet-drop, the marksman can use an elevation hold for a shot at 280 meters on flat ground, even though the actual shot is at a 45° angle over 400 meters.

For angles less than 15° and distances less than 200 m (219 yd), changes in the horizontal distance traveled, and therefore bullet-drop, may be insignificant. (E.g., bullet-drop at 15° is 97% of bullet-drop at 0°.) Although 15° may seem small, marksmen often overestimate elevation changes. For reference, the maximum allowable incline for U.S. Highways is less than 4°, and the steepest roads in the world are less than 18°.

In contrast to the perpendicular component of gravity affecting trajectory, the **component of gravity that is parallel to a bullet's trajectory** affects the bullet's speed. When a bullet flies horizontally, 0% of gravity affects the speed of the bullet; and when a bullet flies vertically, 100% of gravity affects the speed of a bullet. Between horizontally and vertically, when a bullet travels at an inclination, somewhere between 0% and 100% of gravity acts parallel to the trajectory of the bullet.

The parallel component of gravity is rarely considered. This is because gravity only accelerates or decelerates the velocity of the bullet a small fraction up or down. For example, an average high-powered rifle may have muzzle velocities of 1000 m/s (3281 ft/s). Even if air resistance slows a bullet

Cosine and Sine

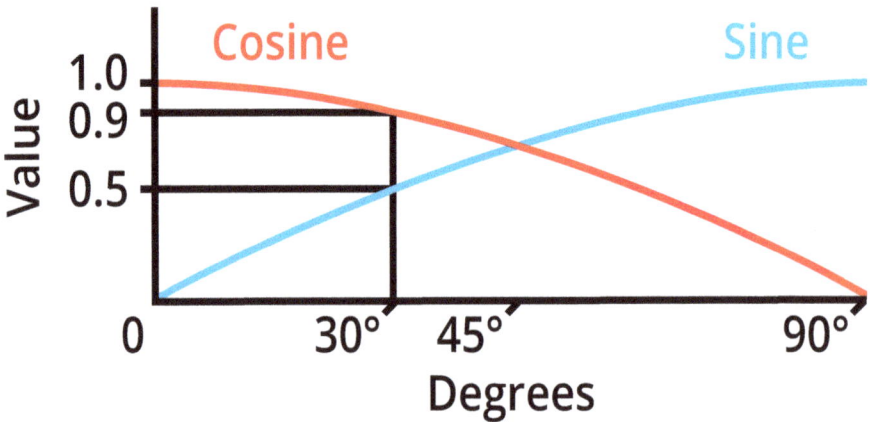

Image 178: In this graph, the X-axis represents the inclination angle, which is the angle between the marksman's line of sight and a horizontal plane (perpendicular to gravity). The Y-axis represents the component value, which is used to calculate how much of the gravitational force acts perpendicular (shown by the red cosine curve) or parallel (shown by the blue sine curve) to the bullet's trajectory at each inclination angle. The two lines are reflections of each other at the 45° mark. This graph includes a vertical line at the 30° mark to show that **cosine changes very slowly at first**. A horizontal line segment is also drawn from the value 0.5 to demonstrate that the component values of sine and cosine sum to more than 1.

down to an average velocity of 500 m/s (1640 ft/s), it would take at most 2 s to impact a target at 1000 m (3281 ft). The approximate average change in velocity can be found at the halfway point, which is 1 s in this example. Gravity accelerates at 9.8 m/s² (32 ft/s²). Therefore, even if a bullet is shot vertically, where 100% of the gravity is parallel to the trajectory, that would only equate to a final velocity change of 9.8 m/s (1 s × 9. 8m/s²) (32 ft/s), which is a 2% change in velocity ((500 m/s) ÷ (9. 8m/s) ≈ 2%).

A 2% change in velocity causes gravity's perpendicular component to deflect the bullet from its trajectory for an additional 2% of the time, resulting in a 2% greater deflection. However, this 2% figure originally came from assuming an exactly vertical shot and 100% parallel gravity. Actual shots would have a much smaller parallel component. To find the parallel component, take the sine (sin) of the inclination. (See Image 178, Pg. 210.) (The perpendicular and parallel components do not add to 100%!)

Shooting Uphill and Downhill

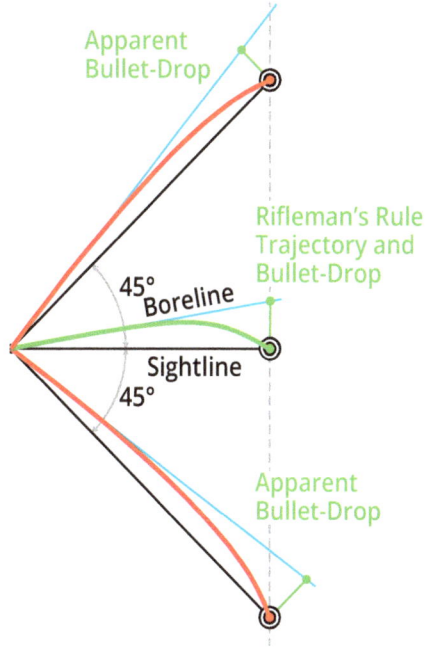

Image 179: The trajectories of bullets shot uphill or downhill are **flatter** than the trajectory of a bullet fired horizontally. This is because gravity becomes more aligned with the boreline. Actual bullet-drop remains relatively consistent regardless of the angle of shooting. However, marksmen **perceive that bullet-drop decreases** because they observe bullet-drop to always be perpendicular to the sightline.

Image 180: To account for the flatter inclined trajectories, marksmen use the **Rifleman's Rule**, which states that the apparent bullet-drop is about the same as the actual bullet-drop of a bullet fired at a target that is the same horizontal distance on flat terrain. (Note, the top and bottom apparent bullet-drops would be much closer in size if the angles between the borelines and sightlines were more realistic (i.e., much smaller).)

▶ sin(0° to 15°)	≤ 0.25	≤ 25% of gravity is parallel
▶ sin(15° to 30°)	≤ 0.5	≤ 50% of gravity is parallel
▶ sin(30° to 40°)	≤ 0.65	≤ 65% of gravity is parallel
▶ sin(45°)	≈ 0.7	= 70% of gravity is parallel
▶ sin(90°)	= 0	= 0% of gravity is parallel

Therefore, if the inclination is less than 30°, then the parallel component can likely only change the bullet-drop by a fraction of a percent.

There are several ways to both measure the inclination to a target and to calculate the cosine component of that inclination. Many phone apps can detect an angle and provide the cosine value. (See Image 182, Pg. 213.) These work by aiming at the target and placing the phone on top of the inclined or declined scope. Similarly, rifles and scopes can have detached tools (i.e., a protractor attached to a level) or attached angle indicators. (See Image 183, Pg. 213.) Many laser rangefinders can also directly measure an angle without involving the rifle at all.

To make conversions easier and more precise, marksmen can use tools to convert the distance to the target to the equivalent flat-target distance with an angle-to-cosine conversion table online, or with an analog tool such as the Mil-dot Master. (See Image 177, Pg. 209.) The fastest but least accurate method of finding an angle to a target is when a marksman trains their intuition on the angle to quickly adjust their point-of-aim. This requires a lot of practice and reference points, as **estimating angles is not intuitive**.

19.b Change in Air Density with Change in Altitude

Air density decreases as altitude increases. (See Image 163, Pg. 190.) This means that bullets traveling upward face less resistance and experience less reduction in speed due to drag. For example, a bullet fired at a 45° angle toward a target 1,000 m (1,090 yd) away would face decreasing resistance as it travels, seeing a maximum density drop of 6% at its apex. Reduced air drag leads to less deceleration of the bullet and decreases the time that gravity has to affect its trajectory, resulting in less bullet-drop per unit distance. In contrast, shooting downward increases air density, which causes more drag and allows gravity to act on the bullet for a longer period. This results in greater bullet-drop per unit distance

That being said, the **change in air density is usually ignored** by everything but the most complicated ballistic calculators. This occurs because the change in velocity caused by varying air density nearly counteracts the change in velocity from gravity's parallel component (See Perpendicular and Parallel Components of Gravity, Pg. 208.) That is, these two effects act in opposition, and they tend to cancel each other out.

Tools That Measure Incline

Image 181: A clinometer is a handheld device that a marksman points at their target to read the incline angle.

Image 182: A level can be placed on the rifle. Many smartphones have access to a level app.

Image 183: This scope mount has an angle indicator (a.k.a., cosine angle indicator) attached to its left side. The the housing and mount rotate around the red line which always stays parallel to gravity. Currently the device shows a 5° decline.

Information
Management Contents

20. Information Gathering   215

Calling Shots   215
Meters   216
Recording   217

21. Calculating   221

Ballistic Calculators   221
Conversion Charts   222

22. Instruction   224

Practice   224
Dry Firing   227
Drills   229
Teaching   230

Information Management

Information is the oxygen of the modern age.
—Ronald Reagan, 40th U.S. President

This section on information management aims to provide you with valuable insights into how you can gather, record, interpret, and learn the information you need to enhance your skills. This section is separate from the others because managing information isn't easily divided by skill level. Both beginner and expert marksmen need to continuously improve their information management skills to join the ranks of the best of the very best.

20. Information Gathering

There are two reasons why a marksman would want to gather information: in order to **reveal past mistakes** and to **predict future performance**. There are multiple types of information to consider gathering, such as information about the equipment, about the environment, and about the marksman's individual performance patterns.

While this section deals with information generally, some information-gathering processes have already been mentioned. For example, zeroing is the process of determining and eliminating the distance between the point-of-aim and the point-of-impact. (See Ensuring Accuracy (Grouping and Zeroing), Pg. 74.) And similarly, gathering information on wind is a huge part of being able to predict a bullet's left and right movement. (See Wind, Pg. 157.) The skills taught in those sections can be generalized to all information gathering.

20.a Calling Shots

"Calling shots" refers to the practice of predicting where a bullet has hit based on the sight picture at the moment of firing. After taking a shot, the marksman mentally notes or "calls" the expected point-of-impact without immediately looking at the target. Later, they check where the bullet actually landed and compare it to their prediction.

To better remember what their point-of-aim was after a shot is taken, a marksman makes a verbal prediction about where their bullet will impact on the target as specifically as possible (e.g., a corner or border). For example, a marksman may think to themselves, "this shot will hit to the right of the bullseye," immediately before and after pulling the trigger. Also, maintaining a pause just after firing (i.e., follow-through) better allows the marksman to concentrate on recalling their point-of-aim and point-of-impact predictions.

Calling shots helps to:

▸ Encourage marksmen to be more deliberate when aligning the sights and aiming within the sight picture.

▸ Minimize flinching by diverting the marksman's attention from the recoil.

▸ Make the marksman better remember potential points-of-impact, which is vital when a bullet completely misses the intended target.

▸ Allow marksmen to develop and verbalize a consistent process into their shooting, making the process smoother and thereby faster overall.

▸ Force marksmen to immediately associate their sight pictures with success or failure, allowing marksmen to better replicate successful sight pictures and avoid repeating any unsuccessful ones.

Calling shots is especially vital in **team settings** where a separate person gives a point-of-aim to the marksman. For example, in marksman-and-spotter scenarios, it is crucial for a marksman to communicate aloud whether any mistaken shots were due to their own error (in any way!), or else due to events beyond their control.

If a marksman knows that the miss was their own fault, they try again with the same point-of-aim. They inform the spotter of this by saying "pulled [a shot]," or they can be more specific by indicating the direction of the error, such as "pulled up" or "pulled 15 cm (6 in) to the 12 o'clock." (The term "pulled" is shooting jargon to mean the trigger finger pulled the entire rifle, unintentionally shifting the point-of-aim. That said, some marksmen use "pulled" to refer to any shot that they caused to be bad.)

If the marksman believes that their point-of-aim was correct, they can provide the spotter with a general assessment, such as saying "good." And if they missed anyway, the fault may lie with the spotter giving the marksman an incorrect point-of-aim. Then the spotter must adjust or double-check the point-of-aim they had given to the marksman. The problem may also be with the equipment; for example, their rifle may have lost its zero.

In any case, if the marksman calls their shots, then the team better understands whose fault the miss was and therefore who must readjust.

20.b Meters

Many kinds of meters independent of the core rifle components have been discussed in this manual, to include those for: time (clocks or chronometer), muzzle velocity (ballistic chronograph), distance (laser rangefinders or telemeters), wind speed (anemometer), temperature (thermometer), humidity (hygrometer), air pressure (barometer), and slope (inclinometer). All of these meters are now commonly used as they have become electronic and compact.

If a marksman is considering buying or using a meter, they must consider a few things. First, they need to decide if the time spent learning how to use the meter and using it regularly is **worth the effort**. In other words, is the time investment beneficial? Ultimately, this decision is up to the individual marksman.

Second, they must weigh the improvement in accuracy or precision against the **cost of the meter**. The marksman is not interested in the meter's data for its own sake, rather what they want to know is exactly how that data can help them improve their skills. To compare different meters, the marksman must determine how much additional precision or accuracy each meter would offer relative to its cost, or even compare a meter against the option of not using a meter at all.

This analysis can quickly become complicated. For example, laser rangefinders greatly increase the accuracy of estimating distance. (See Estimation (Range and Wind) Uncertainty, Pg. 206.) However, laser rangefinders are expensive, standalone devices that can cost hundreds of dollars. In contrast, most hygrometers are built into weather meters, and cost almost no additional money. However, accurately knowing the humidity is unlikely to improve accuracy by more than a fraction of a percent.

20.c Recording

Data is useless if it is forgotten. Therefore, marksmen must develop a way to record and preserve whatever information they find useful or interesting. Common information that a marksman might find useful is wind data, climate data, equipment data, and data on themselves such as the amount of sleep they got the previous night. (Fatigue and lack of sleep can lead to inadvertent errors and ultimately, missed shots.) There are too many kinds of data to list individually, but basically **every section in this book details some kind of useful data that can be recorded.**

Data can be recorded in various ways. The simplest way is binary, which only has two options. Any two mutually exclusive options work, such as: present or absent; hot or cold; energetic or tired; etc. Using a binary system is the fastest and easiest way to record data, and can be done with a checkbox. However, binary is not very exact, so most data is recorded with a specific number; for example, temperature is recorded with a number. Numbers are the best way to record data if computers are involved. A very technical approach is to create data tables. Here, each shot can be represented as a column, while the rows can consist of variables like elevation, windage, and the coordinates on the target.

In addition to binary and numerical systems, data can also be recorded using graphical representations. A marksman attempts to essentially draw a picture of what they see or detect. For example, for wind, a marksman may use arrows to show direction. The most common graphical representation is a paper target, where the paper itself becomes a record once shooting is finished. (See Image 187, Pg. 219.) It is also becoming far more common for people to video record marksmen as they shoot to detect problematic micro-movements, such as flinching from anticipating recoil. Video recording (especially in slow motion) is a great tool for improving one's skills. (See Image 190, Pg. 221.)

Traditionally, marksmen would compile their recordings in a **DOPE (Data On Previous Engagements)** book. (See Image 185, Pg. 219.) Such a book is very helpful because it can be formated with labels and grids to make data organization both fast and easy. However, DOPE books are not strictly necessary, and using a random piece of paper is still better than no recording at all. (See Image 184, Pg. 219.)

While the DOPE book is a general record, there are more specialized logs that the shooting community commonly also uses. For example, in a **shot-by-shot log**, marksmen record the date, time, weather conditions, ammunition specifics, sight settings, and the exact position of each shot on the target. Marksmen can add commentary next to each bullet marking, such as the elevation and windage used or if any mistakes were evident. Some marksmen also record environmental conditions such as wind speed and direction, temperature and humidity, altitude, light conditions, and even barometric pressure. A **barrel log** in contrast, is basically an account of the number and type of rounds fired through a barrel to determine if the barrel has degraded enough to require replacement.

In theory, these detailed records assist marksmen in analyzing their performance, tracking the effectiveness of adjustments to their equipment or technique, and adapting to different environmental conditions. However, the records must actually be used to be "useful." The most obvious way to use records is to find ways to improve. In contrast, many marksmen just find recording to be an enjoyable exercise, and that is also a great reason to record. However, **without enjoyment or use, records are a waste of time**.

A marksman can better determine if they want to record data by starting slow and building up. For example, a marksman can start by only recording

Recording Information

Image 184: This marksman records information on a used box of ammunition. He can transfer the information to a computer later.

Image 185: This marksman records on blank charts, printed on weatherproof paper and stored in a durable binder. **They were prepared**.

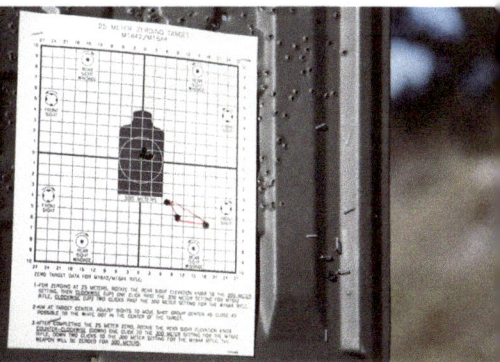

Image 186: While a marksman is shooting, an assistant or spotter can record information for later analysis.

Image 187: The easiest information to keep is used targets. Marksmen can simply keep them in a binder.

the impact location, and then add one data point to the record book each shooting session, instead of trying to record 20 data points from the start.

Tools that assist in recording are becoming more advanced by the day. One interesting analog tool, known as the Plot-o-Matic, involves using clear plastic sheets (e.g., Plexiglas or Lexan) placed over a target replica to compare their precision across different points-of-aim. Marksmen use one sheet per sight setting and overlay all the sheets, aligning each subsequent sheet based on the adjustments made to windage and elevation relative to the first shot group. When looking through the sheets, all shots on each layer converge into one shot-grouping. Digital apps, such as Shooting Companion: Plotter, perform the same function, but digitally.

Ballistic Calculators

Image 188: A U.S. Air Force Tech Sergeant uses a Kestrel-brand weather meter to take ground wind readings. Cheshnegirovo Air Base, Bulgaria, 12 May 2021. Readings. Modern ballistics calculators can be **downloaded as apps on a phone**.

Image 189: A U.S. Marine with 2nd Marine Division, checks his equipment before firing. Camp Lejeune, NC, 08 Mar 2021. He is holding a Kestrel-brand weather meter with a **built-in ballistic calculator**. These devices read the wind and climate data directly, and only require the user to input their rifle-system information.

Image 190: One Missouri Army National Guardsman shows another their shooting position on a phone. **Taking pictures and videos** of a marksman is a great way to analyze the shooting process and communicate any improvements.

21. Calculating

Any information that a marksman gathers must be converted into a usable form by performing calculations. The calculations can be done digitally with a computer or a ballistic calculator, or they can be done manually with the help of a conversion chart. Either way, a marksman must plan and prepare their method of calculation before a shooting session begins.

21.a Ballistic Calculators

Ballistic calculators predict the trajectory of bullets from the point of firing. Thereafter, **the user can receive an elevation hold from the calculator** that compensates for bullet-drop at whatever distance the target is along a bullet's trajectory. Depending on the calculator, the user may also receive a windage hold and an expected energy transfer from a bullet's impact. Therefore, ballistic calculators are indispensable tools for precision shooting beyond 300 m or yd.

To get these outputs, users must input several variables into the calculator, which may include but are not limited to: bullet weight, shape, muzzle velocity, and ballistic coefficient (a measure of air resistance efficiency); the target's distance and inclination; and environmental conditions such as wind speed, wind direction, temperature, humidity, and altitude.

Each ballistic calculator is programmed with proprietary formulas, so some brands are better than others. That being said, the technology is constantly improving, and all reputable brands are capable of accurately calculating well past 1000 m or yd.

Because the "calculator" is a program, many calculators are downloadable apps on phones. (See Image 188, Pg. 220.) Others are built into weather meters. (See Image 189, Pg. 220.) Downloaded app calculators are very cheap, as the marksman can use the phone they already own or instead can buy a disposable phone to solely serve as a calculator. However, marksmen still need to access reliable climate and wind information. This too can be found online. However, some marksmen prefer to not emit radio signals while they shoot. These marksmen require a portable weather meter, and so they may as well purchase one with a built-in ballistic calculator.

21.b Conversion Charts

In this manual, there has been a lot of talk about math and formulas. However, the math and formulas are only for preparation and for the sake of training. On the actual day of any important shooting, math is not performed. Instead, premade conversion charts are used to eliminate the chance of making errors. A conversion chart is very simple to use. It either has a single output derived from either one input (one-dimensional) (See Image 191, Pg. 223.) or two inputs (two-dimensional) (See Image 152, Pg. 174.). Marksmen simply select their inputs on the X and Y-axes, and follow the columns or rows to the resulting cell contents.

In fact, choosing printed conversion charts over electronics is not a matter of being old-fashioned; **it is far easier to reference inert, weatherproof sheets than to click tiny buttons in wet, cold, or dim conditions**.

The most in-depth charts already discussed in this manual are wind tables. (See Image 139, Pg. 160.) (See Image 140, Pg. 161.) (See Image 152, Pg. 174.) (See Image 153, Pg. 175.). Marksmen print out or write down wind tables, and for quick reference, keep the wind tables right next to their shooting position. (See Image 189, Pg. 220.)

However, charts have also been presented in other ways. For example a reticle is a chart of sorts that exists within a rifle (See Image 9, Pg. 19.), and when one is used alongside a Mildot Master (See Image 177, Pg. 209.), a marksman can determine their elevation hold without any electronics. Another example would be the chart of common errors that converts group characteristics into an area of improvement for the marksman. (See Image 191, Pg. 223.) A final example would be a chart that can convert the

Common Errors

Pattern of impacts on a target	Possible Reasons for Error
Horizontal error	Bad wind calls Canted rifle Bad rifle grip or trigger pull
Low-right error (or low-left error for left-handed marksmen)	Pulling the trigger too hard Anticipating recoil Unstable support-hand position Riflestock not firmly pressed against the shoulder Slipping trigger-hand elbow
Vertical error	Poor breathing techniques while firing Inconsistent cheek weld Poorly manufactured or maintained ammunition Changes in temperature, pressure, or humidity
Shots below center	Improper eye relief with parallax error Riflestock shifting in the shoulder pocket Unstable shooting position or sling
Erratic shots around the point-of-aim (Accurate but not precise)	Flinching due to noise or recoil Bucking in response to anticipated recoil Jerking the trigger with abrupt force Over-muscling the rifle by holding it too firmly No follow-through (holding position after a shot)
Compact group not at the point-of-aim (Precise but not accurate)	Incorrectly zeroed scope Failure to compensate for wind Improper point-of-aim Incorrect data input into ballistic calculator Incorrect distance estimation Improper eye relief causing parallax error
Shots scattered all over	Simultaneous vertical and horizontal issues Unstable trigger pull Inconsistent eye, cheek, or body position Focusing on the target instead of on the reticle Incorrect or inconsistent parallax adjustments Bad gear (e.g., loose scope, dirty rifle, bad ammo)

Image 191: A non-exhaustive list of common errors and their potential sources.

movement of vegetation in the wind into an estimated wind speed. (See Image 158, Pg. 183.) **All of these charts are often used alongside electronic calculators, which are unable to make such abstract conversions.**

Image 192: A recon Marine with Force Reconnaissance Company, 13th Marine Expeditionary Unit, **examines his notepad to calculate** proper wind call adjustments. Kaneohe Bay, HI, 23 Aug 2013. He also has his elevation holds in his scope cap. (See Image 120, Pg. 139.)

22. Instruction

Learning the basics of shooting may come naturally to some; however to become an expert marksman, training and practice are necessary. Just as with learning any skill, certain key principles apply, such as receiving feedback and isolating smaller skills. This section shows how those general skills of instruction apply to the specific discipline of long-range shooting.

22.a Practice

Effective practice takes time and cannot be achieved in just one session. It involves consistent and repetitive actions. To develop a good habit, an action must be repeated correctly at least ten times, and even more times if it is meant to replace a bad habit. Since building new habits takes time, ideally practice is done at a slower and more deliberate pace. For example, during a practice session, marksmen focus on taking their time to ensure they are in a good shooting position, rather than rushing into it.

Because practice must be done slowly, over an extended period, the first step to improving any skill through practice is to create a **structured training plan**. To create a training plan that is customized to a marksman or a group of marksmen, the first step is to determine what exactly needs improving. For example, if a marksman mostly has vertical errors in their

Example Drill Composition

Side of body to shoot with			
Dominant	95%	Non-dominant	5%
Knowledge of distance to target			
Known Distance	60%	Unknown Distance	40%
Using a support or not			
With support	50%	Without support	50%
Lighting			
Daylight	90%	Night	10%
Precipitation			
Clear skies	60%	Raining	40%

Image 193: This table shows an example drill composition.

shots, their problem is more likely to be caused by poor breathing techniques rather than canting the rifle. (See Image 191, Pg. 223.) When multiple errors are occurring, optimized training plans focus on areas where the learner is least skilled, as these areas typically offer the most room for improvement.

The skills that a marksman chooses to improve are ideally as specific as possible. For example, a marksman can improve the fastest by concentrating on mastering one particular rifle rather than shooting in general.

Some skills that need improvement may be obvious, while others might not be as clear. Additionally, the mistakes and details a marksman notices can change over time. Therefore, marksmen must continually update their training plans. Tracking progress is essential for identifying patterns and areas that need improvement. Shooting with a friend and exchanging feedback can also be beneficial, as individuals are often unaware of their own weaknesses.

Most marksmen create training plans that work on a diverse mix of skills, with some skills receiving more focus and time than others. While practicing one skill during a session is the most efficient use of time in the short term, it can lead to boredom. In contrast, happy learning is better learning, and varying practice routines keeps each session engaging. Therefore, it is important to do whatever makes a marksman eager to practice again and again.

A good planning tool for structuring a training plan is a **drill composition table**. (See Image 193, Pg. 225.) A drill composition table categorizes the various conditions a marksman might practice under, and assigns a percentage to each category. These percentages represent the ideal proportion of practice time that is ideally dedicated to each condition. By following the table, a marksman can balance their training and ensure that no skill is over or under-emphasized during practice.

Marksmen must assign a higher percentage of practice time both to skills that are important to master and those that are fast to learn. This greatly changes between different disciplines of shooting. For example, a tournament marksman might not need to practice shooting at night, while a hunter may not need to practice using a tripod.

To use a drill composition table, a marksman records each shooting session by adding a tally in the appropriate row and cell that describes the session. When planning future sessions, the marksman can then review these tallies to ensure their practice aligns with their designated percentages. If the tally ratios don't match their ideal percentages, the marksman can adjust upcoming sessions to focus more on under-practiced skills. This approach ensures that overall training maintains the intended balance.

Many marksmen often mistakenly prioritize precision over other crucial skills, such as shooting speed and adaptability. Take deer hunting for example; where a deer's heart at 100 m or yd allows for a roughly 0.5 mil radius of leeway from the center. Consequently, while shooting with 0.25 mil of precision is acceptable, it is not essential for successfully killing the deer. The more valuable skill to enhance would be reducing the time between spotting the deer and smoothly pulling the trigger.

In contrast, marksmen often under-prioritize practicing with external factors, such as rain or dusk. While removing these factors can be helpful for basic training, it doesn't accurately reflect real outdoor environments. For example, many marksmen train on ranges with minimal wind, allowing them to clearly see the effects of their adjustments without wind interference. However, when they hunt or compete in areas where wind is present, they may find themselves underprepared. To successfully make important shots in challenging environments, marksmen must practice in those same conditions.

Practice isn't limited to dedicated training sessions alone. Many marksmen also incorporate marksmanship exercises into their regular shooting activities. For example, many marksmen develop a routine that includes dry firing before all live-firing sessions (except for emergencies). (See Dry Firing, Pg. 227.) This helps calm them down and allows them to practice skills that

Image 194: Multinational Soldiers from seven different countries participate in the Urban Sniper Course conducted by the International Special Training Center. Hammelburg, Germany, 04 May 2015. When teaching, shooting, and teaching how to shoot, it is vital to **set up robust communication** methods beforehand.

need improvement without the added complexity of live ammunition. During live firing sessions, marksmen may record their mistakes in a logbook or with video, enabling them to review and analyze their performance later. These recordings can help marksmen determine the most effective way to proceed and any possible adjustments needed to their individual training plans.

22.b Dry Firing

Dry firing involves pulling the trigger of an unloaded firearm to practice as if it were loaded. (This does not harm modern centerfire rifles, unlike some rimfire weapons.) Following the rules of firearm safety is still paramount! (See Firearm Safety, Pg. 24.) By removing ammunition, marksmen can focus on their shooting for extended periods without worrying about either the cost or the danger of using live ammunition.

Of course, live firing is the gold standard for practice because it is the most realistic simulation of what a marksman is actually practicing for. However, **even if live-firing training were unlimited, marksmen would still practice dry firing**. This is because dry firing does not distract a marksman in the way that live firing does; that is, most importantly, dry firing has no recoil. Therefore, without being distracted by the explosion of gunpowder, a marksman can entirely focus on fixing their bad habits, such as flinching, anticipation, and using too much force on the trigger (a.k.a. "trigger jerk").

More specifically, a clear sign of any effective shooting is when there is no deviation in the sight picture. However, during live firing, recoil always causes some deviation, making it difficult to determine whether a deviation is due to the marksman or the recoil. Dry firing eliminates external forces, ensuring that any deviation is solely caused by the marksman. This feedback enables marksmen to quickly identify and correct their mistakes.

To further enhance visibility of mistakes, marksmen can use a coin and a pencil. By placing a pencil in the muzzle and balancing a coin on top, marksmen can test their ability to keep the coin in place when pulling the trigger. If the coin remains balanced, it signifies that the marksman did not shift their rifle or sight picture while pulling the trigger. (See Image 198, Pg. 231.)

Just as for any other skill, **frequent practice leads to improved competency**. Dry firing can be done much more often and in more locations than live firing can, making it a great option for marksmen who cannot always live-fire. Another advantage of dry firing is its time efficiency. Marksmen can engage in multiple dry-firing sessions, up to fifty times, within a short period of 20 minutes without experiencing recoil fatigue. This time efficiency also makes dry firing an effective warm-up in high-stress situations. Target marksmen, hunters, and military snipers can dry fire at their targets when they assume their position, ensuring that their first live shot is as accurate as possible, potentially avoiding a cold-bore shot. (See Cold-Bore Shooting, Pg. 73.)

Perfect practice makes perfect, and a good setup for dry-firing practice is essential for realistically simulating the entire shooting process to get the best results from practice. For example, just as with live firing, it is crucial to establish a proper shooting position, maintain a sight picture, and control the trigger squeeze. Incorporating laser targets, laser guns, and dummy rounds into dry-fire training can make it more effective and enjoyable. Even acts such as replacing magazines, using the bolt, and resetting the trigger can be included in dry-firing practice. This **more complete immersion** allows the brain to adapt to the entire process, whether it is a dry or live-fire session, and helps the marksman avoid common mistakes that arise when transitioning from dry firing to live firing, such as changing a hand position or reloading.

In contrast, negligent or careless dry-fire practice can lead to poor fundamental skills and may be worse than no practice at all. Safety must always be a top priority, and all firearms must be treated as if they were loaded during dry firing. It is recommended to remove all ammunition from the entire shooting area if possible and to be extra cautious when switching between live and dry firing. Most mistakes occur during this transition.

Image 195: An assistant radio operator with Battalion Landing Team 3/5, 11th Marine Expeditionary Unit, dry-fires a Mark 13 sniper rifle. **Dry firing** is evident because there is no magazine. **The best practice is as real as possible**. Therefore, he is replicating the shooting scenario he is practicing for as closely as possible, including wearing full armor. USS John P. Murtha, Pacific Ocean, 08 May 2019.

22.c Drills

A drill is a repetitive exercise focused on a specific task. Drills are highly effective for improving shooting skills because shooting is a composite of several small, distinct tasks. There are numerous shooting drills, and new ones are constantly being developed. This variety allows students to focus on and enhance the specific skills they need the most.

The most classic drill is the **ball and dummy drill**.(See Image 196, Pg. 231.) In this drill, the marksman assumes their shooting position with no ammunition or magazines. Out of the marksman's sight, an instructor or partner secretly loads the marksman's detachable magazine with a mix of live and dummy rounds so that the marksman cannot know the order of the rounds. The marksman then fires each round, treating all rounds as if they were live. This drill helps identify any flinching or anticipation issues, as any movement of the sight picture when "firing" a dummy round can only be self-caused due to a reaction to the recoil, flash, sound, or smoke.

A good drill that can improve shooting time is the **five-shot metronome drill**. (See Image 197, Pg. 231.) In this drill, the marksman uses a recording that emits a sound at regular time intervals, such as every 10 seconds. The marksman fires at each interval. If every shot is accurate and on time, the interval is shortened. Eventually, when the interval becomes too short to make accurate shots, the marksman gains insight into what is holding back their speed. This allows the marksman to focus on the problem area and improve.

The **manual-recoil drill** is another effective drill that can help correct a poor response to recoil. In this drill, the marksman dry-fires while a partner grabs the stock of the rifle and pushes it backwards into the marksman's shoulder. The amount of recoil applied by the partner is enough to allow the marksman to identify problems with their shooting position and ability to withstand recoil, but not so strong as to cause pain or injury.

To prepare for real-life scenarios, marksmen benefit from incorporating drills that attempt to mimic their important shooting scenarios. For example, practicing hunters who hunt moving prey can set up multiple targets at different distances ranging from 50 to 400 m (55 to 438 yd) and practice shooting all of them in rapid succession to simulate a moving target. Hunters can also practice shooting from different positions (e.g., prone, kneeling, and standing) to simulate realistic conditions. Engaging in less stable positions, such as shooting at a target 400 m (438 yd) away while standing, can expose flaws in technique, while more stable positions (e.g., shooting at a target 50 m (55 yd) away while prone) can be useful for isolating specific skills to work on.

It is important to recognize that **no drill is effective if it is executed poorly**. For example, shooting drills can only begin after a marksman has properly grouped and zeroed their shots. (See Ensuring Accuracy (Grouping and Zeroing), Pg. 74.) Also, rushing a drill can lead to the formation of bad habits, such as rushing important shots or not fully forming good habits. To avoid rushing, it is always best to fully focus on and complete a smaller number of drills rather than attempting to half-heartedly complete a larger number of drills when practice time is limited.

22.d Teaching

Learning alongside another person is invaluable. Collective learning offers fresh perspectives, alternative viewpoints, physical assistance, and most importantly, **moral support**. Shooting with someone who has more experience is even more beneficial because they can elevate a marksman's skill level.

The key difference between a student-teacher relationship and a friendship is accountability. Mastering shooting is a mentally demanding process that requires effort. While a friend may not address a learner's laziness, a teacher must. A teacher reminds their student to call their shots, record data in their data log, and adhere to their training plan.

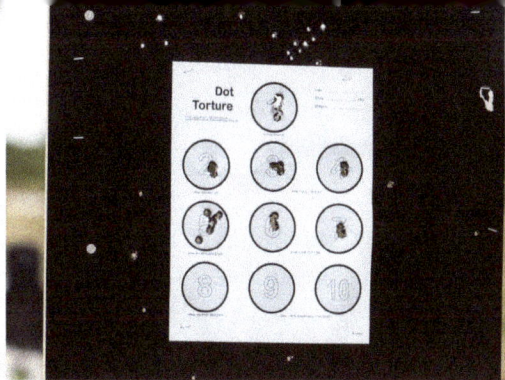

Example Drills

Image 196: The **ball and dummy drill** helps a marksman to overcome the anticipation of recoil. The drill has an assistant randomly load live and dummy (i.e., rubber) rounds into a magazine. The marksman then fires each round. Dummy rounds have no recoil, so if the sight picture moves when a dummy round is "fired," it is the marksman's fault and they must work on not moving the rifle. For safety, dummy rounds must be treated as if they were live rounds.

Image 197: The **metronome drill** helps a marksman to shoot faster and solidify good habits. The drill uses many small targets of identical size, such as on the pictured "Dot Torture" sheet. Marksmen fire a set number of rounds (e.g., 5) at each target within a set amount of time (e.g., 2 minutes). The time decreases after each target. If a marksman misses a single shot, they restart the drill. The goal is to get the time as low as possible without missing any shots.

Image 198: To perform a **coin-balance drill**, a marksman places a pencil in the gun's muzzle and balances a coin on the pencil's end. The objective is to pull the trigger without dropping the coin. This drill can be practiced in any shooting position and offers various levels of difficulty. For example, using a hexagonal pencil is easier than using a round one, and balancing an eraser is easier than balancing a coin. Pencils are ideal because their tapered ends prevent them from slipping into the bore. It is unsafe to use anything that can fall into the bore. Always only point a gun in a safe direction, and never aim at anything that can't be shot.

Teaching is not an innate ability; it must be learned. The most important teaching skill is asking questions of the student, as teachers must confirm their assumptions before suggesting changes. This ensures that the student's actions were indeed mistakes and not consciously made for a valid reason that the teacher may have overlooked. For example, if a student claims to be executing everything correctly, the teacher needs to consider the possibility of equipment error. The teacher may even swap positions with the student or provide the student with a different weapon to determine if the error is reproducible before offering guidance on how to improve. Of course, it is also essential for a student to avoid adopting their teacher's bad habits. Similarly, teachers must avoid instructing above their own skill level; a marksman who can only hit targets at 600 m or yd cannot effectively teach someone how to shoot accurately at 1,000 m or yd and beyond.

Asking questions is also crucial to determine a student's skill level. For example, a teacher can inquire about: the student's shooting history; any past difficulties that they may have faced; changes made to their equipment; or potential underlying issues such as poor eyesight.

It is essential for teachers to be attentive to signs of restlessness and fatigue, as **students in discomfort cannot learn at their best** and may end up developing bad habits. In order to maximize skill retention, individuals must take more frequent breaks when learning a skill compared to when using the skill. This is especially evident when teachers come to a shooting session with an arbitrary list of drills they want to finish by the end of the session. In order to finish the list, teachers often rush students faster than they are able to internalize the skill that the drill intends to teach. Rushed learning is very ineffective, and it is as pointless to move to a subsequent drill when the previous drill was not internalized as it is to build a house on a faulty foundation. Effective teachers only move on once they have confirmed that their students are ready to continue.

Additionally, stress can be very distracting, so it is important for teachers to adopt an **even, encouraging, and fair approach** when working with deadly weapons. That is, while punishment and derision can teach a student how to hit rock with a hammer, they are terrible teachers for complex systems. And if students are forced into stressful situations, they rightfully prioritize safety over learning.

In the long run, the best training plan and teacher aren't defined by how much they can teach in a single day, but by the knowledge learned by a student over time. To achieve this, it's crucial for teachers to keep shooting enjoyable to prevent student burnout. To make shooting fun, some marksmen enjoy

Example Checklist

Gear
Weapon and ammo: ☐ Rifle case with rifle ☐ Ammunition specific to the rifles ☐ Magazines and speed loaders
Range gear: ☐ Ear protection ☐ Eye protection ☐ Targets and stands ☐ Stapler or tape ☐ Range bag ☐ Shooting mat ☐ Spotting scope or binoculars ☐ Cleaning kit ☐ Multi-tool or gun-specific tools ☐ Screwdriver set ☐ Backup batteries ☐ Rangefinder ☐ Pen and notepad ☐ First aid kit ☐ Firearm safety flag ☐ Gloves ☐ Sunscreen and hat ☐ Water and food ☐ Phone and photography or video gear

At the range
Preparation: ☐ Store unloaded firearms in a secure case until use ☐ Confirm the firearm is unloaded before handling ☐ Review the range's safety rules and procedures ☐ Verify range commands and signals ☐ Prepare a checklist of shooting drills
Setup: ☐ Set up targets at appropriate distances ☐ Load magazines with proper rounds ☐ Position shooting mat or bench ☐ Set up any shooting rests or bipods ☐ Prepare to take photos or videos
Shooting: ☐ Start with dry firing ☐ Practice proper shooting stance and grip ☐ Perform controlled shots, focusing on accuracy ☐ Adjust optics/ sights as needed ☐ Regularly check targets for groupings and adjustments ☐ **Perform a prepared checklist of shooting drills**

Post-shooting session
Firearm check: ☐ Unload and clear firearms ☐ Inspect firearms for any wear, malfunctions, or issues ☐ Store firearms in cases
Pack up gear: ☐ Collect all spent casings ☐ Store used targets and dispose of trash ☐ Pack all tools and equipment ☐ Ensure nothing is left behind at the range
Post-shoot cleaning: ☐ Clean firearms thoroughly once home ☐ Clean any tools or gear as needed ☐ Restock ammunition and gear
Post-shoot review: ☐ Review your shooting performance and notes ☐ Plan any adjustments required for your next session

Image 199: Checklists are ideal for complicated, repetitive tasks, and can be as simple or as complicated as a marksman needs them to be.

buying new equipment, while others participate in shooting competitions. No matter the approach, **making learning as fun as possible is key**, even if not every part of the process can always be enjoyable.

Appendices

23. Linear Distance   235
Systems of Measure   235
Meters (m) and Yards (yd)   235
Caliber   236

24. Angular Distance   237
Minute of Angle (MOA)   239
Inches Per Hundred Yards (IPHY)   239
Milliradian (Mil a.k.a. MRAD)   240
Comparing and Converting MOA and Mil   240

25. Other Measurements   241
Speed and Velocity   242
Mass and Weight   243
Force and Pressure   244
Work and Energy   245
Momentum   246

26. Functions and Malfunctions   247
Feeding   247
Chambering   248
Locking   249
Firing   249
Unlocking   250
Extraction   250
Ejection   251
Cocking   252
Standard Malfunction Actions   252

27. Glossary   253

28. Credits   261

Appendices

The difference between ordinary and extraordinary is that little extra.
—Jimmy Johnson, head coach to the Miami Dolphins

23. Linear Distance

Linear distance is a measure of the length between two points in a straight line.

23.a Systems of Measure

There are two main systems-of-measurement used in the world of shooting, the **metric system** (formally known as the "International System of Units (SI)") and the **imperial system** (formally known as the "United States Customary System"). Ballistic charts, scopes, and other shooting accessories often use metric units for easier and more precise adjustments. However, it is common for United States civilians to use the imperial system.

Therefore, this manual presents both systems, usually starting with a unit in metric followed by the same unit converted to imperial within parentheses (e.g. 100 m (109 yd)). Metric units are first because metric is currently the global standard among militaries, including the United States, and it is slowly becoming the global standard for manufacturers as well.

23.b Meters (m) and Yards (yd)

In the metric system, the base unit for linear distance is the meter (m). The metric system follows a pattern where each related unit is a multiple of ten of the adjacent one. For example, 1 kilometer (km) is equal to 1,000 m, 1 m is equal to 100 centimeters (cm), and 1 cm is equal to 10 millimeters (mm).

The imperial units for linear distance include inches (in), feet (ft), yards (yd), and miles (mi). Unlike the metric system, the imperial system does not have a base unit. Because of this, conversions within the imperial system are not as straightforward as within the metric system; 1 mi is equal to 1,760 yd or 5,280 ft, 1 yd is equal to 3 ft, 1 ft is equal to 12 in.

Converting between units has become much easier with the widespread availability of calculators and smartphones. Although mostly non-urgent, in situations where the internet is not accessible, conversions between units can be addressed by printing and using conversion tables. (See Conversion Charts,

Pg. 222.) However, there is a simple trick for converting between meters and yards. 1 m is equivalent to 39.37 in (1.09 yd), while 1 yd is equivalent to 36 in, which means that a meter is approximately 10% longer than a yard. A helpful way to remember this is that the capital letter "M" has more lines and if all the lines were put together to make a straight line, "M" makes a longer straight line than "Y."

Alternatively, for all estimates where a 10% difference is insignificant, meters and yards can be used synonymously. In fact, this manual often uses the term "m or yd" for estimates where the 10% distinction is unimportant or irrelevant and the point being made would be true regardless of the unit being meters or yards.

23.c Caliber

Firearm caliber refers to the internal diameter of a firearm's barrel bore, and it serves as a measure of linear distance. In a rifled barrel, the distance is measured between opposing lands (the high part of rifling) or between opposing grooves (the low part of rifling). Stating the groove measurement is common in cartridge designations originating in the United States, while stating the land measurement is more common elsewhere.

Caliber can be expressed in either inches (in), as part of the imperial system, or in millimeters (mm), as part of the metric system. In the imperial system, caliber is typically denoted as a decimal, followed by the word "caliber" or its abbreviation "cal." For example, a firearm with a ".45 caliber" (explicitly without a preceding zero) possesses a barrel diameter of approximately 0.45 in, and has been designed to fire bullets with a diameter close to 0.45 in. On the other hand, in the metric system, caliber is usually represented by its linear measurement, followed by the abbreviation for millimeter "mm." For example, a 9-mm firearm has a barrel diameter of 9 mm and is designed to fire bullets with a diameter of 9 mm. With that said, some renowned cartridges and bullets are known by both measurements, such as the 5.56 NATO or .223 Remington (5.56 mm equals 0.223 in).

It is important to acknowledge that the nominal caliber can differ from the actual caliber of the firearm. For example, a firearm may be classified as a 30-caliber rifle (i.e., 0.30 in or 7.6 mm), but there can be significant variations in nominal bullet and bore dimensions. Specifically, firearms and projectiles chambered in 303 British are often labeled as ".30-caliber" alongside many U.S. "30-caliber" cartridges, despite utilizing bullets with diameters ranging from .308 to .312 in. (7.82 to 7.92 mm).

24. Angular Distance

Angular distance is a measure of how much an object rotates. For example, if a full rotation is 1 unit of angular distance, then a half-rotation would be 0.5 units of angular distance, and a quarter rotation would be 0.25 units of angular distance. The number of units in a circle is arbitrary and does not have to be 1. For example, if a circle is divided into 4 units of angular distance, then a quarter rotation would cover 1 unit of angular distance. When a circle's circumference is divided into 360 units, each unit is called a "degree." That is, a rotation of 1/360 on a circle is rotation of 1 degree.

A helpful way to understand this concept is by considering slices of pizza. Someone ordering pizza would not say, "3 inches of pizza please," specifying the distance in linear terms. They would instead say, "one slice of pizza please." That is, with 8 slices typically cut into any standard-sized pizza this is one-eighth, or 45°, of the round pizza. Where would one even begin to measure 3 inches of pizza? Using angular distance to measure a circle is unambiguous.

Angular measurements are crucial in ballistics because many concepts of shooting require rotating a specific, quantifiable amount. For example, when adjusting a scope on a rifle, the adjustment effectively rotates the scope's sight picture relative to the rifle. Similarly, when a marksman adjusts their aim to engage a different target, they rotate their body.

A marksman's reticle in their scope serves as their angular measurement tool. Scopes typically employ one of two systems: minutes-of-angle (MOA) or milliradians (mils; mRads). Similar to units of linear distance, these two units can be directly converted. These systems are further explained in the following sections.

One way to understand how reticles measure angular distance is to observe that larger objects farther away can appear the same size in the scope as smaller objects that are closer. That is, as an object moves further from the observer, it appears smaller because it occupies a smaller percentage of the observer's field-of-view. For example, a boulder measuring 4 m (4.4 yd) in width appears to occupy approximately 2° of the marksman's field-of-view when positioned 100 m (109 yd) away. However, if the same boulder were placed 200 m (219 yd) away, it would then occupy approximately 1° of the marksman's field-of-view.

Angular Distance

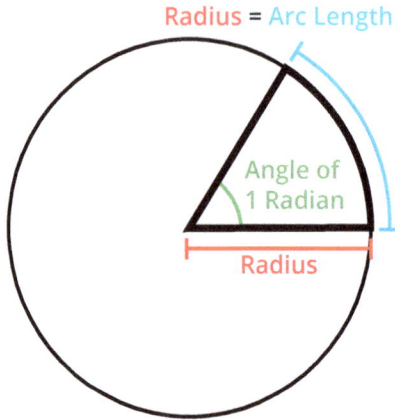

Radius = Arc Length

Image 200: "Angular distance" is a way to say "angle." An angle of 1 radian (equal to 57.3 degrees) is defined by having an arc length that is equal in length to the radius of the circle. 1 radian = 1000 milliradians (a.k.a. "mils" or "mrads").

Linear diameter increases with distance from start
Angle does NOT increase with distance from start

Linear Diameter of 1 MOA →
2.9 cm 5.8 cm 8.7 cm 11.6 cm 14.5 cm 17.4 cm 20.3 cm 23.2 cm 26.1 cm 29 cm
(1.14 in) (2.28 in) (3.42 in) (4.56 in) (5.70 in) (6.85 in) (7.99 in) (9.13 in) (10.2 in) (11.4 in)

Distance From Start →
100m 200m 300m 400m 500m 600m 700m 800m 900m 1000m
(109 yd) (219 yd) (328 yd) (437 yd) (547 yd) (656 yd) (766 yd) (875 yd) (984 yd)(1094 yd)

Linear Diameter of 1 Mil →
10 cm 20 cm 30 cm 40 cm 50 cm 60 cm 70 cm 80 cm 90 cm 100 cm
(3.9 in) (7.8 in) (11.8 in) (15.7 in) (19.6 in) (23.6 in) (27.5 in) (31.4 in) (35.4 in) (39.3 in)

Image 201: Angular units, such as milliradians (mils) and minutes-of-angle (MOA), measure how wide or narrow an angle is. An angle measures a set percent of a circle, so as the circle becomes larger, the linear diameter of an angle may increase but the angle itself remains the same. This image gives the correct numbers for 1 MOA and 1 mil; however, the red triangles are enlarged to an angle of approximately 30 MOA and 30 mils so that they are clearly visible.

24.a Minute of Angle (MOA)

A complete circle consists of 360° or 360 degrees. (See Image 98, Pg. 113.) Under the minute-of-angle measurement system, each degree consists of 60 "minutes," and each minute consists of 60 "seconds." (This would be like a clock with 360 hours.) As a result, a circle is composed of 21,600 minute slices and 1,296,000 second slices. So, a minute-of-angle (MOA), also known as a minute-of-arc, arcminute, and arcmin, represents 1/21,600th of the circle surrounding the observer. Degrees, minutes, and seconds are associated with the imperial system only because they are not an official part of the metric system; however, minutes-of-angle work equally well alongside both systems.

24.b Inches Per Hundred Yards (IPHY)

By pure coincidence, a circle with a radius of 100 yd has a circumference of 22,619 in. This is very close to the aforementioned 21,000 minutes in a circle's circumference. In fact, dividing 21,600 minutes by 22,619 equals approximately 1.047 minutes per inch. Therefore, many marksmen approximate that at 100 yd (91 m), 1 MOA is equivalent to 1 in.

The rounded ratio of 1 minute per inch goes by another name, "inches per hundred yards" (IPHY). IPHY is a third angular measure alongside milliradians and minutes-of-angle. Because 1 MOA represents 1.047 inches per 100 yards, 1 IPHY is about 5% smaller than 1 MOA. There are 22,619 IPHY slices in a circle.

However, because people prefer round numbers, many marksmen and manufacturers have conflated IPHY and MOA. Even the renowned scope manufacturer Leupold makes no mention of this discrepancy between the IPHY and MOA on their official website. They only state, their "riflescopes are available with MOA (1 MOA = 1 inch at 100 yd (2.5cm at 91 m)) adjustments and reticles." This description is incorrect, 1 MOA = 1.047 in at 100 yd.

While Leupold is undoubtedly aware that their information is technically incorrect, it still remains unclear which of the two measurement systems (MOA or IPHY) their scopes actually use. Leupold can continue to sell with ambiguous terms for two reasons. First, the 0.047 (i.e., 5%) difference only matters to extremely precise marksmen. This is because the fractional differences only add up to a significant variation at extremely long distances, 1,000 yd (910 m), for example. The difference between a 40 MOA and a 40 IPHY adjustment at 1,000 yd (910 m) is only 1.88 in (0.047 in × 40).

The second reason is that clicks rarely exactly match the stated measurement anyway. Marksmen must independently determine how much a click on their scope turret actually moves their reticle. For example, one click on a reticle may be advertised as rotating a turret by 1 MOA, but instead actually move the reticle by 0.98 MOA or 1.02 MOA. Part of setting up a scope is determining how much a click on each turret actually moves the reticle.

24.c Milliradian (Mil a.k.a. MRAD)

Under the milliradian measurement system, a circle's circumference is 6.283 (i.e., 2π) radians or 6,283 (i.e., 2000π) milliradians (mils, mRads), where 1 radian is equal to the radius of the circle.

So, if two marksmen fire two rounds at an angular distance of 1 radian, the bullets form an approximate equilateral triangle with the marksmen. Similarly, if the marksmen fire at a distance of 1 mil apart from one another, the distance between the bullets is approximately equal to 1/1,000th of the distance they have traveled. Therefore bullets fired 1 milliradian apart would be 1 yd apart at 1,000 yd, and they would be 1 m apart at 1,000 m.

This seemingly straightforward relationship becomes more complex when marksmen consider that there are different legal definitions for the number of mils in a circle. The U.S. military and NATO often divide a circle into 6,400 mils, whereas Russia and some European countries may divide a circle into 6,000 mils.

Further complicating any conversions, the approximately 2% difference between 6,283 (the circumference of a circle in mils) and the commonly used 6,400 mils, is often within manufacturer tolerances, making it difficult to determine which standard a particular nation or manufacturer uses based solely on equipment. In other words, a scope claiming to adhere to one standard may actually be adhering to another.

Irrespective of the specific standard used, many mil-scopes are manufactured with adjustments in 0.1 (1/10th) mil increments, though some scopes offer .05-mil adjustments. One-tenth of a mil represents an angle that measures exactly 1 cm at 100 m, 20 cm at 2,000 m, 3.6 in at 1,000 yd, etc. Similarly, half a tenth of a mil (i.e., 1/20 mils) would be half that.

24.d Comparing and Converting MOA and Mil

Both MOA and mil measurement systems are equally effective in describing angular distance, with neither being more or less precise than the other. Some advocates note that 1 MOA is smaller than 1 mil; however, this difference is moot since one-tenth of a mil is smaller than 1 MOA. Regardless, once someone is familiar with one system, there is no advantage to switching to the other system unless they need to use equipment that is only available in that alternative system.

With that said, sometimes a marksman still needs to convert between the MOA and mil standards. A common scenario requiring conversion is when marksmen need to input a wind call they learned from an MOA system into their mil scope.

Conversions are often approximated, due to the lack of precision in these measurements. For example, according to the mathematical definitions presented above, a circle contains 21,600 MOA and 6,283 mils, resulting in 3.44 MOA per mil. However, if someone labeled IPHY units as MOA units and used NATO-defined mils instead of mathematical mils, a circle would have 22,619 IPHY and 6,400 NATO mils respectively, resulting in 3.53 "MOA" per "mil." So, in making things less complex, when the exact nature of the units is unknown, **an approximated conversion of 3.5 MOA to 1 mil is fine**.

Despite the ease of conversion by simply multiplying mils by 3.5 to obtain MOA, converting between the two systems can still generate numerous errors. So, it is best to use the unit of measurement favored by the marksman's community. The mil measurement system is supported by more advanced equipment and is therefore preferred by the international community, the U.S. military, and most competitive shooting participants. The primary group still using MOA is North American civilians, although it remains a sizable group. Regardless of the unit of measurement used, it is crucial to employ the same unit for both a scope's turrets and reticle to avoid any conversion errors.

25. Other Measurements

To become an expert marksman, it is helpful to understand the physics of a bullet in motion. To that end, this appendix describes fundamental principles that affect bullet behavior: speed, velocity, mass, weight, force, pressure, work, energy, and momentum.

These topics might seem daunting at first, as they involve substantial scientific concepts. However, readers do not need an advanced degree in physics to understand and apply the knowledge in this book, as only basic concepts are required. That said, for those interested in a deeper exploration of the physics of shooting in more detail, Khan Academy offers free online video courses on the subject.

25.a Speed and Velocity

Movement refers to the change in position of an object over time. This can occur in various ways, such as linear motion along a straight path or rotational motion around a fixed point. Movement has two measures: speed and velocity.

Speed is a measure of how quickly an object moves or the rate at which it covers distance. Formally, speed is calculated as distance over time and is denoted in units like meters per second (m/s), kilometers per hour (km/h), miles per hour (mi/h), or feet per second (ft/s). For example, if a car covers 100 km (62 mi) in 2 hours, its speed is 50 km/h (31 mi/h). Speed is a scalar quantity, meaning it only describes how much an object moves without regard to direction.

▸ *Speed = Change in Distance ÷ Change in Time*

Velocity, on the other hand, is a vector quantity, which means it includes both the magnitude and direction of an object's movement. While speed measures how fast an object is moving, velocity also specifies the direction of its movement. For example, if a car is traveling at 50 km/h eastward, its velocity is 50 km/h east. The distinction between speed and velocity is crucial in scenarios where direction affects the outcome of the movement.

▸ *Velocity = (Change in Distance ÷ Change in Time)* in a *Specified Direction*

In ballistics, speed and velocity are often used interchangeably because the assumed direction is typically from the marksman to the target. However, there are scenarios where a distinction between the two is necessary such as when adding two speeds together. For example, a marksman may fire a bullet at 4,000 km/h (2,490 mi/h) (speed 1) from a truck traveling at 40 km/h (25 mi/h) (speed 2). If both are moving north, the two speeds are added together, and the bullet's speed relative to the ground would be 4,040 km/h (2,510 mi/h). Conversely, if the truck is moving south while the bullet moves north, the truck's speed would be subtracted from the bullet's velocity, which would be 3,960 km/h (2,460 mi/h).

Acceleration is the term used to describe the change in velocity. Like velocity, acceleration is also a vector quantity which must include a direction. Acceleration is typically measured in units such as meters per second squared (m/s^2), kilometers per hour squared (km/h^2), miles per hour squared (mi/h^2), or feet per second squared (ft/s^2). In the case of bullets, gravity acts as an acceleration force, causing them to fall downwards at a rate of 9.8 m/s^2

(32 ft/s²). This means that every second, an object's speed towards the ground becomes 9.8 m/s (32 ft/s) faster.

▶   *Acceleration = (Change in Velocity ÷ Change in Time) in a Specified Direction*

25.b Mass and Weight

Mass and weight are two distinct properties of matter. Mass represents the amount of matter in an object and remains constant regardless of its location in the universe. It is measured in grams (g) or kilograms (kg) in the metric system, and in pounds (lb) in the imperial system. The conversion factor between kilograms and pounds is 0.454 kg per lb and 2.20 lbs per kg.

On the other hand, weight is the gravitational force exerted on an object and depends on the gravitational field the object is in. It is a vector quantity, and it can be calculated using the following formula:

▶   *Weight = Mass × Gravity*

Consequently, an object's weight varies according to the local gravity. For example, an object weighs less on the Moon's surface than on Earth's because the Moon has a weaker gravitational pull.

In fact, the same object can weigh different amounts on Earth because different locations on Earth have different levels of gravity. This occurs because gravity is stronger closer to the Earth's center (i.e., low altitude) and at the poles where centrifugal pseudo-force is not "pushing" objects away from the Earth's surface.

Earth has a maximum variation in its gravitational acceleration of approximately 0.7%. The lowest value is found at Mount Nevado Huascarán in Peru at 9.76 m/s² (384 in/s²), while the highest gravity is at the surface of the Arctic Ocean at 9.83 m/s² (387 in/s²).

Differences in weight can cause serious problems in international trade, where for example a tonne of sugar in Brazil (more gravity) would contain less mass than a tonne of sugar in Canada (less gravity). Therefore, international trade is nominally conducted in units of mass, such as the kilogram. In practice, most scales measure force, or weight, making the distinction moot. However, this is why units of weight and mass are commonly confused with each other.

The appropriate metric unit for weight is the newton (N), which is also the unit of force. However, the pound, the imperial unit for weight, creates further confusion as it can be used to measure both mass and force. The pound-mass (lbm) represents mass, while the pound-force (lbf) measures force. 1 lbm under the influence of Earth's gravitational pull produces 1 lbf.

Nonetheless, this distinction is seldom made, as the metric system is preferred for precise measurements anyway.

Apart from grams and pounds, there is a commonly used third unit of mass known as "grains." This unit is the global industry standard for measuring the mass of gunpowder and bullets. There are 15.43 grains in 1 g and 7,000 grains in 1 lb. However, this is also slowly changing to the metric gram and milligram.

25.c Force and Pressure

In ballistics, the force of a bullet represents its ability to propel itself through air or target material. The unit-of-force is expressed by newtons (N) in the International System of Units (SI). The formula for force is:

▸ *Force = (Mass × Acceleration)*

Since force is determined by multiplying mass by acceleration, the total force of a large object accelerating slowly can be equal to that of a small object accelerating quickly. In cases where an object experiences force, it is due to another object applying force to it. For example, bullets experience force that propels them forward because gas spends its force pushing the bullet.

Additionally, any interaction involving force also involves an equal reaction. So, when one object accelerates another, the second object decelerates (i.e., accelerates in the opposite direction) the first object. Designing bullets involves ensuring that they maintain their force (i.e., accelerated) through air but expend their force upon reaching solid targets.

Pressure is determined by dividing force by area and is measured in pascals (Pa). The formula for pressure is:

▸ *Pressure = (Force ÷ Area)*

Pressure is relevant in several aspects of shooting, particularly in bullet design. For example, a bullet with a smaller tip applies more pressure to a target with the same amount of force. The level of pressure determines whether force causes an object to deform or break. Armor-piercing rounds, for example, retain their pointed shape upon impact, applying greater pressure than fragmentation rounds, which disperse rapidly upon contact with the target (i.e., have a larger area of impact). Therefore, using the same amount of force from the gunpowder but over a smaller area (the tip of an armor-piercing round) allows for easier breaking and penetration of target objects.

The downside of high-pressure rounds is that they are bad at transferring energy. Armor-piercing bullets are likely to shoot right through a target, retaining a substantial amount of their energy, whereas a lower-pressure, fragmentation bullet (e.g., a soft-point or hollow-point bullet) would transfer all of its energy to the target, causing considerably more damage.

25.d Work and Energy

Work is a concept in physics that helps to understand how things move. Work is most often described through its formula:

▸   *Work = Force × Distance*

That is, work is defined as the force that is applied over a distance. For example, when a bullet is fired from a gun, the explosion inside the gun creates a force that pushes the bullet forward. The exploding gas only applies force to the bullet while the bullet is in the barrel, so it can apply its force over a longer distance (i.e., do more work) in a longer barrel than a shorter one.

Energy is a related concept in physics that describes work that can be done, but is not currently being done. In other words, energy is the ability to do work (a.k.a. potential work, or stored work). For example, gunpowder possesses energy because it can do work in the future. In contrast, exploding gases are an example of work being done (i.e., transferring their energy to the bullet).

There are a few equivalent units for both work and energy. For example, in the metric system, one unit of work is a newton-meter (N × m). The newton-meter is equal to the joule (J), which is a unit of energy. (That is, 1 J is equal to the amount of work done when a force of 1 N displaces a mass by 1 m.) The imperial unit for work and energy is the foot-pound (ft × lb), which is calculated by multiplying a foot by the pound-mass.

Work and energy are also equal to mass times velocity squared. To prove this, the previous formula can be rearranged into the following:

▸   *Work = Force × Distance*
▸   *Work = Acceleration × Mass × Distance*
▸   *Work = (Distance ÷ Time²) × Mass × Distance*
▸   *Work = (Distance² ÷ Time²) × Mass*
▸   *Work = Velocity² × Mass*

This has two implications. First, energy and work are directly proportional to the mass of the object. This means that objects with more mass require more energy to move or stop. Second, energy and work are proportional to the square of the velocity. This implies that even a small increase in the velocity of an object results in a much larger increase in the energy contained within it. For example, doubling the speed of a bullet requires four times the kinetic energy to be applied to it, and subsequently, the bullet can do four times the amount of work on the target.

The "conservation of energy" is a key principle stating that when one object does work on another, the exact amount of energy lost by the first object is transferred to the second. This means the total energy remains constant; the total energy before any work is done is equal to the total energy after. Although energy conservation might not always be visible at the macroscopic level, (e.g., some of the energy is converted to heat, or multiple objects and fluids are involved), the total energy always remains the same.

The conservation of energy can be used to infer, for example, the minimum amount of energy that was contained by the gunpowder in a cartridge. For example, if both the mass and the muzzle velocity of a bullet are known, the bullet's energy can be calculated. And due to the conservation of energy, it can be inferred that the gunpowder and rifle must have done an equal amount of work on the bullet to transfer energy to it. Therefore, the energy of the gunpowder must have been at least as much as the energy of the bullet.

In fact, to determine the efficiency of energy transfer from the gunpowder to the bullet, one can divide the energy imparted to the bullet by the total chemical energy contained in the gunpowder. The remaining energy not transferred to the bullet is lost as heat, either to the rifle or the surrounding environment.

25.e Momentum

Momentum is a fundamental concept in physics that describes the motion of an object in terms of its mass and velocity. It's a vector quantity, meaning it has both an amount and a direction. The formula for momentum is:

▸   *Momentum = (Velocity × Mass)*

These same variables can be converted to be expressed in a different way:

▸   *Momentum = (Force × Time)*

Momentum is most often calculated to employ a property of matter called the "principle of conservation of momentum." In physics, the term conservation refers to something which doesn't change. This means that the variable in an equation which represents a conserved quantity is constant over time. It has the same value both before and after an event. So "conservation of momentum means that two objects, without external interference, always have the same total amount of momentum between them, even if one gives momentum to the other.

This principle is very helpful in making certain conclusions. For example, if one object changes momentum, it can be deduced that another object has also changed momentum. Also, because momentum has a direction component, if one object changes direction, it can be deduced that another object also changed direction. In those regards, momentum is related to Newton's third law of motion, which states that any force exerted on one object (over time) requires another object to have also experienced the same force (over time) but in the opposite direction.

The conservation of momentum is evident on a pool table, where the cue ball transfers momentum and energy to the target ball. If the cue ball comes to a complete stop, the momentum of the cue ball before the collision must equal the momentum transferred to the target ball after the collision (excluding friction from the surface of the table).

In that vein, momentum is useful in explaining the concept of recoil. Due to the conservation of momentum, moving a bullet forward at a faster speed also causes the rifle stock to move backward faster. Likewise, moving a bullet with more mass forward also leads to a faster backward movement of the rifle stock.

26. Functions and Malfunctions

A rifle is a machine that translates human input (e.g., pulling a trigger) to mechanical output (e.g., releasing the firing pin). To achieve this, rifles perform a series of functions. The specific steps can vary depending on the rifle's mechanism, as some steps are a matter of preference. However, military rifles generally follow a common list of functions: 1) feeding, 2) chambering, 3) locking, 4) firing, 5) unlocking, 6) extracting, 7) ejecting, and 8) cocking. The following subsections start with feeding because feeding is a firing cycle's starting point when starting with an unloaded rifle. However, since functions do occur in a cycle, any function could technically act as the cycle's starting point.

Since a rifle has a limited number of functions, it can only have a limited number of malfunctions. When attempting to repair a broken rifle, the best initial step is to identify which function is not working correctly. However, this can be challenging as malfunctions often occur simultaneously or commonly cause other malfunctions. Additionally, some malfunctions can be fixed immediately using the solutions provided here. However, if a malfunction is caused by broken or defective rifle parts, the marksman may need to consult a gunsmith for repairs.

26.a Feeding

Feeding is the process of positioning a cartridge centrally within the bolt channel (See Image 4, Pg. 16.). In a bolt-action rifle, this involves manually

New and Used Primers

Image 202: Used primers have divots in them (left). New primers are flat (right). If any sort of divot is present on a live round, then it must be considered a dangerous explosive, since the divot indicates that the primer has already been struck, making it unknown whether the primer is unstable and spontaneously explosive.

inserting a round or pushing the bolt forward. In a semi-automatic rifle, the first round is typically fed by locking the bolt to the rear, loading a round, and releasing the bolt, with subsequent feeding being automatic.

A common malfunction, a failure-to-feed, occurs when a round fails to enter the action. The most common cause of a failure-to-feed is a worn or damaged magazine, and the failure can be resolved by reseating or cleaning the magazine. However ideally, the magazine is replaced whenever there is a failure-to-feed. If the magazine is not the cause of the malfunction, then alternatively, the action may require cleaning and lubrication. However, marksmen must avoid over-lubricating, as it can attract dirt and cause problems in cold conditions. Other potential issues include a weak or broken mainspring and a bolt that binds in the receiver.

During shooting, a failure-to-feed often goes unnoticed by the marksman, who may mistakenly believe that a round is in the chamber. In such cases, a failure-to-feed is indistinguishable from a failure-to-fire. To avoid injury, marksmen must wait for 30 seconds and keep the rifle pointed in a safe direction before attempting to open the bolt, as the primer may have been struck and could have a delayed reaction (a.k.a., a "hang fire").

26.b Chambering

Chambering involves moving a cartridge from its centrally aligned position within the bolt channel to being fully seated in the firing chamber. (See Image 4, Pg. 16.) Feeding and chambering rely on different mechanical parts and can be differentiated by checking for specific failures.

A failure-to-feed indicates that there is no force pushing a round into the chamber, whereas a failure-to-chamber means that a cartridge cannot enter the chamber. A failure-to-chamber may occur if the cartridge itself is damaged, so it is crucial to first inspect and safely discard any damaged cartridges before any shooting session. Additionally, a dirty chamber can cause problems, so it must also be inspected and, if needed, cleared and cleaned before use. Other issues that can arise include a weak or broken mainspring or a bent receiver housing. Repairing or replacing components may require the assistance of a qualified gunsmith.

26.c Locking

Locking refers to the immobilization of the action and sealing of the chamber, preventing any explosive gases from escaping out the back of the rifle. While feeding and chambering require pushing the bolt, locking generally involves a rotation of the bolt and engages separate mechanical parts such as a bolt latch and bolt latch trip, or a bolt and a star chamber.

Issues related to locking or unlocking may arise due to an internal obstruction between the firing pin and the bolt. This is because many actions have a safety mechanism which requires the firing pin to rotate to engage in a locked chamber and enable firing. To fix a firing pin obstruction, a marksman may have to disassemble and clean the rifle bolt. Bolts contain springs and small parts, and so this exercise must always be done in an area where dropped parts can be found (e.g., above a flat floor without gaps.)

That being said, locking and unlocking problems may be caused by root issues such as: a blown or wedged primer; a broken or burred bolt latch; a broken bolt-latch spring; or a bent or improperly seated bolt-spring. All of these issues all best addressed by an expert gunsmith.

An improper lock may also be caused if a marksman rests the rifle on its magazine. In this case, the marksman only needs to adjust their rifle hold to take pressure off of the magazine.

26.d Firing

Firing as a function of a rifle encompasses the steps that cause the release of the sear (part of the trigger assembly), thereby causing the firing pin to strike the primer on a cartridge. Firing is completed when the pressure within the barrel and chamber returns to a neutral state.

Failure-to-fire may occur due to faulty ammunition. Specifically, sometimes the firing pin correctly strikes a primer, but the primer fails to ignite. If this happens, a marksman must inspect and potentially replace all of the ammunition from the same manufacturing lot. (See Image 202, Pg. 248.)

It is also possible that the firing pin or spring itself is faulty or defective. This can be identified by a shallow impression or no impression on the primer after firing. Any defective firing pin, firing mechanism, or trigger components requires the expertise of a gunsmith.

In cases where the primer was struck, but nothing happens, marksmen must wait 30 seconds while keeping the rifle pointed in a safe direction before attempting to open the bolt. This prevents a delayed reaction from causing an injury. Similarly, when unloading, the ejection port must also be pointed in a safe direction in expectation of a delayed reaction. Once the round is ejected, the marksman can inspect the primer. If it was struck, then the ammunition was the cause; if it was not struck, then the rifle was the cause. Regardless, marksmen must never attempt to reuse a round that failed to fire.

During this phase, there can be overly-strong recoil, which must not be confused with a failure-to-fire. Hard recoil may be caused by faulty or heated ammunition, which must be replaced or allowed to cool. (See Temperature, Pg. 188.) Damage to the mainspring, mainspring buffer, or muzzle break can also contribute to this issue, necessitating the replacement of the damaged components by an expert.

26.e Unlocking

Unlocking, the next function or phase in the rifle's cycle, refers to the reversal of the immobilization of the action that occurred during the locking step, allowing the spent cartridge to be extracted. Unlocking failures occur for the same reasons as locking failures. (See Locking, Pg. 249.)

26.f Extraction

Extraction is the process of moving the spent case from the chamber into the action of the firearm. It is the opposite function of chambering, however, it may use a different mechanism in the rifle (e.g., the extractor). Any extraction failures can occur if the extractor is not moving freely in its slot. In that case, the extractor is broken or stuck and it must be removed and cleaned or completely replaced with a new one. However, another common cause of extraction failure is bad or mishandled magazines. Preventing double feeds is a primary reason marksmen do not rest rifles on their magazines.

Extraction failure is usually detected when attempting to chamber another round, resulting in a double feed and a failure-to-chamber. To clear a

Image 203: A **double feed** is when a bolt attempts to feed two bullets into the chamber at the same time. This is caused by a **failure-to-extract** a previous bullet from the rifle. The extractor can fail to extract if it is broken, but more often the magazine is old or being mishandled, or the gun may lack proper lubrication.

stuck casing in the extraction port, a marksman can pull it out with pliers. To clear a stuck casing in the chamber, a marksman can push it out by inserting a cleaning rod into the muzzle and pushing it down into the stuck casing.

If there is a failure-to-fire and an extraction failure, meaning the stuck round is live, a marksman must wait 30 seconds to ensure the round does not spontaneously explode. Then the marksman can remove the bolt, thereby rendering the rifle incapable of firing, and take the rifle to a gunsmith, who can safely remove the stuck, live round. Although there is no risk of explosion by slowly and carefully pushing on a bullet if the ammunition is good, stuck ammunition is by definition bad ammunition. And this manual cannot recommend that an amateur forcefully push on a live round with a metal rod.

26.g Ejection

Ejection as a function of a rifle is the removal of the spent casing from the firearm's action to outside the firearm. A failure-to-eject occurs when a cartridge remains partially in the chamber or becomes jammed in the upper receiver. One type of failure-to-eject is typically referred to as a "stovepipe," and happens when a spent cartridge casing gets stuck in the ejection port. (See Image 204, Pg. 253.) Stovepipes can usually be fixed with standard malfunction actions. (See Standard Malfunction Actions, Pg. 252.) A failure-to-eject could also be caused by a frozen or damaged extractor, ejector, or ejector spring. Often, failures-to-eject are due to friction from dirt or carbon buildup. In semi-automatic rifles, a "fouled" (i.e., has carbon build-up) gas tube can restrict the flow of gas and prevent sufficient force from reaching the extraction and ejection mechanism. These malfunctions that are caused by problematic rifle parts require expert attention to fix.

26.h Cocking

Cocking is the setting and resetting of stored energy within the firearm to allow the firing pin to launch forward once the sear is disengaged by pulling the trigger. Cocking can be caused, for example, by the manual rearward pulling of the bolt in a bolt-action rifle or by the force of redirected gases in a semi-automatic rifle. Cocking failure is likely caused by worn, broken, or missing components of the trigger assembly and sear assembly. A cocking malfunction requires expert attention.

26.i Standard Malfunction Actions

If a rifle stops working during any function, a marksman can save a lot of time by performing a few quick actions that immediately fix most malfunctions. The specific actions depend on the rifle. However, all fixes begin with ensuring that the weapon is pointed in a safe direction and the safety mechanism is activated.

When a standard bolt-action rifle malfunctions, a marksman can:
1) Remove the ammunition source to prevent further feeding.
2) Pull the bolt back to its rear position to address any potential malfunction caused by a stuck or misaligned round.
3) Complete two full cycles of the bolt, ensuring that it moves smoothly, and then leave it in the open position.
4) Reinsert the ammunition source, chamber a round, and attempt to fire

To fix a standard semi-automatic rifle, marksmen first attempt "immediate action," which is a quick procedure to try and get the weapon back into operation without diagnosing the problem in detail. This is often referred to by the acronym SPORTS:

Slap – the bottom of the magazine to ensure that it is properly seated.
Pull – the charging handle to the rear.
Observe – the ejection port to check for any ejection or obstruction.
Release – the charging handle to allow the bolt to go forward.
Tap – the forward assist (if there is one) to ensure the bolt is locked.
Squeeze – the trigger to attempt firing.

If immediate action fails, a marksman can attempt "remedial action" to correct the malfunction:
1) Reattempt immediate action. If it still fails, proceed to the next steps.

Image 204: A **stovepipe** malfunction is a type of firearm malfunction where a spent casing fails to eject properly and gets caught in the ejection port of the firearm. The casing typically sticks out of the ejection port, resembling a "stovepipe," hence the name. Stovepipe malfunctions can occur due to a clogged gas port that blocks gas from operating the ejector, a fault ejection mechanism, or weak gunpowder.

2) Remove the source of feed (e.g., the magazine).

3) Lock the bolt to the rear.

4) Clear the weapon by visually inspecting the chamber, magazine well, and bolt, and remove any obstruction or jammed cartridge.

5) Reinsert the ammunition source, chamber a round, and attempt to fire.

27. Glossary

Accuracy	The degree to which shots are centered on the point-of-aim.
Action	The part of a gun that moves cartridges in and out of position.
Adjustable objective dial	A parallax adjustment dial located on the scope's objective bell (i.e., it's far side).
Aerodynamic	Relating to how air moves around objects, and how objects move within the air.
Air density	How thick the air is. Influenced by station pressure, temperature, and humidity.
Ambidextrous	Capable of using both hands with equal skill.
Ammunition (Ammo)	The projectiles and propellant (a.k.a., gunpowder) used in firearms.
Angle cosine indicator	Device mounted on a rifle to measure and adjust for shooting angles.
Angular distance	Angular measurement between two points.
Angular height	Measurement of an object's apparent height in angular distance units.

Aperture	A small opening that is used for looking through.
Arc	A curved path of travel
Auxiliary	Additional or supplementary components or equipment.
Ballistic calculator	A device that calculates bullet trajectories based on factors such as distance, wind, and temperature.
Ballistic coefficient	A measure of a bullet's aerodynamic drag in the air. Bullets with higher ballistic coefficients resist drag better than a bullet with low BC.
Ballistics	Study of the behavior of projectiles, including trajectory, penetration, and impact.
Barrel crown	The point at which the bore meets the outside. The final point of the bore that touches the bullet.
Barrel	The long, narrow part of a gun that the bullet or projectile travels through when fired.
Benchrest	A steady support for a rifle, usually including a bench or table. Also a type of shooting competition that uses said support.
Bipod	A two-legged support attached to the front of a rifle, allowing for increased stability. The legs are often collapsible and adjustable for varying heights.
Bolt	A part in a firearm that moves behind the chamber and seals the rear of the chamber during firing.
Bolt-action	A type of manually-reloaded firearm for which the marksman operates a bolt to cycle rounds in and out of the chamber.
Bone support	The use of the body's skeletal structure to support the weight of the rifle.
Bore	The internal cavity of a barrel in which bullets travel.
Boreline	The imaginary line that extends from the center of the chamber, straight out of the center of the bore, and on to infinity.
Bore-sighting	Roughly aligning a rifle's bore with the scope's sightline.
Bracket	First, establishing a minimum and maximum to set a range. Then, establishing new minimums and maximums that are closer to the average to make the range smaller.
Brass	Another name for casings, since most casings are made from brass.
Breech	The loading entry point for ammunition. In everyday conversation "breech" and "chamber" could be used interchangeably. Technically, the breech is where the bolt rides, but forward of the ejection port.
Bullet-drop	The vertical, linear distance that a bullet falls due to gravity over distance.
Bullet-drop compensator	A kind of reticle in scopes that spaces hashmarks according to a set increment of linear distance instead of angular distance.
Butt, buttstock	Rear portion of the rifle stock that rests against the marksman's shoulder, providing support and stability.
Caliber	Internal diameter of a bore, expressed in inches or millimeters.
Cant	A rotation of the rifle or scope off the vertical axis (directly up or down).
Cartridge	A complete unit of ammunition containing a casing, a primer, gunpowder, and a bullet.
Casing, Case	The part of a cartridge that contains the other parts (i.e., the primer, gunpowder, and bullet).
Center average	The vertical and horizontal center of all the points in a grouping.

Centerline of the bore	See boreline.
Chamber	Part of a firearm where the cartridge is inserted and locked into for firing.
Cheek piece	A part of a stock that supports the marksman's cheek, aiding eye alignment with the sight(s) while firing.
Cheek weld	Proper placement of the marksman's cheek on the rifle stock for consistent sight alignment.
Centerfire	A type of ammunition where the primer is located in the center of the cartridge base.
Clicks	Units of adjustment for firearm sights, both iron and optical. These sights have rotating turrets that click when adjusted by one increment.
Climate	Characteristics of the air, including density (pressure, temperature, humidity) that affect bullet trajectory.
Cold-bore	Refers to a firearm barrel that hasn't been recently fired and is therefore cool.
Cone-of-fire	The dispersion pattern of bullets fired from a firearm, where the point of the cone is at the muzzle, and the base is at the target.
Conical	In the shape of a cone.
Cowitness	Using the point-of-aim of one sight on a gun to set the point-of-aim on a second sight on the same gun.
Crosshairs	The horizontal and vertical lines that respectively extend the full width and height on the reticle and intersect in the center.
Crosswind deflection	Deflection due to the portion of the wind's power that is perpendicular to the trajectory of the bullet.
Crown	The meeting point between a barrel's muzzle and its bore. The crown can be slightly recessed or relieved to protect the forward edge of the rifling from damage, which can affect accuracy.
Cycle	To operate the action of a firearm.
Deflect	To push an object from one trajectory to another trajectory.
Depth-of-field	Range of distance in which objects appear sharply focused.
Dispersion	The spaced-out pattern of individual points.
DOPE (Data On Previous Engagement) book	Record book used by marksmen to log specific details of each shooting session for future reference and adjustment.
Double feed	The simultaneous attempted feeding of two or more rounds into a chamber.
Drag	The portion of air resistance that slows a bullet down.
Dry fire	The act of cocking, aiming, and pulling the trigger of an unloaded rifle for the purpose of practicing marksmanship fundamentals.
Duplex reticle	A scope reticle design featuring coarse crosshairs at the outside that narrow down to fine lines at the center.
Eddy	Unpredictable wind patterns, typically created by obstacles such as trees or buildings blocking the wind's forward path causing swirling currents of air.
Ejection	Removal of spent casing from the firearm's action after firing.
Ejection port	Opening on the rifle through which spent cartridges are ejected after firing.
Elevation, Elevation hold	The vertical measure of a sight or scope in angular measurements.

Environment factors	External conditions such as wind, gravity, and temperature that affect the flight path of a bullet.
Equivalent crosswind	The speed of a hypothetical wind perpendicular to the trajectory of the bullet that acts on the bullet with equal force to the actual wind.
Exponential	The dependent (Y) variable changes faster than the independent (X) variable.
Exterior ballistics	The study and calculations related to factors affecting a bullet's flight after it leaves the barrel.
Extraction	Removal of a spent casing from the chamber into the firearm's bolt channel.
Extractor	A part of a firearm that removes the spent cartridge after firing.
Eye relief	The distance from the ocular lens of the scope to the marksman's eye.
Failure-to-chamber	Cartridge unable to enter the firing chamber.
Failure-to-eject	Spent cartridge not fully ejected from the firearm.
Failure-to-feed	Cartridge fails to enter the action for firing.
Felt-recoil	The subjective force of recoil felt by a marksman. Actual recoil feels different to different people for various reasons.
Fine-tune	To make precise adjustments or modifications.
Firing	Release of firing pin to strike the primer of a cartridge, causing an explosion.
Firing pin	Component in a firearm that strikes the primer of a cartridge to initiate ignition.
Fliers	Outliers in a shot group that deviate significantly from the main pattern.
Flinching	Involuntary movement or reflex.
Focus	To adjust the scope to make the reticle and sight picture sharply visible.
Follow-through	Maintaining focus and technique after firing the shot to ensure accuracy.
Forestock	The front part of a rifle stock. It may or may not connect to the barrel depending on the design of the firearm.
Forward assist	A mechanism on some rifles, such as the AR-15, that is used to ensure proper chambering of a round by pushing the bolt forward.
Functions check	The performance check of each function of a firearm in sequence to ensure a firearm operates properly.
Functions	The mechanical steps that a firearm cycles through to operate.
Gas-operated	Refers to an automatic (i.e., self-loading) firearm that uses the expanding force of the propellant's gas to cycle a weapon's action.
Grime	Dirt or residue that has accumulated and adhered to surfaces of both the inside and the outside of a weapon or object.
Grip, rifle grip	The part of the firearm that the trigger hand holds, and the manner in which the marksman holds that part.
Group, grouping	A series of shots fired with the same point-of-aim. Five-shot groups are standard.
Group size	The size of the spread of impacts in a group. Can be the center average or the diameter.
Gunsmith	Skilled professional specializing in the design, construction, repair, and modification of firearms.
Gust	A sudden and temporary increase in wind speed.

Handguard	See forestock.
Handling	The act of touching or manipulating firearms or ammunition, requiring caution and respect.
Hashmark	A small marking, such as a line or dot, used to indicate distance on a reticle.
Height-above-bore	The vertical distance between the boreline and the sightline where the scope is mounted.
Hold	A vertical or horizontal adjustment in angular units to the rifle that allows the marksman to use a point on the reticle other than the center as their point-of-aim.
Holdover	Either another name for height-over-bore, or the hold used to compensate for it.
Imperial system	The nickname for the United States Customary System, primarily used in the United States and based on inches, feet, pounds, etc.
Inclination	The angle between a straight line and the horizontal plane perpendicular to gravity.
Inferior mirage	Mirage that makes objects appear lower than they are.
Interior ballistics	The study of the forces inside the cartridge and barrel during firing.
Jacketed bullet	Bullet with a metal casing (the jacket) covering the core.
Joists	Horizontal supporting beams in structures, mentioned as a safe direction to point a rifle.
Kentucky windage	An estimated and imprecise aiming adjustment to compensate for wind effect or target motion.
Laser rangefinder	Electronic device using laser technology to precisely measure distance to a target.
Lateral throwoff	Random deviations from the boreline caused by equipment.
Lift	The component of aerodynamic force that causes a bullet to deflect (i.e., move perpendicularly to its direction of travel).
Linear	A change in one variable causes a consistent and constant change in another variable.
Linear average	The center point between the two farthest points in a group.
Linear distance	The straight-line distance between two points.
Line-of-sight	An imaginary line extending from the marksman's eye through the rifle's sights and onto an aiming point on a target.
Locking	Immobilization of the cartridge within the action to prepare for firing.
Loophole	An opening within a barrier that a marksman fires through.
Magazines (Mag)	The part of a rifle that stores ammunition. It may or may not be detachable.
Magazine well	The cavity in a firearm's receiver where a detachable magazine is inserted and held in place.
Magnification	Optical feature on scopes that enlarges target view for precise aiming.
Malfunctions	A failure to perform a function of a firearm's operation.
Mechanical center	Midpoint of a turret's adjustment range on a scope.
Metric system	The nickname for the International System of Units, a decimal system of measurement based on meters, liters, kilograms, etc.
Mil, mils	See milliradians.
Milliradians (mils)	A unit of angular measurement. Depending on the definition, there are between 6,000 and 6,400 milliradians in a circle.

Mildot reticle	A crosshair design with circular (dot) hashmarks along its crosshairs.
Milling	Measuring the angular height of an object in mils or MOA through a reticle to calculate the distance to that target.
Minute, Minute-of-angle (MOA) – A unit of angular measurement. Usually a minute is 1/60 of a degree. There are 360 degrees in a circle. A different definition has ~22,620 MOA in a circle.	
Mirage	A wavy optical distortion caused by variations in air pressure that cause light to move through the air in slightly different directions.
Muzzle	The barrel's end, from which the bullet exits upon firing.
Muzzle jump	Upward movement of the firearm's muzzle upon firing.
Muzzle velocity	The speed at which a bullet leaves the muzzle of the firearm.
Natural point-of-aim (NPA) – Position where the rifle aligns naturally when the marksman is relaxed.	
Objective lens	The lens at the front of a scope.
Ocular lens	The lens of a scope closest to the marksman's eye.
Optics	Devices related to light and lenses, including scopes and rangefinders.
Parallax	The apparent shift in the position of the reticle or sights relative to the target when the marksman's eye moves.
Pinwheel	A small rotating propeller that uses the power of the wind to rotate.
Plumb	Aligned to gravity.
Plumb line	A string with a freely hanging weight attached to and pulling on the bottom that is perfectly aligned to gravity.
Point-blank range	Distance from the muzzle at which a firearm can hit a target of a specified height without adjusting the aim to compensate for bullet-drop.
Point-of-aim	The exact location a marksman intends to hit on a target, and the location on a reticle they use to aim at that location
Point-of-impact	Where the bullet strikes the target.
Power	Power can refer to either the energy of the bullet or the magnification level of the scope sight.
Precision	The degree to which shots cluster closely together, indicating consistency and control.
Probabilistic shooting	Concept acknowledging that every shot has inherent random factors affecting trajectory, resulting in a dispersion pattern rather than a single trajectory.
Projectile	An object propelled through the air, such as a bullet or artillery shell.
Range	The distance between two points.
Range, Shooting range – A designated place where live-fire weapons training is conducted.	
Rangefinder	Device used to measure the distance to a target.
Recoil	Backward movement of the firearm caused by the explosion of gunpowder.
Repeatable	The concept that an adjustment in one direction exactly equals an adjustment in the opposite direction.
Rest	A support for the rifle that a rifle can sit in, supporting both the front and the back of the rifle. Rests can range from specially-designed devices to sandbags or improvised supports.

Reticle	A pattern of crosshairs and hashmarks in a scope that is used for aiming and finding angular measurements.
Rifle system	The set of equipment that allows a projectile to impact at the marksman's point-of-aim, including a rifle, optics, ammunition, and auxiliary equipment.
Rifling	The grooves cut into the barrel's bore that induce a spinning motion to the bullet, improving its stability during flight.
Rimfire	A type of ammunition where the primer is located within the rim (i.e., side) of the cartridge base.
Round	See cartridge.
Safety	Mechanism on firearms to prevent the firing pin from striking the primer and causing ignition or the cartridge.
Safety selector	Mechanism on a firearm that activates or deactivates the safety.
Sandsock	A fabric bag (e.g., a sock) filled with a grain (e.g., sand or pellets) used as a shooting support to stabilize a firearm.
Scope, riflescope	An optical sight that uses lenses to enable a marksman to see a magnified sight picture without parallax.
Scope adjustment	Fine-tuning the scope's turrets to adjust the reticle's location within the scope and thereby change the reticle's point-of-aim.
Scope mount	Device to attach a scope to a rifle.
Scope tube	Outer housing containing the lenses.
Semi-automatic	Firearm (or the action therein) which fires one round per trigger pull, automatically chambering the next round.
Shooting positions	Various body positions used for shooting, including prone, kneeling, and standing.
Side focus turret	Turret on the left side of a scope that adjusts for parallax and target focus.
Sight	A device that a marksman uses to aim.
Sightline	See line-of-sight.
Sight picture	The view that a marksman sees when the point-of-aim is aligned to the point-of-impact. For a scope, it is the image seen through the scope.
Silencer	A common term for a "suppressor," a device designed to reduce the noise of a gunshot.
Simple mirage	See mirage.
Slack	The distance that a trigger can be pulled before it engages the sear and discharges the firearm.
Sling	A long strip of fabric or leather that connects the forestock to the marksman's upper arm, allowing the marksman to aim more precisely.
Spotter	A partner to a marksman who feeds the marksman information on the situation, target, and optimal holds.
Station pressure	Actual air pressure at a specific location and elevation.
Stock, Rifle stock	Bottom part of the rifle that the marksman holds, including the buttstock and forestock.
Stock weld	See cheek weld.
Strap	A long strip of fabric or leather attached to a firearm that allows it to be carried. It is not useful for aiming.
Streamer flags, Streamers – Simple strips of cloth or plastic, often brightly colored, that flutter and thereby show wind direction.	

Sub-MOA	The capability of a rifle to consistently shoot groups with a diameter of less than 1 MOA in diameter.
Subtension	The distance between two hashmarks.
Superior mirage	Mirage where that makes objects appear higher than they are.
Supersonic	Any speed that is faster than the speed of sound.
Support	An inanimate object that helps to support the weight of a rifle.
Suppressor	Device attached to a muzzle to reduce the noise and recoil of the firearm.
Target focus	Whether the target can be seen clearly or is blurry.
Target identification	Recognizing and identifying specific targets for accurate shooting.
Terminal ballistics	The study of a projectile's behavior upon impact with a target, including penetration, expansion, and energy transfer.
Trajectory	The curved path followed by a projectile in flight, affected by factors such as gravity and wind.
Trigger control	Technique of smoothly pulling the trigger to avoid disturbing aim.
Trigger finger	The finger used by the marksman to pull the trigger, usually the index finger.
Trigger group, Trigger	assembly – The complete set of components within a firearm that includes the trigger, sear, and other parts that cause a firing pin to suddenly strike the primer and fire the rifle.
Trigger guard	A loop or enclosure that surrounds the trigger, protecting it from negligent discharge or external interference.
Turret	A rotating mechanism that adjusts part of the rifle incrementally. Each incremental turn makes a click.
Turret cap	A cover that protects the adjustment turrets on a riflescope.
Unload	To remove ammunition from the chamber or magazine of a firearm, ensuring it's not ready to fire.
Vane	A horizontal stick with a tail and pointer mounted on a vertical pivot, where the tail catches the wind to direct the pointer into the direction of the wind.
Weathervaning	Bullet aligning with incoming wind direction due to spin stabilization, akin to how a weathervane aligns with the wind.
Wind reading	Assessing wind speed and direction to adjust aim and ensure accurate shots.
Windsock	Cone-shaped fabric tubes inflating at specific speeds to indicate wind direction and intensity.
Wind value	Fractional representation of the equivalent crosswind component relative to the bullet's trajectory angle.
Windage	The horizontal adjustment to the aim of the firearm to compensate for wind drift.
Windage turret	The turret on a scope that allows for horizontal reticle alignment.
Zero	The alignment of the scope's point-of-aim with the bullet's point-of-impact at a specific distance for a specific rifle and scope.
Zero distance	The specific range at which a rifle and its scope are set to hit the target where the center of the reticle points.
Zero point	The two points at which the sightline intersects the trajectory of the bullet.
Zero stop	A lock that can be set to prevent a scope turret from turning into negative elevation (i.e., the minimum turn sets the reticle to its zero).

28. Credits

The explicit and implied contents herein do not imply or constitute endorsement by the U.S. DOD or any of its branches.

Front Cover: John Mark, U.S. Army SGT Thoman Johnson, Mitch Kezar
Back Cover 1: U.S. Air Force A1C Steven Cardo
Back Cover 2: U.S.M.C. GYSGT Robert B. Brown Jr.
Back Cover 3: Courtesy Photo
Back Cover 4: U.S.M.C. CPL Sean Potter
TOC Image 1: U.S. Army VIS Dee Crawford
TOC Image 2: U.S.M.C LCPL Christopher J. Nunn
TOC Image 3: U.S. Air Force SRA Renee Nicole Finona
TOC Image 4: U.S. Army SGT Patrik Orcutt
TOC Image 5: U.S. Space Force A1C Joshua Fontenot
Intro TOC: U.S. SOCOM SSGT Angelita Lawrence
Setting Up TOC: U.S. Army SPC Nyatan Bol
Beginner TOC: U.S. Army SGT ShaTyra Reed
Intermediate TOC: U.S.M.C CPL Sean Potter
Expert TOC: U.S. Army 1LT Benjamin Haulenbeek
Info TOC: U.S. Air Force A1C Breanna Carter
Appendix TOC: Markus Rauchenberger
Image 1: Ragnhild Kjeldse
Image 2: Sonaz
Image 4: U.S.M.C CPL Bethanie Ryan
Image 6: ArmourerIAA
Image 7: Francis Flinch
Image 8: Captaindan
Image 11: U.S. Army SSG John Mark
Image 12: U.S.M.C LCPL Samuel Fletcher
Image 13: U.S. Navy MC2 Deven B. King
Image 14: U.S. Army SPC Craig Jensen
Image 17: U.S. Air Force SRA Kristen Heller
Image 18: K. Kassens
Image 20 et al: K. Kassens
Image 21: U.S. Navy MC1 Ryan G. Wilber
Image 22: U.S. Army MAJ Michelle Lunato
Image 23 et al: U.S.M.C LCPL Antonio Garcia. U.S.M.C LCPL Christopher Mendoza. U.S.M.C LCPL Ryan Young. U.S. Army SGT Connor Mendez
Image 25: U.S.M.C CPL Isabella Ortega
Image 26: Sumbria Vikramaditya
Image 27: Justin Connaher
Image 30: P. Mateus
Image 35: Cayambe
Image 36: Rhododendrites
Image 37: Mattes
Image 38: U.S. Army SGT Khylee Woodford
Image 43: U.S. Army SGT Bill Boecker
Image 44: U.S. Army SPC Alisha Grezlik
Image 45: U.S. Army SPC Jacob Krone
Image 46: Justin Connaher
Image 48: U.S. Army PFC Christopher Brecht
Image 50: U.S.M.C LCPL Juan Carpanzan
Image 51: U.S. Army SPC Thomas Crough
Image 55: U.S.M.C LCPL Brennon Taylor
Image 58: U.S. Air Force SSGT Perry Aston
Image 61: U.S.M.C LCPL Brienna Tuck
Image 62: U.S. Army SGT Nicodemus Taylor
Image 64: U.S.M.C LCPL Jacqueline Smith
Image 65: U.S. Army SPC Shannon Westpfahl
Image 66: U.S. Army N.G. SGT Remi Milslagle
Image 67: U.S.M.C LCPL Jacqueline Smith
Image 69: U.S.M.C LCPL Taylor Cooper
Image 70: U.S.M.C PFC Kevan Dunlop
Image 71: U.S. Army CPT Charles Emmons
Image 72: U.S. Air Force SSGT Donald R. Allen
Image 73: U.S.M.C LCPL Vaniah Temple
Image 74: U.S.M.C LCPL Aaron S. Patterson
Image 75: U.S.M.C LCPL Jacqueline Smith
Image 76: U.S.M.C SGT Alicia R. Leaders
Image 77: U.S. Army PFC Jesus Menchaca
Image 78: U.S. Army SPC Kelvin Johnson Jr
Image 79: U.S.M.C CPL Sean Potter
Image 80: CPT Thomas Cieslak
Image 81: U.S.M.C CPL Sean Potter

Image 82: U.S.M.C CPL Alexander Mitchell
Image 83: U.S. Army SGT Thoman Johnson
Image 84: U.S.M.C CPL Sean Potter
Image 85: U.S. Army SPC Alisha Grezlik
Image 86: Noah Wolf
Image 87: K. Kassens
Image 88: U.S. Army SPC Alisha Grezlik
Image 89: U.S.M.C SSG Matt. Orr
Image 89: U.S. Army SPC Ethan Scofield
Image 91: U.S. Army SGT Patrik Orcutt
Image 92: U.S. Army SPC Billy Brothers
Image 93: U.S. Air Force SRA Curtis Beach
Image 94: U.S. Army SPC Alisha Grezlik
Image 95: U.S. Navy PO1 Jesse Monford
Image 96: U.S. Army SGT Matthew Lucibello
Image 101: Justin Connaher
Image 102: U.S.M.C SGT Allison DeVries
Image 108: U.S. Army SSGT Teddy Wade
Image 109: U.S. Army CAPT Peter Smedberg
Image 110: U.S.M.C GYSGT Robert B. Brown Jr.
Image 113: Nathan Boor
Image 117: Hugdum
Image 118: Smartiejl
Image 120: AliveFreeHappy
Image 128: Adamrpatel97
Image 129: Gurdeepdali
Image 133: Cal Zant
Image 134: U.S. Army SPC Taylor Shaffe
Image 135: U.S. Army MSG Michel Sauret
Image 136: U.S. Army SPC Michelle C. Lawrence
Image 137: Alejandro Pena
Image 142: U.S. Army 1LT Benjamin Haulenbeek
Image 147: Dan Periard
Image 148: Dan Periard
Image 151: Iowa Environmental Mesonet
Image 159: U.S.M.C CPL Elliott A. Flood-Johnson
Image 160: U.S.N.G AMN Ashley Williams
Image 161: U.S. Air N.G. SRA Kregg York
Image 164: Sebastian Saarloos
Image 165: U.S. Army SGT Jarred Woods
Image 166: U.S. Army SPC Matthew Marcellus
Image 167: U-ichiro Murakami
Image 168: Juris Sennikovs
Image 170: Robert L. McCoy
Image 171: Bryan Litz
Image 174: Bryan Litz
Image 175: Bryan Litz
Image 176: U.S. Army SPC Jason Johnston
Image 177: U.S.M.C CPL Logan Kyle
Image 181: U.S. Air N.G. SRA Caleb Vance
Image 182: Rufiyaa
Image 183: John Mark
Image 184: U.S. Army SGT Steven Lewis
Image 185: U.S.M.C LCPL Christopher Mendoza
Image 186: U.S.M.C CPL Andrew Kuppers
Image 187: U.S. Army SPC Tracy McKithern
Image 188: U.S. Air Force MSGT David W. Carbajal
Image 189: U.S.M.C CPL Corey Mathews
Image 190: Missouri National Guard SPC Trevor Wilson
Image 192: U.S.M.C LCPL Matthew Bragg
Image 194: U.S. Army SPC Dee Crawford
Image 195: U.S.M.C SSGT Donald Holbert
Image 196: U.S. Air Force A1C Scott Warner
Image 197: Timothy L. Hale
Image 202: Courtesy Photo
Image 203: U.S. Army SGT Michael Hunnisett
Image 204: U.S. Army SGT Peter A. Ford

Please leave a review. Thanks!

Positive reviews from awesome people like you help other people to benefit from the valuable instructions in this manual. Could you take 60 seconds to share your thoughts?

Thank you in advance for helping the community!

If you liked this book, consider getting a copy of *Small Unit Tactics*.

Or to learn urban tactics, consider getting a copy of *Small Unit Raids*.

ADVANCED
LONG
RANGE
SHOOTING

Learn vital shooting skills from professionals who instruct foreign snipers.

Tell people what you think? Thanks!

Reviews from amazing people like you help other marksmen discover these insights, and make the entire community better and smarter.

Thank you in advance for your help and time!

Become one of the most precise and accurate marksmen on the planet.

Multivariable Shooting

1. Moving Participants   9
Moving Target   10
Moving Marksman   18

2. Multiple Participants   22
Working with a Spotter   22
Multiple Marksmen   28
Multiple Targets   29
Multiple Marksmen and Targets   32

3. Restriction-Based Positioning   34
Non-Standard Shooting Positions   34
Non-Dominant Hand   39
Non-Vertical (Canted) Rifle   39

4. Disadvantaged Conditions   41
Stress and Fatigue   41
Time Pressure   42
Nighttime   45
Cold   50
Shooting through Barriers   54

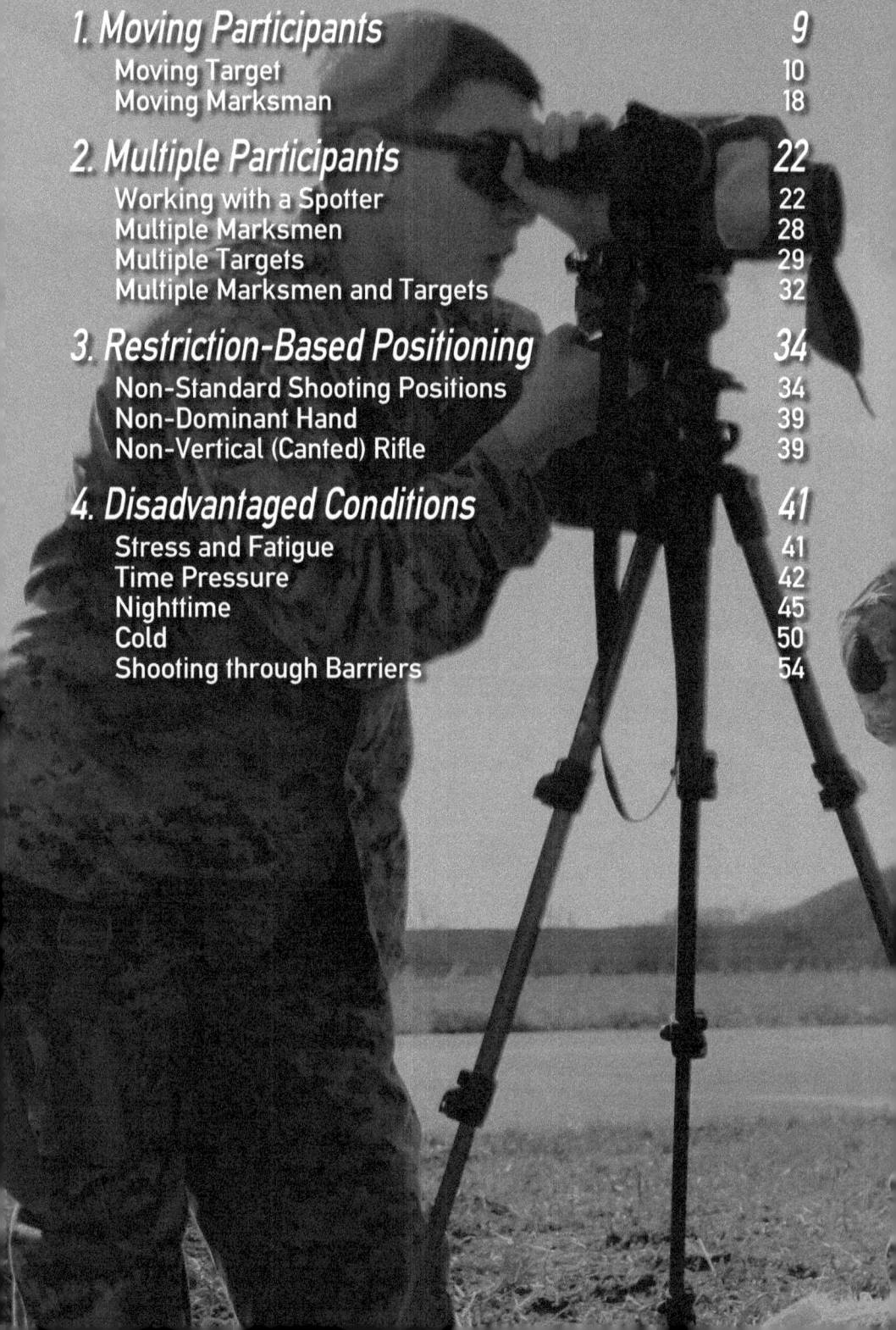

Advanced Effects, 600 Meters and Beyond

5. Trajectory Information 59

Determining Bullet Trajectory 60
Determining Bullet Impact 64
Making Adjustments 67
Trajectory Validation for Ballistic Calculators (Truing) 68

6. Bullet Spin 72

Barrel Twist 73
Spin Stabilization (Precession) 74
Bullet Orienting to Vector (Weathervaning, Tractability) 79
Optimal Spin Rates 80
Vector Orienting to Bullet (Lift Force) 83
Precession-Induced 90-Degree Turning 84
Aerodynamic (Muzzle) Jump 86
Spin Drift 89

7. More Air Effects 92

Headwinds and Tailwinds (Parallel Winds) 92
Transonic Range (Supersonic to Subsonic Shift) 94

8. Acceleration on a Rotating Body 101

Latitudinal (North-South) Coriolis Effect 101
Longitudinal (East-West) Coriolis Effect 104
Altitude (Up-Down) Eotvos Effect 107

De Minimis Effects

9. Local Gravity   111

10. Thermal Effects   113

11. Magnus Force   116

12. Coning and Nutation   118

13. Additional Effects   118

Appendices

14. Glossary   123

15. Credits   127

Multivariable Shooting

1. Moving Participants   9
Moving Target   10
Moving Marksman   18

2. Multiple Participants   22
Working with a Spotter   22
Multiple Marksmen   28
Multiple Targets   29
Multiple Marksmen and Targets   32

3. Restriction-Based Positioning   34
Non-Standard Shooting Positions   34
Non-Dominant Hand   39
Non-Vertical (Canted) Rifle   39

4. Disadvantaged Conditions   41
Stress and Fatigue   41
Time Pressure   42
Nighttime   45
Cold   50
Shooting through Barriers   54

Multivariable Shooting

Practice.
—Simo Häyhä, the deadliest sniper in history with over 500 confirmed
*   kills, when asked how he became such a good marksman.*

Almost all shooting is performed as a single shot from a stationary marksman to a (relatively) stationary target under daylight conditions. Normally, the only two major considerations (or variables) a marksman must consider are the distance to the target and the wind.

However, when these **marksmen incorporate additional factors**, they are engaging in multivariable shooting. Such additional factors may include: how others act and react, movement of themselves and targets, restricted positions, and abnormally bad conditions.

Some hobbyists seek the challenge of multivariable shooting, while hunters often want to expand their skills and available targets. In contrast, occupational marksmen (i.e., military and police) sometimes have no choice but to take multivariable shots because they must complete a mission.

1. Moving Participants

Sometimes, a marksman is aiming at a moving target. Sometimes, the marksmen themselves are moving. And in rare instances, both the target and the marksman are moving.

In any case, accounting for movement is related to two techniques already explained in *Long Range Shooting: An Illustrated Manual*: **elevation and windage**. "Elevation" is the vertical adjustment a marksman must make with their rifle, typically measured using the reticle, to ensure bullets reach distant targets. When a target moves closer or farther away, it changes the distance that a bullet must travel. To compensate for the changed distance, the marksman compensates by adjusting the elevation setting higher or lower to alter the trajectories of their bullets.

"Windage" is the left-or-right distance that a marksman must move their rifle to account for wind pushing the bullet sideways (and any other sideways consideration). If movement causes the target to be to the side of the point-of-aim, the marksman uses a windage hold or dial just as if wind pushed the bullet to the side. In fact, when a target is moving in the wind, both windage holds are simply added together.

1.a Moving Target

A moving target can be a deer in the woods sauntering around, an enemy patrolling, or a target on a moving rail system. When considering moving targets, the first consideration is simply to **wait for a target to stop moving**. Although it seems obvious, many marksmen get caught up in the moment and engage too quickly. In fact, most hunters deliberately avoid shooting at moving targets because they consider maiming an animal to be unethical and are unwilling to accept the slight decrease in accuracy that comes with shooting a moving target. Therefore, these hunters wait for a deer to stop or slow down, maximizing their chances of achieving a quick and accurate kill. **A full stop is not necessary**, and simply waiting for the target to slow down or change direction can be enough to greatly increase accuracy.

If the target does not slow down, the second thing that a marksman must determine is how long their bullet's flight time would be. Bullets take time to go from the rifle to the target, and during that time the target moves. Therefore, by determining their bullet's flight time, the marksman is also considering the target's movement time. In other words, if the marksman places the point-of-aim directly on the target, the target would move out of the way as the bullet flies. Therefore, the marksman must fire while their point-of-aim is in front of the target so that both the target and the bullet simultaneously intersect their movements at the point-of-aim.

This sideways distance that the target moves while the bullet is airborne is called the "**lead**" (rhymes with "seed"). To determine the lead, a marksman must know the sideways speed of the target and their bullet's flight time (i.e., the amount of time that the target has to move). For example, if a 400-m-or-yd target is trotting at a pace of 0.5 m or yd/s and a bullet takes 2 s to travel 0.5 m or yd to a target, the marksman must aim 1 m or yd ($0.5 \times 2 = 1$) in front of the target for the bullet and the target to intersect at the point-of-aim.

A bullet's flight time can be calculated using two different methods. The simplest method is to **fire a bullet and time it**. If a marksman does this at various distances, they can prepare a chart that matches target distances to flight times. These charts can record the flight time as a function of distance (e.g., 0.45 s at 500 m). Many of these tables are available online or preprogrammed into ballistic calculators. Then, once the actual target's distance is known, the marksman can look up that target distance in their table to find the corresponding bullet flight time.

To create these tables oneself, a marksman can go to a long-distance range with a recording device. They shoot at metal targets spaced 100 m or yd

Image 1: Marines set and prepare Robotic Moving Targets to shoot at. Quantico, VA, 24 Sep 2013.

apart. Then, at home, they determine the travel time by measuring the time difference in the audio recording between the shot and the impact. If the target is too far to hear, the same process can be used with a video recording. A little bit of extra time must be added to the final number to account for the time that it takes for a bullet to exit the muzzle after the trigger is pulled.

The second method uses calculations to determine the travel time. That is, a marksman **multiplies the average velocity of the bullet by the distance** that the bullet travels. The simplest way to "calculate" average velocity is to substitute in muzzle velocity. This method works over short distances since velocity does not change much at the start of its trajectory. For example, if the muzzle velocity of a bullet is 800 m/s, then the marksman can be relatively certain that a bullet would take about ½ s to travel 300 to 400 m.

Once a target is far away, bullets slow down significantly. At that point, substitution no longer works. Instead, a marksman can calculate the average velocity by using a ballistic calculator. A good ballistic calculator can give the bullet velocity at various points in the trajectory. By averaging together a few points, the marksman can get a good approximation of the average velocity. For example, although not exact, a marksman can then take a mean of: the muzzle velocity, the mid-distance velocity, and the velocity at the target.

Once the bullet's flight time is calculated, the marksman must determine the target's **lateral speed**. That is, if a target is moving diagonally, only the

Movement Value Chart

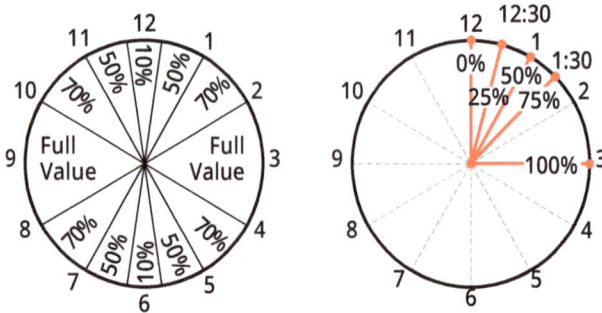

Image 2: **Movement value** is based on where the target is moving to. (Movement values and wind values are identical since they are both based on the same trigonometry. However, instead of the wind pushing the bullet back to the target, the target moves into the path of the bullet.)

The simplest movement value system is the **clock system**, where the marksman points at 12 o'clock. A target moving parallel has no value, while one moving sideways has a full value. On the left is a full system with only 4 values: 100%, 70%, 50%, and 10%. This system becomes less precise as the target strays from the 12 or 6 o'clock, because the percentages change quickly there. A filled in table can be useful for showing more numbers. (See Image 3, Pg. 13.)

---

side-to-side component is useful since a target is unlikely to outrun a bullet when moving closer or farther from the marksman. The only possible exception to that may be vehicles where it may be required to determine their front-back speed. In that case though, the left-right and front-back components of movement are still separated because they are separately accounted for with windage and elevation respectively. (In fact, a shot at a moving helicopter would, in theory, additionally require determining vertical velocity.)

Lateral velocity can be measured with a reticle's windage hashmarks. For example, a marksman can look through their scope and measure how far across the reticle the target traveled in 1 s. Or the marksman can determine how long it took the target to traverse 0.5 mils. Either method measures the angular distance the target traveled per unit of time.

Getting the angular velocity is very helpful because it can be multiplied by the bullet's flight time to directly calculate the windage hold. For example, if a target is moving right-to-left at 0.5 mil/s, and the flight time is 0.2 s, then the lead is 0.1 mils (0.5 mil/s × 0.2 s = 0.1 mils).

The other methods for getting the target's angular velocity require knowing the linear velocity and converting it to an angular velocity. The

Lateral Movement Value

Target Direction		Target Speed m/s (ft/s)									
Degrees	O'clock	1 (3.3)	2 (6.6)	3 (9.8)	4 (13)	5 (16)	6 (20)	7 (23)	8 (26)	9 (30)	10 (33)
0°	12:00, 6:00	0 (0)	0 (0)	0 (0)	0 (0)	0 (0)	0 (0)	0 (0)	0 (0)	0 (0)	0 (0)
10°		0.2 (0.7)	0.3 (1)	0.5 (1.6)	0.7 (2.3)	0.9 (3)	1 (3.3)	1.2 (3.9)	1.4 (4.6)	1.6 (5.2)	1.7 (5.6)
15°	11:30, 12:30, 5:30, 6:30	0.3 (1)	0.5 (1.6)	0.8 (2.6)	1 (3.3)	1.3 (4.3)	1.6 (5.2)	1.8 (5.9)	2.1 (6.9)	2.3 (7.5)	2.6 (8.5)
20°		0.3 (1)	0.7 (2.3)	1 (3.3)	1.4 (4.6)	1.7 (5.6)	2.1 (6.9)	2.4 (7.9)	2.7 (8.9)	3.1 (10)	3.4 (11)
30°	11:00, 1:00, 5:00, 7:00	0.5 (1.6)	1 (3.3)	1.5 (4.9)	2 (6.6)	2.5 (8.2)	3 (9.8)	3.5 (12)	4 (13)	4.5 (15)	5 (16)
40°		0.6 (2)	1.3 (4.3)	1.9 (6.2)	2.6 (8.5)	3.2 (11)	3.9 (13)	4.5 (15)	5.1 (17)	5.8 (19)	6.4 (21)
45°	10:30, 1:30, 4:30, 7:30	0.7 (2.3)	1.4 (4.6)	2.1 (6.9)	2.8 (9.2)	3.5 (12)	4.2 (14)	4.9 (16)	5.7 (19)	6.4 (21)	7.1 (23)
50°		0.8 (2.6)	1.5 (4.9)	2.3 (7.5)	3.1 (10)	3.8 (13)	4.6 (15)	5.4 (18)	6.1 (20)	6.9 (23)	7.7 (25)
60°	10:00, 2:00, 4:00, 8:00	0.9 (3)	1.7 (5.6)	2.6 (8.5)	3.5 (12)	4.3 (14)	5.2 (17)	6.1 (20)	6.9 (23)	7.8 (26)	8.7 (29)
70°		0.9 (3)	1.9 (6.2)	2.8 (9.2)	3.8 (13)	4.7 (15)	5.6 (18)	6.6 (22)	7.5 (25)	8.5 (28)	9.4 (31)
75°	2:30, 3:30, 8:30, 9:30	1 (3.3)	1.9 (6.2)	2.9 (9.5)	3.9 (13)	4.8 (16)	5.8 (19)	6.8 (22)	7.7 (25)	8.7 (29)	9.7 (32)
80°		1 (3.3)	2 (6.6)	3 (9.8)	3.9 (13)	4.9 (16)	5.9 (19)	6.9 (23)	7.9 (26)	8.9 (29)	9.8 (32)
90°	3:00, 9:00	1 (3.3)	2 (6.6)	3 (9.8)	4 (13)	5 (16)	6 (20)	7 (23)	8 (26)	9 (30)	10 (33)

Image 3: This table takes the target's actual movement speed (X-axis) and the target's direction (Y-axis), and outputs the **lateral movement velocity (cell contents)**. The value is found by multiplying the actual movement velocity by the sine of the degree when the direction the marksman faces is defined to be 0°. This is an expanded version of the above image.

linear velocity of a target can be estimated from prior knowledge, or it can be measured precisely in the moment. An example of an estimate based on prior knowledge is that people typically walk at about 1 m/s (3.3 ft/s), run at 2 m/s (6.6 ft/s), and sprint at 4 m/s (13 ft/s). (The average individual is often in poor shape.) A healthy adult male can sprint at 7 m/s (23 ft/s), while Usain Bolt set a world record of 12.42 m/s (40.75 ft/s).) An experienced marksman can refine these numbers by differentiating between different gaits and different groups of people. For example, a man may jog at 1.5 m/s (4.9 ft/s) while a woman may typically be 0.25 m/s (0.8 ft/s) slower than a man due to their shorter stature; therefore, women may jog at 1.25 m/s (4.1 ft/s).

Speed can also be estimated based on body width. A healthy man has a chest depth of about ¼ to ⅓ m (10 to 13 in), so if they travel 6 body-widths per second, they are traveling at 2 m/s (6.6 ft/s). Body-widths can even be a measurement in and of themselves. For example, if the marksman knows that a bullet would take 1 s to reach the target, they can aim 6 body-widths ahead.

Finally for this method, linear, lateral velocity is converted into angular velocity. The formulas for mils and MOA to **calculate an object's angular velocity** *(T Angular V)* with a known distance to target *(D to T)* and an estimated linear velocity in m or yd *(T Linear V)* are:

▸ *(T Angular V in Mils) = ((T Linear V) × 1000) ÷ (D to T)*
▸ *(T Angular V in MOA) = ((T Linear V) × 3438) ÷ (D to T)*

Once a marksman knows the target's actual velocity, they must then **derive the component of that velocity that is perpendicular** (i.e., lateral) to the trajectory of the bullet. The simplest method of deriving lateral velocity from actual velocity is to place the target into one of three categories: full turn (100% of actual velocity), half turn (70% of actual velocity), and no turn (0% of actual velocity). (See Image 2, Pg. 12.)

How far a man or animal is turned can be estimated by the number of arms or legs that are visible. That is, if only one arm is visible then they have a full turn; if both arms are visible but one shoulder is a little hidden, then they have a half turn; and if both shoulders are clearly visible then the target has no turn. Of course, a marksman can create as many categories and corresponding lead multipliers as they can reasonably employ.

Putting this together with an example, assume that a target at 350 m is walking diagonally at 1 m/s. Because they are moving diagonally (i.e., half turn), the turn multiplier is 70%, so they have 0.7 m/s of sideways movement. Using the above formula to find the target's angular, lateral velocity:

▸ *(T Angular V in Mils) = ((T Linear V) × 1000) ÷ (D to T)*
▸ *(T Angular V in Mils)* = ((0.7 m/s) × 1000) ÷ (350 m) = **2 mils/s**

The resulting angular velocity is then multiplied by the bullet's flight time to determine the lead. Assume a bullet has an **average** velocity of 1000 m/s. The formula for flight time is:

▸ *(Bullet Time) = (D to T) ÷ (Bullet V)*
▸ *(Bullet Time)* = 350 m ÷ 1000 m/s = **0.35 s**

The final windage hold is determined by this formula:

▸ *(Windage in Mils) = (T Angular V in Mils) × (Bullet Time)*
▸ *(Windage in Mils)* = 2 mils/s × 0.35 s = **0.7 mils**

Movement to Windage in Mils (MOA)

Target Lateral Velocity m/s (ft/s)	Bullet Average Velocity m/s (ft/s)						
	400 (1312)	500 (1640)	600 (1969)	700 (2297)	800 (2625)	900 (2953)	1000 (3281)
1 (3)	3 (11)	2 (7)	2 (7)	1 (4)	1 (4)	1 (4)	1 (4)
2 (7)	5 (18)	4 (14)	3 (11)	3 (11)	3 (11)	2 (7)	2 (7)
3 (10)	8 (28)	6 (21)	5 (18)	4 (14)	4 (14)	3 (11)	3 (11)
4 (13)	10 (35)	8 (28)	7 (25)	6 (21)	5 (18)	4 (14)	4 (14)
5 (16)	13 (46)	10 (35)	8 (28)	7 (25)	6 (21)	6 (21)	5 (18)
6 (20)	15 (53)	12 (42)	10 (35)	9 (32)	8 (28)	7 (25)	6 (21)
7 (23)	18 (63)	14 (49)	12 (42)	10 (35)	9 (32)	8 (28)	7 (25)
8 (26)	20 (70)	16 (56)	13 (46)	11 (39)	10 (35)	9 (32)	8 (28)
9 (30)	23 (81)	18 (63)	15 (53)	13 (46)	11 (39)	10 (35)	9 (32)
10 (33)	25 (88)	20 (70)	17 (60)	14 (49)	13 (46)	11 (39)	10 (35)

Image 4: This table takes the Bullet Average Velocity (X-axis) and the Target Lateral Velocity (Y-axis), and **outputs the windage hold or dial (cell contents) in mils and MOA.** Having different bullet velocities for one rifle is important because average velocity falls as a function of distance. Therefore, marksmen would want to keep an additional table that shows Bullet Average Velocity as a function of target distance. Note that without a fast bullet, a slow target, or a robot pulling the trigger, even relatively slow-moving targets are impractical to hit at long distances. The windage hold from this table is simply added to the windage hold for wind.

However, because distance to the target increases a bullet's flight time and decreases angular velocity, these two formulas can be simplified:

▸ *(T Angular V in Mils)* × *(Bullet Time)* =
   (((T Linear V) × *1000)* ÷ *(D to T))* × *((D to T)* ÷ *(Bullet V))* =
   ((T Linear V) × *1000)* ÷ *(Bullet V))* = *(Windage in Mils)*
   (Windage in Mils) = 1000 × ((T Lateral Linear V) ÷ (Bullet V))
   (Windage in MOA) = 3438 × ((T Lateral Linear V) ÷ (Bullet V))

In other words, distance to target *(D to T)* cancels out. Put another way, if a bullet takes twice as long to arrive, the target has twice the amount of time to arrive at the point-of-aim and the lead is twice as large. The result is:

▸ **(Windage in Mils)** = **1000** × **((T Lateral Linear V)** ÷ **(Bullet V))**

The turn value can be included as a separate variable as well:

▸ *(Windage in Mils)* = 1000 × *((T Actual Linear V)* × *(Value))* ÷ *(Bullet V))*

▸ *(Windage in Mils)* = (1000 mils × 1 m/s × 0.7) ÷ 1000 m/s = **0.7 mils**

Note that **the units for target velocity and bullet velocity do not matter** because they cancel each other out, but the units do have to be the same (e.g., both are m/s, or both are mi/h, but one cannot be m/s while other is mi/h).

To prepare to shoot at moving targets, a marksman could create tables with precomputed values, such as: X-axis = *(T Lateral Linear V)*, Y-axis = *(Bullet V)*, cell contents = *(Windage in Mils)*. Similar tables are used to account for wind speed very often. (See Image 4, Pg. 15.)

Leads can quickly become impractical as targets get farther because the margin of error shrinks with distance. For example, for a man with a chest depth of ⅓ m (1 ft) who walks at 1 m/s (3 ft/s), their chest only exists in a particular spot for ⅓ of a second. Shooting with this level of time accuracy is relatively easy when the target is only 10 m (30 ft) away and 1 m (3 ft) represents a large angle. However, as the target gets farther from the marksman, the angular width of a target decreases, and the opportunity for something to change during the flight time increases. Therefore, the range for shooting at moving targets is typically limited from 50 m or yd where a lead is useful, to 400 m or yd where a lead is practical.

Once the lead is determined it is added to the elevation and windage holds or dials to determine the final point-of-aim. Then, a marksman can either "track" or "ambush" the target. "Tracking" is to follow the target with the point-of-aim, moving the rifle as the target moves. "Ambushing" is to place the point-of-aim in front of the target and wait for the target to move into position. (See Image 7, Pg. 17.)

To **ambush** a target, a marksman aims at a point in front of the target's direction of movement. Then, the marksman fires when the target is away from the point-of-aim by the lead distance. Thereby, the target moves into the path of the bullet.

Ambushing is the easier technique because the marksman can stay still while shooting. However, marksmen who use the ambush technique tend to jerk their triggers because their timings must be very precise. Further, winds may have changed while a marksman was waiting, changing the windage.

Shooting a Moving Target

Image 5: A running deer is one of the most common moving targets. White-tailed have been recorded sprinting at speeds of 60 km/h (40 mi/h) and sustaining speeds of 50 km/h (30 mi/h) over distances of 5-6 km (3-4 mi).

Points-of-aim for a moving target as distance increases

(300 m or yd zero)

- Heart
- Lungs
- Liver

100 200 300 400 500
Distance, m or yd

Image 6: As targets get farther from the marksman, a bullet drops more and takes longer to reach a target. Because bullets drop exponentially more with distance, the marksman must aim higher and higher. However, because the angle is constant, the horizontal offset increases linearly. Therefore, as targets get farther, the marksman's point-of-aim follows this upward curve as a function of distance.

Ambush, Trapping

Tracking

Image 7: A marksman can either hold their aim and wait for the target to move into position (i.e., **ambushing**) , or they can move with the target (i.e., **tracking**).

The **tracking** method moves the point-of-aim with the target so that it is one lead distance ahead of the target at all times. This technique allows the marksman more opportunities to fire, and they can be more patient. However, it is a more difficult technique because it requires moving the rifle while precisely aiming. While tracking, the marksman does not stop to shoot; the marksman pulls the trigger while continuing to move the rifle with the target and re-engages the sight picture to where they expect the target to be.

Tracking with a reticle in a perfectly straight line while maintaining a lead, windage, and elevation hold is not easy. In fact, focusing too much on the target and not the reticle is a major issue with marksmen who use the tracking method. Another danger of the tracking method is that what is behind the target changes, and so the marksman must pre-clear the entire potential background behind the target to ensure they don't shoot something unintentionally. A third issue is that if the rifle is moved too much, the marksman may move themself into a bad shooting posture.

To address some of these issues, a marksman can improve their tracking by practicing moving the reticle in a straight line. And they can position their rifle on a single pivot point such as a sandbag (instead, for example, a tripod or bipod). Further, the marksman must practice pivoting their entire body if the necessary swing is more than a few degrees.

A great method is to combine the two techniques. A marksman can begin with an ambush because it is better for maintaining a stable position. But the marksman is then prepared to immediately track the target if they miss their window of opportunity to ambush.

1.b Moving Marksman

The concept of a lead was explained in the previous section as a way to hit a moving target. (See Moving Target, Pg. 10.) That lead is a positive lead since it requires aiming in front of the target's direction-of-movement.

In contrast, when the marksman is moving, they must apply a negative lead, which requires **aiming behind the target**. A negative lead is required because when a marksman is moving, their rifle and ammunition move with them. Therefore, a fired bullet retains the marksman's sideways momentum, causing it to move sideways from the initial point-of-aim.

The concepts of positive and negative leads make intuitive sense when combined. That is, if both the target and marksman were to move in sync, then the negative and positive leads would cancel each other out and the marksman could fire without using any lead at all.

Shooting from a Truck

Image 8: In this scenario, the bench has been lowered to create a shooting support. Island of Crete, Greece, 12 Mar 2018.

Image 9: While a **gimbal stabilizer** could be used, if a marksman can plan for a gimbal they can usually plan a way to shoot without being in motion.

Image 10: Italian special operations forces engage targets from a moving truck. They spread their legs to **find stabilizing footholds**. Crete, Greece, 18 Mar 2018.

Calculating a negative lead is the same as calculating a positive lead, except the resulting hold is in the opposite direction of movement. In fact, it is quite a bit easier since often a marksman precisely knows their own movement speed, unlike having to guess the speed of a moving target. Therefore, it is even possible to shoot out the back of a fast-moving vehicle. (Again however, the easiest way to shoot from a moving position is to simply stop moving.) For example, the final formula for a lead was this:

▸ *(Windage in Mils) = 1000 × ((T Actual Linear V) × (Value)) ÷ (Bullet V))*
(T Linear V) can be replaced with *-(M Linear V)* for negative marksman linear velocity. That gives the formula:

▸ *(Windage in Mils) = 1000 × -(M Actual Linear V) × (Value) ÷ (Bullet V)*

For example, assuming that a vehicle is moving laterally at 30 k/h (18 mi/h), or -8.3 m/s (-9 yd/s) after converting so that the vehicle velocity and the bullet velocity use the same units, and assuming that the bullet is traveling at 1,000 m/s (1093 yd/s), then the formula looks like this:

▸   *(Windage in Mils)* = 1,000 mils × (-8.3 m/s × 1.0) ÷ 1000 m/s = **-8.3 mils**

Because the bullet moves sideways at the speed of the vehicle, **even moderate vehicular speeds can quickly require a large amount of windage.**

The most difficult aspect of shooting while moving is **muzzle vibration.** There are a variety of methods that have been devised to stabilize a rifle on a moving platform: bag supports, single straps, and spider straps are a few examples. A gimbal may be used, but the scenario where a rifle gimbal would both be useful enough to plan for, but the shooting task couldn't be reconfigured to not involve vehicular movement is unclear (See Image 9, Pg. 19.) Whatever the support used, a rifle cannot be stabilized directly against a vehicle frame because vibrations easily transfer from metal to metal.

The **bag support method** is the fastest to establish and provides a stable shooting platform. A soldier can utilize a variety of bags, such as: a rucksack, an assault pack, or a luggage bag. The more firm cushioning, the better. It is crucial to ensure that the bag itself is secured firmly to the interior floorboard of the platform to guarantee stability and accuracy.

The **single strap method** is also reasonably quick to establish. First, a marksman needs a readily available, inelastic line such as 550 cord, a rifle sling, or even a roll of duct tape if desperate. Then, the marksman secures either end of the line to the left and right sides of the vehicle frame, creating a horizontal shooting platform. The line must have some degree of slack so that the rifle can hang in the center of the line. To stabilize the rifle, a marksman leans the rifle into or away from the line, so ideally the line is strong enough to support many times the force of an adult man. (See Image 11, Pg. 21.) (See Image 12, Pg. 21.)

The **spider strap method** is the most complex method, but also the most stable. This technique uses four attachment points instead of the two used with the single strap method. To make the spider strap, a marksman attached two or more lines to the top left, top right, bottom left, and bottom right corners of the vehicle's opening (e.g., a doorway). The lines converge at the center, where they are held together by a carabiner, tape, or another locking device. A cushion can also be attached to the center to help dampen vibration. (See Image 13, Pg. 21.) (See Image 14, Pg. 21.)

To reiterate, the best way to improve accuracy while shooting and moving is to reduce vibration. To do this, the marksman can position themself on a

Shooting from a Helicopter

Image 11: A Soldier from Alaska trains with a M107 .50-Caliber Long Range Sniper Rifle in a UH-60 Black Hawk. Note the string is wrapped around the rifle. Joint Base Elmendorf-Richardson, AK, 07 Mar 2014.

Image 12: A U.S. Army Special Forces Sniper engages a floating target from a Greek Navy Bell UH-1 Iroquois (Huey) helicopter. He is pulling his line slightly back for increased stability. Souda Bay, Greece, 21 Apr 2021.

Image 13: A U.S. Marine corporal practices maneuvering a M110 semi-automatic sniper system inside a UH-1Y Venom helicopter. Camp Lejeune, NC, 12 Mar 2015.

Image 14: A U.S. Marine Corps lance corporal scout sniper engages a target with an M110 semi-automatic sniper system while aboard a UH-1Y Huey. Okinawa, Japan, 14 Jun 2022.

cushion. They can also use the scope on the lowest magnification setting so the vibration does not appear as significant in the sight picture.

Shooting from a moving vehicle at any significant distance (e.g., beyond 100 m or yd) has a high chance of requiring multiple shots to hit the target. To that end, a semi-automatic rifle is much faster to reload than a bolt-action.

Finally, **shooting from a moving platform is a team effort** between the marksman, driver, and possibly a spotter. If the marksman needs a different speed or direction, they can always ask the driver to change course or hold still. In fact, a vehicle can hold multiple snipers, a sniper team leader, spotters, and a driver to create a long-range shooting vehicle. Having multiple marksmen aiming at the same target is a sure way to increase the odds of at least one successful hit.

2. Multiple Participants

A typical shooting interaction has one marksman and one target. However, there are many instances in which there are more people or targets involved in a single scenario. For example, if a marksman has too much work, they can offset some of that work onto an assistant called a "spotter." In other scenarios, there may be multiple targets, multiple marksmen, or even multiple targets and multiple marksmen. These complex scenarios may involve tens of people who all must **coordinate their actions with high precision**.

2.a Working with a Spotter

A two-person long-range shooting team is composed of a marksman and a spotter. The marksman is the man behind the gun, and their job is simple; the **marksman only needs to aim at the target and pull the trigger**.

In contrast, the spotter's job is more complex. They serve as the vital **information supplier for the marksman**. The spotter gathers information such as the wind speed and direction, and also the impact location of prior shots. With this information, they use the various charts, tables, and the ballistic calculator they are holding to generate elevation and windage holds for the marksman. Finally, for occupational marksmen with command teams, **the spotter is the point-of-contact** and handles all external communications.

Teamwork is highly effective because dividing the tasks among multiple people is faster than having one person handle all the work. This increased speed not only allows each shot to be taken more quickly but also improves accuracy. Increased accuracy occurs because dynamic factors in shooting, such as wind or moving targets, can be addressed before they change a lot.

Notably, given two experienced marksmen, the more skilled marksman is the spotter because they can give more accurate information faster. Aiming and pulling the trigger is considered to be the more straightforward function. That is, while being a marksman does require skill, the job requires fewer total skills than being an effective spotter requires.

Having a spotter enables the team to use additional optical equipment. Because spotters normally do not shoot, they can utilize binoculars or spotting scopes instead of riflescopes. Both devices can be held closer to the eye than riflescopes because riflescopes must be mounted far from the marksman's eye to prevent recoil impact. Thereby, compared to riflescopes, **binoculars see a wider area at the same resolution and spotting scopes see the same area at a higher resolution.** (See Image 15, Pg. 23.)

Working with a Spotter

Image 15: Two Marines from the 3rd Marine Regiment form a standard **sniper-spotter team** with a spotting scope and an M110 sniper rifle. The most common position for a spotter is right next to the marksman in order to facilitate easy communication. Fort Bragg, NC, 22 Mar 2021.

Image 16: A joint US and UK sniper team engages targets from a truck to elevate the sniper's vantage point while protecting the spotter. So long as **communication is reliable**, the spotter and marksman can be in entirely different positions. Camp Asazi, Latvia, 16 Nov 2021.

To begin shooting, the spotter and the marksman must agree on their communication and verbage. Communicating information in the fastest, least ambiguous way is vital to successful teamwork. The most important parts of communication start at target identification and end with the marksman shooting the target.

During shooting, **the first step in any situation is to find and identify the target**. Once the spotter determines what the next target is, they must be able to direct the marksman onto that target. Many methods are available; for example, a common method in a well-planned area is the 3D's: a direction, a distance, and a descriptor. For example, "278 degrees at 500 m, male with red shirt and jeans." Another, rougher method is for the spotter to use clock degrees relative to the direction the marksman is facing: "Your 1 o'clock at approximately 500 m, male with red shirt and jeans." A spotter can also direct a marksman from an obvious landmark in the area: "From the intersection look 50 m to the East, male with red shirt and jeans." Any target description must be so specific and distinctive that it only applies to a single possible target! The marksman responds once they find the target: "male with red shirt and blue jeans, ready."

After confirming the target, the spotter reminds the marksman to complete a **basic checklist** to eliminate any chance that a vital step could be forgotten. (The demographic most likely to accidentally skip vital steps are experienced teams that become complacent!) An example is, "laze, parallax, mil":

Laze – Use a laser rangefinder to determine the exact distance to the target.
Parallax – Adjust the scope's parallax setting to match that distance.
Mil – Measure the height of the target in mils or MOA.

Once the target and its characteristics are identified, the spotter **calculates the elevation and windage holds** using their tools, such as charts and calculators. (Premade tables are manufactured using ballistic calculators prior to the shooting event.) If technology is unavailable, the spotter can ask the marksman for their estimate and average it with their own, as averaged estimates are usually more accurate than any individual's estimate.

Then, the spotter gives winds and elevation holds to the marksman. A hold refers to distance on the marksman's reticle, so for example, a spotter may say, "Left 2, Up 6.8" to tell the marksman to move the center-of-the reticle left 2 mils and up 6.8 mils (i.e., the target is centered in the bottom right quadrant 2 mils to the right and 6.8 mils down from the reticle's center). **The marksman always confirms**, "Left 2, Up 6.8, Ready." Regardless of the number of inputs, the call to the marksman is always the same: a single left-or-right number and a single up number. A spotter does not give feedback in clicks, nor do they give guesstimates for the marksman to interpret. That said, a spotter must not hesitate to update their instructions if something changes that requires a new call, such as a gust. **Spotting is a continuous process.**

The spotter and marksman speak the fewest number of words possible to maximize verbal clarity. Some spotters even prefer to only give one direction and number at a time. However, it can be a good idea to add the word "point" before the direction, for example, "point left 2." This additional word ensures that the marksman understands that the given direction is where the marksman must point the center of the crosshairs (i.e., to the left), and that the marksman does not instead mistakenly put the target in the reticle's left quadrant (i.e., pointing the crosshairs to the right). While some hunters work with spotters that give information in less compressed form, such as giving the wind speed itself instead of a wind call, this is only to entertain the marksman and is not a maximally effective setup.

Spotters base their corrections on the impacts of previous shots. Therefore, a marksman must either follow the spotter's instructions precisely or refrain from acting altogether (for example, declining for safety or pausing for verification). In other words, **the marksman must not modify the spotter's directives**. Any deviation by the marksman causes the spotter to adjust based on their original, ignored instructions rather than the marksman's actual shooting, leading to baseless adjustments. For instance, if the spotter calls "left 2.2 mils" but the marksman only adjusts left 1.1 mils, subsequent instructions from the spotter would be based on the 2.2-mils instruction instead of the reality.

Image 17: A Marine sniper attached to the 24th Marine Expeditionary Unit **receives information from his spotter** behind him as he shoots at a target in the ocean during a crew-served weapons shoot. USS Nassau (LHA 4), 12 Feb 2010.

Sample spotter-sniper interaction:

Spotter – identifies the target.

"From building 4, go 9 o'clock 10 mils. One male, red shirt, blue jeans."

Marksman – confirms the target.

"One male, red shirt, blue jeans, standing next to a white Hilux truck."

Spotter – runs through the checklist.

"Check parallax, laze, and mil."

Marksman – confirms ready; gives the requested information.

"Ready. Height, 2.0 mils."

Spotter – gives the elevation hold.

"Point up 6.2 mils."

Marksman – confirms the hold.

"Pointing up 6.2 mils. Ready."

Spotter – gives the windage hold.

"Point left 0.5 mils."

Marksman – shoots.

Shoots.

Once a bullet impacts, the spotter incorporates the information into a new call. They do not inform the marksman of anything other than "hit" or "miss."

Reporting unnecessary information back to the marksman can only distract them. Even worse, telling the marksman where they missed allows for conflicting opinions in the team on what a good adjustment would be.

A miss can occur due to either incorrect instructions from the spotter or poor performance from the marksman. If the marksman knows that they shot poorly, they must inform the spotter so that the spotter knows their instructions do not need to be changed. Then the marksman can try again using the same windage and elevation holds.

A marksman explicitly indicates the direction of their mistake if they are confident. For example, "bad shot, 6 o'clock." A marksman may also indicate that they shot well with a phrase such as "good."

If neither the spotter nor the marksman cannot identify where an impact occurred, the spotter can give the marksman the same holds to repeat the shot. (When bullets leave no signs of impact, they typically are going high.) A spotter may also use the same hold again to verify that the marksman is shooting consistently. If the same holds lead to different impacts, the marksman and their rifle must both be inspected.

If a spotter is working with different marksmen at different times, they must consider that every marksman responds to information differently. How elevation and windage holds affect shots depends on the marksman's rifle, scope, batch of bullets, and interpretation of commands. Therefore, even if two marksmen appear identical, the same hold command can lead to different results.

Spotters are particularly relevant to the military and police because those professions place additional administrative duties on a marksman that are external to actual shooting. Such duties include: communicating with higher command, maintaining situational awareness, terrain navigation, and recording events. If long-term observation is required, the marksman and spotter can switch positions to reduce their eye strain. Military teams can also carry diverse loadouts, with a team carrying a bolt action rifle for sniping and a semiautomatic rifle for the spotter as a medium-range backup weapon.

All that being said, modern technology is obsoleting spotters because modern technology makes it easier for a marksman to gather their own information. For example, shooting competitions use cameras at the target to determine impact. Ballistic calculators, better reticles, and more aerodynamic bullets increase the chances of first-shot impacts at distances that would have been extremely unlikely in past years. And communication is easier with headsets, cellphones, and digital readouts. For example, in the past a command team had to relay orders to multiple units by handheld radio, and those units then had to verbally report information back up the chain-of-command. Now, marksmen can receive commands through an earpiece, and transmit their points-of-aim and sight picture in real time by using scopes with a broadcasting video feed.

This influx of technology has not completely depreciated the role of a spotter in the military and police, but rather, the new technology has changed the spotter's role. Sniper-spotter teams are still teams of two; however, neither marksman is a dedicated spotter and both can either shoot from their own rifles independently, or they can work together to shoot one rifle more accurately. This flexibility doubles the potential firepower of the team. (See Image 18, Pg. 28.)

Technology also allows the spotter to be in a **separate location** from the marksman. (See Image 16, Pg. 23.) Being in separate positions can be useful because it allows the marksman to shoot over an obstacle to hit a target that they cannot see. To perform this shot, the spotter picks a proxy target that is close to the actual target, such as a bright rock, and describes it to the marksman. The marksman-spotter team shoots the proxy target until they can hit it consistently. The spotter then proceeds to determine the angular distance between the proxy target and the actual target. The spotter tells the marksman to always aim at the proxy target and dial that angular distance into their scope. As the marksman always aims at the same spot (i.e., the proxy target) changing the dial by the angular distance to the proxy target makes the bullet impact on the proxy target. If the first shot does not impact correctly, the spotter can guide the marksman to impact the actual target by having them dial their scope while continuing to aim at the same location. This process is similar to how a forward observer would guide artillery.

All that said, being in separate locations is usually a bad idea, as it increases the vulnerability of the marksman. A primary role of a spotter is **situational awareness** to ensure the safety of the marksman while they shoot.

Although the most common military teams only employ one marksman and one spotter per team to conserve manpower, competition shooting may not be constrained to a two-man team. One marksman can assign multiple spotters to observe and report on different environmental factors individually, such as one on bullet impact and one on wind changes. Similarly, a single spotter can set up two spotting scopes to switch between, one to observe the target and one to observe mirage. In fact, a competition marksman can use another competitor as a quasi "spotter." The first marksman can wait for their competition (of equal or greater skill) to shoot first, and then use their impacts to better understand the wind conditions.

Multiple Marksmen

Image 18: A U.S. Army Special Forces sniper team fires upon a floating target at sea. While spotting scopes can see targets in more detail, often **another scope can perform the same function**. Using two rifles allows the spotter to be a backup marksman. Souda Bay, Greece, 20 Apr 2021.

Image 19: **One spotter facilitates two competitors** in the USASOC International Sniper Competition. And vice versa, multiple spotters can facilitate one marksman. For example, one determining wind speed and another looking for impacts. Fort Bragg, NC, 22 Mar 2021.

2.b Multiple Marksmen

A marksman only has one chance to hit a living target before that target moves to cover, be it an animal or an enemy. However, this limitation can be overcome by using multiple marksmen who all shoot at the target to **increase the odds** that at least one bullet impacts correctly. Marksmen do not have to be collocated if they can still communicate via radios.

Every marksman involved in the simultaneous engagement must initially verify that their target is in sight and that they are ready to execute the shot. This confirmation is typically delivered through verbal notification. However, modern camera technology may allow the marksman to wirelessly send their scope's sight picture to other devices. This technology allows a spotter or team leader to directly see whether the target is in sight without asking.

To **facilitate synchronization of shooting**, a team leader or spotter often counts down, such as by saying, "Standby. Ready. Fire." Or a team may use a reverse order countdown, "Three. Two. One." Although shooting on the "one" is possible, even that duration of sound is often considered too imprecise. Therefore, snipers can be instructed to shoot at the precise sounding of the "T" in "Two." A third option is to use an electronic audio recording with three beeps. Ideally, all the shots from every marksman are fired so closely in time that they sound like a single shot. Marksmen can practice their coordination with dry-firing, so that the clicks of their firing pins sound like one click.

Whatever trigger is used, marksmen must never predict the countdown. If the trigger word is not said at the end (e.g., "fire" or "two") it is likely because there was an interruption and the team leader does not want the shots to be fired. In fact, some teams utilize **abort-codewords** specifically to avoid inadvertent shooting.

Another common pitfall to avoid is jerking the trigger. Because marksmen are so focused on precise timing, they may jerk their trigger at the last moment. It is important to remember that both no shot and a delayed shot on target are preferable to any missed shot.

2.c Multiple Targets

Sometimes, a single marksman has to shoot at multiple targets. For example, a military sniper may be responsible for suppressing one side of a building with many windows. A tournament competitor may have to hit multiple targets in quick succession. Or a hunter may spot multiple animals that they want to engage to bring home as much meat as possible.

The first step in planning to shoot multiple targets is to **prioritize and order the targets**. This process always begins with defining the area of engagement. For example, the marksman may be responsible for a building or a ground sector. Hunters may want to monitor one field more closely, or multiple fields more broadly. For complicated terrain, the marksman can create a map or chart with specific distances listed. In contrast for simpler sessions, the marksman may simply use estimates and a rangefinder in the moment.

Within their decided area, the marksman or marksman-spotter team **systematically reviews** the area for potential targets. For example, when scanning the side of a building, every opening (i.e., window, door, or hole) is examined from the top left to the top right, and then each floor below in succession. A common method of labeling for buildings is to label the top left opening "1A," with letters going to the left and numbers going down. The top floor is floor "1" because the bottom floors are not always visible. (See Image 20, Pg. 31.) When a potential target is identified, it is noted.

The second step in shooting between multiple targets is **creating an order of priority**. There are multiple methods a marksman can use. Target marksmen often diagram their targets and shoot them in the order of closest to farthest. Using a diagram enables the marksman-spotter team to refine their communication and get in the groove of shooting as they move along to more difficult shooting.

Military snipers often engage targets by eliminating the most serious threats first. "Threat" can be defined by either danger or reactivity. That means that a sniper would first engage the targets that are the most dangerous and the most likely to react to prior shots. This prioritization both minimizes the danger and ensures that as few targets flee as possible. If an extreme threat is identified mid-shooting, a military marksman may even pause to reassess their shooting order to take immediate action against that extreme threat.

Hunters with multiple targets have the goal of taking home as much prey as possible. To that end, "a bird in the hand is worth two in the bush." In other words, hunters take down the easiest targets first. That said, if the herd's lead doe is both available and obvious, shooting them first can sometimes create confusion in the herd, potentially creating a brief window to shoot again. Similarly, shooting a deer on the fringe of the herd may also delay a complete scattering. However in all other cases, attempting to shoot some animals to influence the fleeing direction of the remaining animals can be an exercise in futility. Finally, trophy hunters always prioritize the trophy animal.

Once the marksman has determined where their targets are and in which order they would shoot them, the **marksman must get into position**. The chosen shooting position must be able to accommodate transition between targets. For example, the prone position limits lateral mobility and is not ideal for shooting short-range targets over a broad area. In contrast, a kneeling position is not stable enough to engage multiple targets far away. The standing position with a support tripod is a good choice for engaging multiple targets at a long range because it provides support and stability while also providing room to observe and transition. However, instead of locking the rifle into the tripod, the marksman can rest the rifle on a sandbag situated on top of the tripod to enable easy adjustments.

When shooting and transitioning between targets, a marksman must be **methodical**. Some military marksmen fire two rounds at every enemy to ensure that at least one hits the target. Other military marksmen fire one round at every enemy as quickly as possible to put down as many threats as quickly as possible. Afterwards, they return to each threat to reassess and shoot again. Hunters in contrast, may fire only one round and re-engage their sight picture before firing a second shot in order to preserve as much meat as possible. Target marksmen engage in whatever method their competition prescribes.

Reducing the time between transitions is difficult and takes practice. One method of reducing time is to maintain a tempo. Keeping a tempo between shots encourages the entire movement to become methodical (e.g., moving the left leg, then the right leg, then the hips, etc.) Keeping tempo also encourages a marksman to not rush their shots, which can cause the marksman to jerk their trigger. Of course, using a semi-automatic rifle and practicing shooting are also both imperative for becoming faster.

Window and Door Coordinates

Image 20: When marksmen and spotters communicate about buildings, they must be able to **quickly identify windows and doors**. This is done by starting at the top left corner because the bottom of the building may not be visible. If multiple wings, faces, or frames are visible, they too must always be specified.

Image 21: In this example, the highlighted area could be fully identified with: "East wing, East face, window 1C, bottom frame." (Doors and windows are counted separately.) Alternatively, to **prevent confusion** about wings on this four-story building, the area can also be identified with: "East face, window 3C, bottom frame."

2.d Multiple Marksmen and Targets

Rarely, multiple targets must be taken out simultaneously, leaving no time for any target to react. This exact scenario occurred when Captain Phillips was held hostage in a lifeboat by three Somali pirates. Navy Seals had to eliminate every pirate before they could react and take revenge by killing Captain Phillips. In the movie reproduction of this event, the Seals took three simultaneous shots to confirm three kills. In real life, it is unclear whether the Seals were as perfect in their synchrony, but the scenario and successful rescue of Captain Phillips were the same nonetheless.

Coordinating multiple marksmen to shoot multiple targets simultaneously is not much different from coordinating multiple marksmen to shoot one target. Therefore, the individual instructions from a spotter to their marksman are the same. (See Multiple Marksmen, Pg. 28.) And assigning multiple targets to multiple marksmen is done the same way as assigning them to a single marksman. (See Multiple Targets, Pg. 29.)

The new aspect of coordinating multiple marksmen and targets is the introduction of a team leader. **Having a single, centralized team leader is vital to ensure proper coordination among various, dispersed teams.** Team leaders must ensure that each spotter is entirely sure of their assigned target using simple, precise language. This clarity can be created by using only a single reference point for all the shooting teams. For example, on a building, that reference point could be the top left corner of the building. Then, each shooting team is assigned different coordinates in relation to that single reference point.

Also, because each target may appear or disappear at any time, each team must send up reports on their readiness to fire on the target whenever their situation changes. Traditionally, these reports would be sent up in a **round-robin report.** That is, the team leader would ask for a status report from each spotter, and the spotters in order would report their identity and status, such as "Team 1, ready." or "Team 1, no go." In contrast, modern spotters can send a video feed of their sight picture to the team leader so the team leader can view the status of all marksmen simultaneously.

Once the team leader is satisfied, he can issue the preplanned trigger word, such as "engage." Or he can order the spotters to wait on whatever other trigger has been preplanned, such as a specific start time.

Positioning is also vital. The team leader must weigh the benefits of disbursing the shooting team to achieve better angles of fire against the costs of worse communication, security, and responsiveness.

Image 22: On February 4, 1976, near the Djibouti–Somalia border, a group of independence militants held a busload of children hostage. The GIGN, a French counter-terrorism unit, deployed to the scene and introduced a new tactic known as "**simultaneous shooting**." After careful coordination, each sniper targeted a different hostage-taker inside the bus, waiting on a precise, silent three-second countdown triggered by the command "zero." When they fired in unison, only a single shot was heard. All four captors fell instantly, and the children emerged unharmed.

Caption: Djibouti 1976: 1. Somalian MG 42. 2. Somalian border post. 3. Bus with the thirty child hostages. 4. Border barrier. 5. GIGN firing position. 6. French GNA fort. 7. Command post of General Brasart. 8. Legionnaires' starting point. 9. Armored car starting point (800 meters).

3. Restriction-Based Positioning

Marksmen want to be as accurate and precise as possible. So they tend to always use the same, optimal shooting positions. However, sometimes there is a physical restriction that prevents being optimal and the marksmen must adapt and make the best out of a suboptimal position. These are some of the more common scenarios.

3.a Non-Standard Shooting Positions

Marksmen are often trained to shoot from standard positions, such as prone, kneeling, and standing, as these provide the most stable and predictable platforms. However, **real-world scenarios are not in ideal conditions**, and marksmen must adapt to a variety of non-standard shooting positions.

Some of the more common non-standard shooting positions are sitting, squatting, and lying on one's back (supine position). Key features of non-standard positions include: unstable platforms, restricted space, and odd angles. Whatever the case, the end result is the same: there are fewer points-of-contact with the ground, less bone support (i.e., more muscular support) to the rifle, and less resistance to recoil.

Therefore, the first step in shooting from a restricted position is to **maximize the rifle's points-of-contact with the ground**. This means placing the rifle on a supporting surface whenever possible, even if the marksman themself must be in an awkward position. Of course, normal supports such as sandbags, bipods, and tripods are always useful. However if they are used, then the position is a standard supported position and not "non-standard."

Because locations that result in restricted positions are often characterized by unstable or uneven surfaces, such as rocky terrain or vehicle hoods, marksmen often cannot maintain a sight picture in these restricted positions for a long time. However, marksmen can extend the longevity of their sight picture by using objects in their surroundings as a support, such as local rocks and dirt, to elevate themselves or their rifles. Of course, physical training and endurance training can also both help a marksman to stay in an unexpected position for a longer amount of time.

Image 23: This marksman supports his rifle with a sandsock. **Rifle support** and stability are the foundations of every shooting position.

Image 24: This marksman is modifying a standing position. However he puts **no weight on the rifle** because forcing a rifle decreases its precision.

Image 25: Marksmen can shoot off **any stable surface**. It is important to not get tunnel vision, and incorporate unconventional scenarios in training.

Image 26: Sometimes when thinking of novel scenarios, it is helpful to look to **historical examples**. Many marksmen have shot from rooves.

Image 27: The **environment** may allow for positions that otherwise would be unstable. This slope makes sitting a much more stable position.

Image 28: Similar non-standard positions can be used in **both urban and rural** environments. This rock and a roof require similar positions.

Image 29: There are various ways of getting **as low as possible**, while still maintaining front-support for the rifle. This is a kneeling crouch.

Image 30: To get low, this marksman sits down and uses **the crux of his elbow** to support the rifle. He could also get low using the bipod.

Image 31: To get low, this marksman **lays on his back**. This is possible because his bulletproof vest and the tree are providing back support.

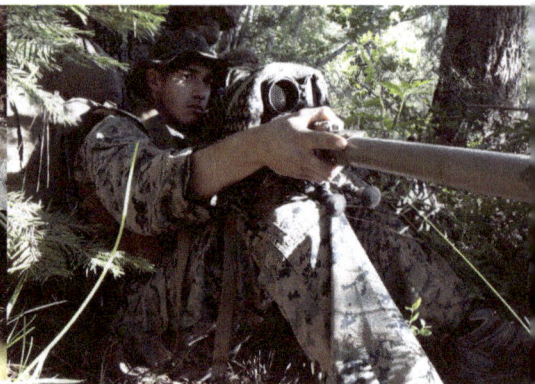

Image 32: Good **back support** can allow for a position that is almost as stable as a prone position. Trees and walls are common back supports.

Image 33: To match the height of the window, this marksman has to go lower than a crouch, but higher than a prone. So he **swings his legs**.

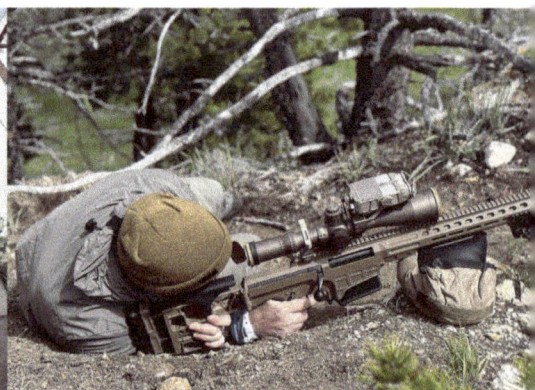

Image 34: If there is not enough space behind the marksman, they may have to completely **swing their body** to the side to get low.

Image 35: Often, **cover dictates** a marksman's position. This marksman, however, would be safer standing and using his left side to shoot.

Image 36: To compensate for his unavoidably poor shooting position, this marksman has **wrapped his sling** around the ladder to **lock in his rifle**.

Image 37: Although this is training, the marksman has achieved **concealment but not cover** because plastic barriers are not bulletproof.

Image 38: Only the **engine block** is bulletproof, meaning the marksman on the left may be safer and more accurate in the prone.

Image 39: This is a fun trick shot. But there is **no reason** to have a marksman kneeling in a bare field, devoid of a real barrier or a support.

Image 40: Many times, marksmen practice positions because they are **fun and interesting** rather than optimal.

Non-Dominant Hand

Image 42: The most common scenario which requires using a non-dominant hand is when **cover is only available from one side**. In this diagram, the center figure shows what happens when a right-handed marksman fires from right-side cover. They must either find left-side cover (left figure), or switch hands (right figure).

Image 41: A U.S. Marine platoon commander with Special Purpose Marine Air-Ground Task-Force Crisis-Response Africa fires an M4 carbine around a high barrier using his **non-dominant hand** to maintain cover. He must be careful to not block the ejection port and have a hot casing get caught. Germany, 09 Apr 2016.

3.b Non-Dominant Hand

Ambidextrous shooting is a skill that many police and military snipers benefit from mastering. Training to fire a rifle from both sides of the body prepares marksmen to be able to **use cover on both their right and left sides**. In contrast, attempting to fire from cover that is on the side of the support hand leave's the marksman's torso exposed.

The most significant challenge when shooting from the less dominant shoulder is correctly positioning and maintaining the rifle against the shoulder throughout the firing process to mitigate excessive recoil. Also, maintaining correct eye relief with the non-dominant eye can be tricky. Both of these issues can be overcome with practice. **Practice is in fact the only answer**, as there is no trick or shortcut to becoming better at off-handed shooting.

It's advisable to start practicing non-dominant-side shooting from the prone position before gradually advancing to more challenging stances because it's typically easier to fire from the less dominant shoulder while in a prone position. Similarly, marksmen can **practice with dry-firing** before transitioning to live-fire practice to avoid wasting ammunition. Eventually, the goal is to be proficient at shooting from a standing, unsupported position from either side.

3.c Non-Vertical (Canted) Rifle

Rifles are held with the reticle perfectly vertically whenever possible. Introducing cant (i.e., sideways rotation) to a rifle causes bullets to travel in a complex three-dimensional trajectory that is difficult to predict or account for. However sometimes, a marksman has no other choice because they are shooting from a very restricted space, and both the scope and the barrel cannot point out of the opening at the same time.

If a rifle must be canted, a marksman is better served by rotating their rifle **either a full 90 degrees (i.e., the scope is beside the barrel) or only 45 degrees (i.e., the scope is diagonal to the barrel).** Doing this allows a marksman to more accurately adjust their elevation and windage holds. If the rifle is oriented at 90 degrees, the windage dial is adjusted with the elevation hold and vice versa, essentially switching the roles of the windage and elevation turrets. If the rifle is held at a 45-degree angle, the elevation dial setting is divided between the windage and elevation dials. Each dial receives about 70% (the sine and cosine of 45 degrees) of the elevation hold.

Every other angle requires more precise math to account for bullet-drop. Performing those calculations involves applying the sine of the cant times

Non-Vertical (Canted) Rifle

Image 43: A U.S. Marine Corps team leader with Special Purpose Marine Air-Ground Task-Force Crisis-Response Africa fires an M4 carbine through a nine hole barrier during a combat marksmanship range. Germany, 09 Apr 2016. Sometimes there is only a narrow slot from which to both aim and shoot through. In these cases, canting the rifle is necessary.

Image 44: Other times, **the position dictates canting**. This position may occur when shooting below a vehicle in a compressed space. On the other hand, it is unclear why such a rare scenario would be part of training. Germany, 09 Apr 2016.

Image 45: This Soldier is from the 11th Armored Cavalry Regiment. A more common, though still rare situation would be when a marksman must take a shot while pushing as far back as possible into **sloped cover**. Fort Irwin, CA, 04 Apr 2017.

the elevation to the elevation turret, and the cosine of the cant times the elevation to the windage. However, at that point, **it would be better for the marksman to just improve or change their shooting position to uncant their rifle**.

If a marksman somehow knows that they need to shoot through a very narrow opening, they can also address the cant of their rifle during the zeroing process. That is, a rifle can be zeroed with a cant. However, that zero point would then be invalid for the rifle in a vertical position, meaning the rifle could only be accurately be shot with a cant present, and only at a single distance (i.e., the zero distance). Instead, **a rifle can be "zeroed" with no zero point**. That means that the scope is adjusted to not account for the holdover, and is instead set so that the sightline is parallel to the boreline. This arrangement means that when aiming, the error introduced by cant cannot be greater than twice the holdover, which is a relatively small error when shooting at very long distances.

4. Disadvantaged Conditions

Basic shooting instructions assume that the conditions of shooting are optimal. The marksman is relaxed and not pressed for time. The sun is bright and warm. However, there are often circumstances that make a shooting session stressful, time-limited, dark, and cold.

4.a Stress and Fatigue

Physical stress can negatively impact both the accuracy and precision of shooting. An elevated heart rate, rapid breathing, and tremors or shaking in the hands make aiming more difficult. Then, high stress may lead to a rushed firing sequence, resulting in poor aiming practices. Finally, stress can cause a marksman to jerk their trigger more often, leading to inaccurate shots.

Fatigue, in addition to stress, can compromise critical cognitive functions such as high-level decisionmaking and the ability to perform calculations. For example, a tired marksman may choose to lay in the prone because it is more comfortable, despite another position providing a superior point-of-aim. Also, the precise calculations of factors such as wind, distance, and trajectory become more difficult.

Most importantly, **stress and fatigue can disrupt a marksman's established routine**. A marksman may skip specific steps in their routine because their memory recall is not as strong. Deviating from or rushing through any step in a standard procedure can significantly degrade performance in not just that step but also in every following step.

There have been a few scientific studies specifically done on the effects that physical stress and fatigue have on shooting, and they consistently show that the effects are significant but also minor and short-lived. For example, one study found that ingesting 300 mg of caffeine (a form of artificially induced stress) slightly decreased shooting performance from the standing position but not the prone. Another found that marksmen experienced a significant return to their baseline skill level after just three minutes of concentrated rest.

There are a few implications that can be learned from these findings. First, **taking a moment to deeply inhale and exhale** can eliminate most of the negative effects of stress and fatigue. That is because most of the errors can be traced back to rushing and not a decline in core performance. Another takeaway is that **forming habits and routines is important**, as stress and fatigue make marksmen rely more heavily on actions that can be performed without much active thinking or decisionmaking.

Practicing how to combat the effects of stress and fatigue is often misunderstood. The"skills" of calmness are to rely on previously learned skills and generally to slow down. That is, there is no special or different shooting technique that is optimal only for stressed and fatigued marksmen. Therefore, **making training stressful is a waste of time**, and a better use of time would be further ingraining core skills so that they appear when stress is induced.

That said, military and law enforcement snipers often do rigorous physical training and stress inoculation exercises. Introducing stress during shooting helps evaluate how marksmen might perform under pressure in real situations. (It's important to distinguish between training and testing, as they serve different purposes.) To test the stress reaction of these marksmen, they perform high-intensity physical activities followed by immediate shooting. Stress is not added to improve shooting accuracy, but to test how quickly the marksmen can calm themselves. Being calm allows a marksman to rely on their regular, practiced training routines even when stressed. Many marksmen achieve calm by focusing on and controlling their breathing.

4.b Time Pressure

There are many reasons why a marksman may be shooting under a strict time limit. For example, marksmen in competitions only receive a limited amount of time to complete their shots. For hunters or military, targets are often only briefly visible. For example, perhaps a deer may be eating grass, and the hunter can only see the deer's head when it raises it to look around. Or perhaps a military sniper only sees their target when they look out a window. There are also soft time "limits." For example, a marksman can improve their accuracy by only shooting during the portion of their time when environmental conditions, such as wind, are optimal.

The harsh reality is that sometimes a time limit simply does not give enough time, and a marksman misses their shot. However far more often, marksmen overcompensate and rush through their shooting when in fact, they still had remaining time to shoot in a slower, more controlled manner. The best marksmen learn to not miss their windows of opportunity by

Image 46: Two Soldiers, one Australian and one American, compete in a sniper competition. Rushing to get into position causes rushed and inaccurate shooting. It is more important to pause and remember that "**slow is smooth, and smooth is fast**" than to risk missing a shot by rushing. Camp Taji, Iraq, 29 Dec 2019.

following the **same deliberate process every time they shoot** to build up muscle memory. That is because quickness is the natural result of control and consistency. This idea is captured in sayings such as "slow is smooth, smooth is fast" and "consistency is key." In contrast, the aversion to forcing speed is captured in the sayings: "you can't miss fast enough" and "fast is fine, but accuracy is final." **The worst marksmen are those that shoot as fast as possible, without proper control.**

One of the best ways to improve muscle memory is to become more familiar with one's own equipment; for example, learning the specific way to manipulate the bolt of a bolt-action rifle. Marksmen who are familiar with their bolt can work the bolt quickly and quietly without looking at it. An even more advanced technique is for a marksman to time the moving of their bolt to their inhales and exhales. Another example of muscle memory is a deliberate trigger pull. Practiced marksmen know precisely how much force pulling the trigger requires and **are not surprised when the trigger releases**.

The shooting position also helps prepare a marksman for time pressure. For example, a marksman in the prone position cannot as efficiently change their natural point-of-aim (NPA) as a marksman in the standing position can change their NPA.

Image 47: A member of the Marine Corps Shooting team fires his rifle during the Quantified Performance Match. **The best way to prepare for time-sensitive shooting is to practice it.** Here, the marksman has his partner hold a beeping clock to enforce time standards. Quantico, VA, 24 Apr 2022.

Preparation is always necessary to shoot well under pressure. This rule applies not only to creating habits but also to the environment one is shooting in. For example, ammunition for follow-on shots must be placed within easy reach. Elevation holds for various possible target distances must be precalculated on a reference chart. The expected wind is logged and written down. A marksman can run through an entire scenario before it happens and perform every available action (save shooting) in advance of the actual event to save time.

Observing environmental factors can also help determine the true time limit for taking a shot. For example, if a target is about to stop moving, the marksman would know they have more time to aim carefully. If a marksman is relying on mirage to read the wind speed, they would benefit from noting how the clouds are moving and when they are likely to block the mirage. That is, environmental awareness can help such a marksman to anticipate periods when wind readings may become unavailable.

It's just as important to plan what not to do as it is to plan what to do. For example, if a marksman has a spotter to track previous shots, there's no need for the marksman to do it themself.

4.c Nighttime

Humans can't see well in the dark. It takes 30 minutes for the eyes and the brain to fully adjust to seeing at night. During this period, night vision gradually "comes alive," and everything seems to brighten. It's akin to the moonlight being turned up and the stars shining brighter.

The full moon can shine surprisingly brightly once one's eyes have acclimated to the dark. Under strong moonlight, a marksman with their naked eye can shoot pronounced targets within about 200 m or yd or less, with distances of less than 150 m or yd being ideal. Even a half-moon can illuminate some targets if they are large and prominent enough. In fact, hunters may align their schedules with the brightest phases of the moon if they plan on shooting at night.

That said, impacting targets over 200 m or yd away at night becomes very unlikely with eyesight alone. That is because **at night, eyes work differently** and lose much of their ability to see detail. This loss happens because low light does not activate some cells in the eye, disabling a portion of human sight. Specifically the eyes' cone cells, which handle central vision, color vision, and fine detail, do not work in darkness. That means that sight only occurs with the eyes' rod cells, which detect black-and-white vision and movement in the periphery (i.e., outside the center). As a result, although a marksman may sense a target's presence, aiming precisely at night becomes much more difficult. Eye fatigue also quickly occurs at night, after just a few minutes of intense watching. Therefore, some marksmen alternate eyes or plan for resting periods.

Another change in eyesight is **pupil dilation**. During direct daylight, pupils constrict down to 2 mm (⅓ in) in diameter, whereas at night they dilate up to 8 mm ($^1/_{12}$ in) (pupil size changes less with age). Pupil dilation dictates how large the most useful objective lens on a scope can be. A bigger objective lens creates a bigger exit pupil, (the exit pupil is the diameter of the light beam that passes through the back of a scope's eyepiece).

The size of a scope's exit pupil is determined by dividing the diameter of its objective lens by its magnification. For example, a scope with a 32-mm objective lens and 4x magnification has an exit pupil of 8 mm (32 mm ÷ 4). This 8 mm exit pupil matches the typical size of a human eye's pupil in low-light conditions, allowing the scope to fully illuminate the marksman's eye with light when the exit pupil is properly aligned. (To accommodate eye movement, good settings often have an exit pupil that is slightly larger than the size of the observer's eye's pupil.)

In contrast, a scope with 24x magnification and the same 32-mm objective lens produces a 1.33-mm exit pupil (32 mm ÷ 24). This much smaller exit pupil is insufficient to fully illuminate an 8-mm pupil in low light; therefore, the sight picture would be dimmer than the sight picture of a scope with a larger objective lens. This is why low-light scopes with high magnification settings can have truly giant objective lenses. This also means that it is not possible to judge the effectiveness of a nighttime scope in a store; real-world night conditions and fully adjusted night vision are required.

Some marksmen use flashlights, car headlights, or spotlights to add light to areas to night-shoot. Of course, adding light is effective at increasing the ability to see. However, if target shooting, operating in low-light conditions is adding unnecessary danger. And when hunting or on a mission, shining bright lights alerts the target to the marksman's presence and location. Therefore, adding light is usually not a good option.

Instead of relying on moonlight or manmade lighting, a marksman can use **night-vision devices that read the existing light and then amplify it** so they can see it. These devices can either be integrated into the scope itself or be a separate clip-on device that goes in front of or behind the scope. Most night vision devices are clip-ons. Because a clip-on night-vision device does not disturb a scope's mounting, the scope and rifle can maintain their zero both with and without the device. In contrast, marksmen with a scope that has integrated night-vision capabilities typically use a different rifle for daytime shooting with a daytime scope to avoid having to unmount and remount scopes. And even if mounting a clip-on device does shift the zero (a.k.a., a "zero shift"), the clip-on usually shifts the zero the same repeatable amount every time, so the marksman can simply dial or undial the zero-shift into the scope. The downsides of clip-ons compared to dedicated night scopes are the added weight, the added cost, and that both devices must be compatible.

Night vision devices generally use one of two methods: **image intensification (II) or thermal imaging (TI).** II devices see and amplify visible light. In contrast, TI devices see infrared light and display it as visible light. In other words, an II device displays how bright an object is; e.g., a cold white rock will show brighter than a warm black cat. (See Image 51, Pg. 47.) In contrast, a TI device displays how warm an object is because warmer objects emit more infrared light; e.g., a cold white rock will show darker than a warm black cat. (See Image 50, Pg. 47.) Some modern devices combine both II and TI into one image. Due to the niche and complicated nature of night vision devices, it is always necessary to read their manual. And all night vision devices require batteries, so carrying extra batteries is a must.

Image 48: A nighttime U.S. Air Force KC-135 Stratotanker. Al Udeid Air Base, Qatar, 09 May 2021.

Image 49: A daytime Nebraska KC-135 Stratotanker. Pardubice Air Base, Czech Republic, 14 Sep 2020.

Image 50: A thermal image of a KC-135 Stratotanker aircraft. MacDill Air Force Base, FA, 27 Feb 2019.

Image 51: An intensified image of a KC-135 Stratotanker aircraft. Al Udeid Air Base, Qatar, 02 Sep 2017.

Night vision devices display their output on a screen. Because screens have limited resolution, **when a night vision scope is magnified, no additional information is revealed**. However, more information may be perceived by the marksman since it is easier to focus on an enlarged area. And many night-vision devices are built with very high resolution specifically to allow for magnification.

An II device intakes light and converts and multiplies it into electrons that hit the back screen. That screen then displays an image to the marksman. Because II devices only intensify light, they rely on a difference in light between the target and the background. That is, II devices cannot illuminate; they can only intensify. Therefore, on plains under full moonlight, marksmen using even relatively old II devices can easily see and shoot large, reflective targets at 1,000 m or yd and beyond. However, if a marksman were shooting at a target inside a triple canopy forest under a new moon, the II device's effective range would be much shorter. That said, technology is continuously improving, and images are becoming crisper with each generation of II technology.

The advertised ranges of night vision devices can be deceiving. That is because night vision devices advertise the range at which a person can perceive the presence or absence of a target, and not the range at which a target can be accurately aimed at. For example, one high-end II device claims

Light Signature

Image 52: A U.S. Marine sergeant sets security. Night vision devices can **create a light signature** that can be detected in complete darkness. USS Essex, Pacific Ocean, 02 Sep 2021.

Image 53: A Marine with 3rd Battalion, 3rd Marines conducts a live-fire sniper range. Of course, the most obvious light signature is the firing of the rifle. Okinawa, Japan, 16 Aug 2022.

that it can "identify" a human at 1800 m but only recognize that human's characteristics at 450 m. In other words, with a target at 1800 m, the observer may only know whether the target is present or absent, which is not very useful for aiming nor shooting.

Although II devices are not primarily or explicitly thermal or infrared devices, most can interpret infrared light. Therefore, **marksmen often pair infrared illuminators (i.e., lights, spotlights, or floodlights) with their II device** to shine light onto an area that unequipped enemies and animals cannot see. Boar hunting with an II device and an infrared spotlight is relatively common. In contrast for the military, since II devices have become cheap, most militaries do not use infrared illuminators since they give away the marksman's location. Target marksmen also do not use illuminators since they could just shoot during the day if they wanted to add illumination.

Another potent combination for long-range shooting is to use an infrared laser as an aiming device. (See Image 54, Pg. 49.) The laser is attached to the rifle and runs parallel to the boreline. Thereby, when shooting at night, even if the target is difficult to see, the marksman can shine the laser onto the target to align their boreline and illuminate the target. Because the dot must reflect off of something to be seen, when the laser dot is seen on the target that means the boreline is pointing at the target.

If the target is relatively far away, the marksman can also apply windage and elevation dials to the laser to aim more precisely, just as they would for the scope during the daytime. The laser dial does not need to be seen because the dialing is confirmed through the scope. In contrast, the scope itself is rarely dialed at night for the simple reason that the dial markings cannot easily be seen in the dark.

Lasers Sights

Image 54: Paratroopers from the 82nd Airborne Division train using **infrared lasers and night vision**. Laser sights allow a marksman to illuminate their point-of-aim. Good laser sights can be dialed for elevation and windage hold, and have low light divergence so the laser dot stays small at long distances. Ramadi, Iraq, 26 Oct 2009.

Image 55: Unlike passive sights (e.g., a scope), laser sights are rarely mounted exactly vertically above the bore. In fact, **some are mounted to the side of the barrel**. Therefore, a marksman must know how far the actual point-of-aim is from the left or the right of the laser dot. East China Sea, China, 23 Sep 2020.

Another issue is that a laser increases diameter with distance, so at longer distances the dot becomes bigger and less precise as a point-of-aim. (This is called "laser divergence.") Therefore, it's best to zero a laser at 200 m or yd or less and then refine it at a second distance if necessary.

Although ideally a laser is mounted on the top of a rifle to horizontally center the laser with the barrel, there often is not enough mounting space to put one. Therefore, when zeroing a laser sight, **it is important to consider the laser's sideways-offset from the barrel**. For example, if the laser is placed 5 cm (2 in) to the right of the barrel, the marksman must always place the laser dot 5 cm (2 in) to the right of the actual point-of-aim. That is, the bullet and laser would travel in parallel and never intersect, similar to shooting with a canted rifle. (See Non-Vertical (Canted) Rifle, Pg. 39.) This offset prevents the laser from traveling sideways to meet the bullet, which would only occur perfectly at the zero distance.

An important factor to remember when using lasers in combat situations is their bidirectional nature. This means that while a soldier can see the laser, the target can also trace the laser back to the soldier, exposing the marksman's position. Another problem arises when multiple marksmen, all using laser sights, aim at the same target. Distinguishing one laser dot among many is difficult.

Thermal imaging (TI) devices only read infrared light. TI devices are heavier, bulkier, more expensive, and have less battery life than image intensification (II) devices. Although II devices also read infrared light, specialized TI devices are far more sensitive and accurate. Also, by not reading visible light, they can better reveal information that is only broadcasted through the infrared spectrum.

Just as the sun emits visible light (i.e., sunlight), **all objects naturally emit infrared light**. The amount of infrared radiation emitted is proportional to an object's temperature, and living organisms are typically warmer than their surroundings at night. Therefore, at night, humans and warm-blooded animals shine like beacons in the infrared spectrum. Even during the day, they continue to shine brightly in the infrared as long as their body temperatures exceed that of the environment, despite any high-intensity visible light. In contrast, if the surroundings become significantly hotter than normal body temperature, humans and animals still appear in the infrared spectrum, but relatively darker against the hotter background. When the environment is around body temperature, humans and animals become indistinguishable against the background.

Infrared light has different interactions with materials than visible light has. For example, materials that are transparent to visible light are often not transparent to infrared light. **Glass is opaque to infrared light**, and a TI device can only determine the glass's temperature and not the temperature or infrared image of whatever is behind it. In contrast, **dust clouds, smoke, and fog are all much more transparent to infrared light than visible light**. For another example, metal, wood, and paper targets are generally the same temperature as the environment and so can be difficult to distinguish in the infrared spectrum.

At night, finding targets in the first place is one of the most difficult steps in the shooting process. Therefore, a quality set of binoculars is crucial for target acquisition. Using binoculars with a minimum 40-mm objective after the eyes have adjusted to the darkness and moonlight can offer an astonishingly clear view without any artificial light. A very effective setup for hunting is to use a pair of II binoculars for detecting targets moving at night, then switching to a thermal scope to shoot.

4.d Cold

Cold can not only make for an intensely uncomfortable experience, but it can also directly change or degrade performance. Specifically, **the cold causes numerous performance issues** through changes in: material plasticity, thermal contraction, the water vapor capacity of air, battery performance, and a marksman's body temperature. Therefore, a marksman must consider the temperature of their shooting location during preparation.

Cold-Weather Clothing

Image 56: A U.S. Marine lance corporal tries to warm up his hands. Being cold is not just a matter of pain. Cold can cause uncontrollable shaking that ruins aiming. A marksman in serious pain or shaking from cold is a **failure of leadership** to plan and supervise. Sekiyama, Japan, 14 Mar 2017.

Image 57: Shooting in the cold requires a **full suit** of layered, waterproof clothing, boots, gloves, and headgear. On the other hand, excessive sweating is also bad. Sunglasses prevent snow blindness, and hiking sticks can double as shooting sticks. MT, 09 Feb 2023.

Changes to plasticity are straightforward: materials become more rigid and less viscous. Rubber materials such as gaskets are more brittle at lower temperatures and so are more likely to tear. Standard lubricants may thicken or freeze, hindering a rifle's performance. Switching to a lightweight, cold-weather-specific lubricant ensures smooth and reliable operation of moving parts.

Thermal contraction (or thermal expansion) is when a material changes physical size as the temperature changes. **Thermal contraction becomes a problem when there are multiple materials used closely together** in high precision. When it comes to scopes for example, metal housings dissipate heat more quickly than the glass lenses they enclose and therefore contract at a faster rate. This discrepancy in contraction rates can put tremendous pressure on the glass, leading to cracking in either the lens or the housing if the temperature shifts too rapidly. Thermal contraction may also pose a problem if a rifle is zeroed in a high-temperature environment but then fired in colder conditions because the rifle's overall shape changes slightly, leading to a corresponding small shift in the point-of-aim. Wooden stocks are the most prone to having this issue.

The third issue is that as air cools, it retains less water vapor. That means that if air inside of a scope is humid and cannot escape fast enough, cooling it down leads to water condensing inside of the scope. (See Image 58, Pg. 52.)

Image 58: This photo is of the shopping center Lauttis in Lauttasaari, Helsinki, Finland. It was photographed through a **fogged-up camera lens** because the rapid transition from exiting cold outdoor weather to entering warm indoor weather caused water vapor in the warm air to condense onto the outside of the camera. A similar effect can occur in reverse, **when a warm camera enters a cold environment and the warm air in the camera suddenly cools and condenses to water on the inside.** While exterior condensation can just be wiped off, internal condensation is much harder to fix, and often just requires time to acclimatize.

The problems caused by thermal contraction and water vapor capacity can mostly be resolved by allowing scopes to **slowly acclimate** to the ambient temperature. Transporting a rifle in the bed of a truck or otherwise exposing it to the cold over an extended period before shooting are ways to do this.

Electronic devices such as rangefinders and scopes experience **reduced battery efficiency** in cold weather. This reduced efficiency is caused by cold temperatures slowing down chemical reactions. To mitigate this problem, spare batteries are best kept inside a coat, close to one's body, where they can remain warm. Attaching hand warmers to electronic devices can also help.

Colder gunpowder does not expand as much as hotter gunpowder because its chemical reaction releases a fixed amount of energy into the system. When the propellant starts at a lower temperature, more of the energy is consumed bringing the gunpowder up to its burning point, leaving less energy for propulsion. As a result, **colder ammunition produces a lower muzzle velocity than hotter ammunition.**

This energy loss is only significant when a rifle is zeroed at a very hot temperature and then shot at a very cold temperature (or vice versa). Therefore, ammunition is ideally always shot at the temperatures in which zeroing occurred. This can be done either by zeroing in the same climate as one shoots, or by keeping warming up or cooling down the ammunition itself.

Radiation and Temperature Cycle

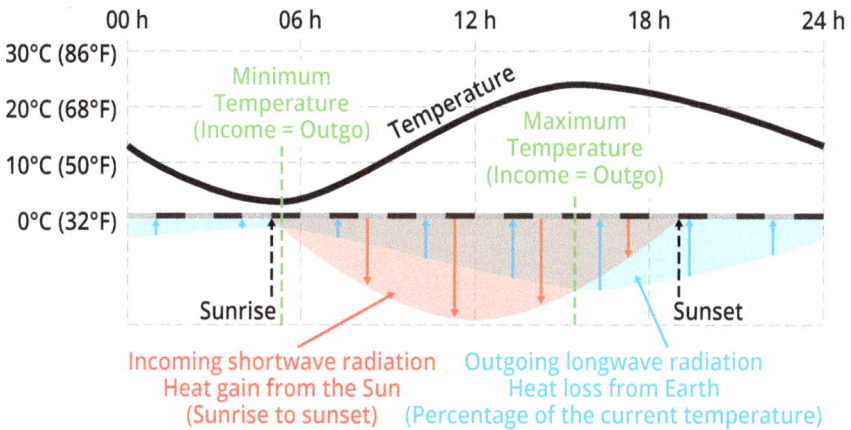

Image 59: Over the course of the day, **the temperature can significantly change** depending on the location and climate. In the example, the temperature ranges from 2°C (35.6°F) just after sunrise to 25°C (77°F) at around 16:00 (4 PM). Water retains heat and so smooths out daily temperature changes. Therefore, **inland, mountainous, and desert regions are prone to large temperature swings**.

Beyond equipment degradation, **cold temperatures also greatly affect the marksman's ability to shoot**. A cold body temperature leads to body tremors and shaking that make aiming difficult. Cold fingers go numb, reducing the feedback a marksman receives from pulling the trigger. The pain of cold is just generally distracting and can reduce focus and concentration.

Therefore, proper attire is vital for comfort and performance in cold weather. Dressing in layers that offer wind resistance and insulation helps maintain core body temperature. The human body constricts blood vessels in the limbs when the core is cold, so **warming the core is the best way to warm up one's arms and fingers.** However, the most effective way to "warm" hands is to not let them get cold. Utilizing hybrid glove-mitten combinations with hand warmers allows for both warmth and functionality. Whatever attire is used, a marksman must practice and dry-fire in that specific attire to find any significant interactions between their attire and equipment.

Insulating the body from the ground is also critical, especially when shooting from a prone position on snow or ice, which can rapidly drain body heat. A proper shooting mat, or even some spare pieces of cardboard, can serve as an effective barrier against frozen ground.

Dehydration is a common occurrence in cold weather because people dislike drinking very cold water when they are already cold. This a serious problem, as serious dehydration for extended periods can cause performance issues on par with being legally drunk. To address this, marksmen can warm up their liquids either with external heating or by keeping containers inside their coat to let their body heat up the liquid.

Managing condensation from breathing is important to prevent fogging or frost on optics. Exhaled breath must be directed away from equipment, and adequate air circulation must be allowed for. Historically, some snipers have reportedly put snow in their mouths to reduce the visibility of their exhale, but this is excessive in a modern conflict where isolated sniper teams operate at significantly farther distances. Instead, marksmen can simply breath through the nose, which directs air down and not forward.

Finally, **cold weather affects the environment**. For example, cold air is denser than warm air, increasing aerodynamic drag on bullets and causing them to lose velocity more rapidly. This loss of velocity can result in lower points of impact over long distances. Accounting for this is as simple as inputting the temperature into a ballistic calculator.

After shooting in cold weather, a common issue is the formation of condensation on the rifle when it is brought indoors and begins to warm up. Condensation occurs when warm air quickly cools upon contacting the cold rifle, causing excess water vapor to be released and condensed into liquid water. This water condensation can rust a rifle's metal parts. To avoid condensation on the rifle's surface, marksmen can wrap their rifles in towels which both absorb excess water and insulate them to slow the warming process and prevent condensation from forming in the first place.

4.e Shooting through Barriers

On rare occasions, a long-range marksman may want to hit something on the other side of a barrier. The most rigorously studied example is shooting through glass, so that is the primary subject of this section. That said, the conclusions would equally apply to other situations, such as shooting an animal hiding behind a thick bramble patch.

How well a bullet penetrates a barrier depends on two things: the energy it carries and its robustness. **A bullet that has more energy and is more robust is better able to penetrate a barrier and accurately impact a target on the other side.**

The energy contained in a bullet is proportional to the mass, but proportional to the square of the velocity. Therefore, faster bullets are more likely to penetrate and less likely to be deflected. This can be somewhat misleading though, as the mass of a bullet somewhat limits its potential

Shooting through Barriers

Image 60: An Italian soldier shoots through glass with a SIG Sauer SSg2000. **Non-laminated glass shatters**, so a second marksman can then shoot through the opening. Hohenfels, Germany, 15 Jun 2015.

Image 61: Snipers engage 100m-targets through glass panels. Without prebreaking the glass, **the target must be relatively close behind the glass**. Hohenfels, Germany, 20 May 2017.

Image 62: A Marine explodes a **water charge** with a detonating (det) cord to clear the glass from a window. Camp Lejeune, NC, 16 Jul 2013.

Image 63: Once the glass shatters, it can send fragments backwards, so **eye protection is always necessary**. Camp Lejeune, NC, 16 Jul 2013.

Impossible - Perfect Reflection or Perforation

Actual - Deformation and Deflection

Image 64: Bullets do not travel in a straight line through barriers.

velocity; therefore, mass and velocity typically increase in tandem. Similarly, the diameter, or caliber, of a bullet correlates well with its mass. Therefore, marksmen and manufacturers often use the caliber as a proxy measurement for a bullet's energy, even though it is not a particularly good proxy. That said, while specifics depend heavily on the specific bullets compared, in general, a .30-caliber bullet is much more likely to accurately penetrate a barrier than a .22-caliber bullet is.

A much better measurement of a bullet's energy is to calculate the actual energy by measuring the bullet's velocity at the point of impact using a ballistic calculator and multiplying it by the bullet's mass. Alternatively, a marksman could compare the grains of powder each bullet uses, although this does not account for the energy that is lost to air resistance during the bullet's flight.

The material composition of the bullet also affects its penetrative ability. Every type of bullet is designed to exist on a spectrum of over or under-penetration. Obviously, armor-penetrating bullets perform the best at penetrating a barrier. (In fact, they may not stop after penetrating the target, so knowing what is behind the target is critical.) On the opposite end of the spectrum, fragmentation bullets are designed to fragment immediately upon impact. While they may technically break through a barrier and exit the other side, the "bullet" would be nothing more than a spray of (still dangerous) fragments. Between those two extremes, hollow-point bullets fragment a lot, full-metal-jacket bullets fragment somewhat, and bonded bullets fragment a little.

The most significant factor when shooting at a smooth surface is the angle of that surface. **Bullets deflect off surfaces that are too angled.** For glass, shots are best taken if the surface is less than 15° from perpendicular. Anything higher than 15° requires extreme power from high-powered cartridges. For example, a .30-caliber bullet can usually shoot through glass at a much greater angle than a .22-caliber bullet can. Regardless, shots at window-glass at more than 45° from perpendicular are always too unreliable to be taken.

When a shot is taken at (non-laminated) glass or other material, the bullet exits the other side alongside fragments (a.k.a. "spall"). (See Image 60, Pg. 55.)This spall ranges from the size of dust to the size of the bullet itself and even larger. **The danger of spall is that it may cause collateral damage,** so it is important to consider the surroundings of any target behind a barrier. Even small wood chips and glass dust can be extremely hazardous to the eyes of anyone in the area. Maiming a deer or a hostage is unacceptable.

The exact exit of a bullet shot through a barrier is not predictable; instead, marksmen predict that **the bullet exits somewhere within a cone**. This cone has been the most well-studied for glass. For a bullet shot at a perpendicular pane of glass, the cone of the possible bullet trajectories begins at the glass and expands on the other side at a rate of about 0.5 to 2 cm (0.3 to 0.8 in) in diameter per 30 cm (12 in) of distance traveled. The cone of all possible fragments and glass dust is significantly wider, at about 10 cm (4 in) in diameter per 30 cm (12 in). When bullets pass through an angled barrier, they are deflected within a cone that is more perpendicular to the barrier, causing them to deviate from a straight trajectory. (See Image 64, Pg. 55.)Consequently, when shooting through a barrier, it is ideal to have a target be as close to the barrier as possible, and shoot as perpendicular to the glass as possible. (See Image 61, Pg. 55.)

Of course, the range of possible trajectories is only so wide because it must be comprehensive of all of the possible bullets, glass types, and glass thicknesses. When a marksman plans on shooting with a specific setup and through a specific barrier (including other material types), they can get far more precise predictions. For example, for every 30 cm (12 in) that a bullet may travel, it may deflect in a particular direction by 1 cm (.04 in), give or take ⅓ cm (.012 in) (i.e., the standard deviation may be only one-third of the mean prediction).

The most effective way to shoot "through" a barrier is to remove the barrier entirely. In fact, this is the only method that works when the target is far behind the barrier. In the case of glass, there are two common techniques for immediately removing the barrier. The first is to have two marksmen fire in quick succession; the first shot breaks the glass, removing the vast majority of the resistance for the second shot. The second technique is to attach an explosive, such as a blasting cap, to the glass and fire the blasting cap just before firing the rifle. (See Image 63, Pg. 55.) For either technique, eye protection is a necessity.

The second most effective method is to practice with the exact material one plans on shooting through. For example, elite military units may purchase and shoot the exact type of vehicle they plan to shoot through in an actual mission. And hunters may test the ability of their ammunition to shoot through brush and bramble while staying intact.

Advanced Effects

5. Trajectory Information — 59

Determining Bullet Trajectory — 60
Determining Bullet Impact — 64
Making Adjustments — 67
Trajectory Validation for Ballistic Calculators (Truing) — 68

6. Bullet Spin — 72

Barrel Twist — 73
Spin Stabilization (Precession) — 74
Bullet Orienting to Vector (Weathervaning, Tractability) — 79
Optimal Spin Rates — 80
Vector Orienting to Bullet (Lift Force) — 83
Precession-Induced 90-Degree Turning — 84
Aerodynamic (Muzzle) Jump — 86
Spin Drift — 89

7. More Air Effects — 92

Headwinds and Tailwinds (Parallel Winds) — 92
Transonic Range (Supersonic to Subsonic Shift) — 94

8. Acceleration on a Rotating Body — 101

Latitudinal (North-South) Coriolis Effect — 101
Longitudinal (East-West) Coriolis Effect — 104
Altitude (Up-Down) Eotvos Effect — 107

Advanced Effects, 600 Meters and Beyond

Problems worthy of attack prove their worth by fighting back.
—Piet Hein, Danish scientist and mathematician

A natural cutoff in shooting equipment occurs at various distances. For example, at about 300 m or yd, the naked human eye can no longer distinguish targets, making high-powered scopes necessary. A similar cutoff exists at 600 m or yd, where **tight part-tolerances** (i.e., low empty space between parts) become necessary to shoot accurately. These tight-tolerance parts are rigidly held in place by the other parts, and so enable more consistent shooting.

The reason loose-tolerance parts are used before 600 m or yd is because they make rifles more durable. The space between parts is where dirt and grime can be shoved when such a rifle fires, allowing that dirty weapon to continue to be mostly accurate. In contrast, a tight-tolerance weapon without that extra space available, may sustain internal damage upon firing as any excess dirt (especially quartz sand) is pushed through the chamber and bore. This principle explains why the maximum effective range of the standard-issue U.S. Army M4 is designed to only be 600 m. If the range were any longer, the rifle would be less reliable in harsh wartime environments.

There is another functional range-cutoff at 1000 m or yd, which is why shooting at that distance is called extended or **extreme long range shooting** (ELR). First and foremost, the cutoff is at "1000" because that is a nice round number. However more practically, this is the rough distance at which many **additional effects begin to significantly affect the trajectories** of bullets. These effects include parallel winds, transitioning to subsonic bullet speeds, spin drift, and the Coriolis effects.

5. Trajectory Information

As bullets travel farther distances, they must travel in more arced trajectories to hit their targets. In addition, there are more environmental factors that can alter a bullet's trajectory as it travels. Therefore, it becomes more difficult to predict where a bullet goes as it travels farther from the marksman. For a marksman to be as accurate as possible then, getting feedback on how a bullet travels after being fired is critical. Specifically, a marksman can use various methods to determine how a bullet travels, where it impacts, and how to incorporate that information into both their following shots and their ballistic calculators.

5.a Determining Bullet Trajectory

An observer may be able to see the location of a bullet as it travels using one of four methods: glint, tracer rounds, bullet trace (a.k.a. bullet wake), and vapor trail. (See Image 65, Pg. 61.) (Bullet trace is unrelated to "tracers," which are phosphorus bullets that light up.)

Glint is the most straightforward of the four; it is a reflection of sunshine on the bullet itself. Glint happens in rare instances when the sun shines on the bullet but not the surrounding area. For example, a 20-degree sun at sunset often may reflect off a bullet's copper jacket, producing a small, bright reflection that is visible to the naked eye. However, glint is not very helpful because it is so small and usually only appears in the part of the trajectory for which the sun is not blocked.

A "**tracer round**" is a type of ammunition that contains a small pyrotechnic charge that burns after being fired to create a visible trail of chemically produced light along the bullet's entire trajectory. This bright light allows anyone and everyone to easily track the bullet's trajectory. Modern marksmen never use these rounds because the enemy can see the origin of the tracer (i.e., the marksman's position), endangering the marksman to counterattack. However, historical marksmen would use tracer rounds to point out targets for other, stronger weapons such as tanks. And modern machine gunners use tracer rounds to see where their bullets are landing, since otherwise they may not be able to know.

In contrast to glint and tracer rounds, bullet trace and vapor tail are very useful indicators. They both appear for the entire length of a bullet's trajectory (under the right climate conditions) and do not give away nearly as much information to potential enemies.

Specifically, **bullet trace** is a form of mirage unrelated to tracer rounds. Mirage is a pattern of distorted light that occurs when light passes through pockets of air with different densities, bending the light in random directions. (How light travels through air depends on the air's density.) Thereby, bullet trace, like any other mirage, appears as a wavy pattern in the air.

Bullets create this wave of pressure and mirage in a cone shape as they travel and push air out of their way at supersonic speeds. This change in pressure, however, only changes the refractive index of the air (i.e., how air

Image 65: An F-35A Lightning II demonstration-team pilot and commander takes off. This jet shows "**trace**" (in the form of mirage), **vapor trail,** and **glint** in the same image. The mirage is created from the heated air coming out of the jet engine. While the mirage in bullet-trace is caused by mechanical force, it looks the same as this mirage. The vapor trail is formed at the tips of each wing and appears as white streaks or clouds. The glint appears on the top of the cockpit and at the top of the right wing as a shiny reflection. Hill Air Force Base, UT, 12 Feb 2020.

distorts light) very little, and so light has to travel through a lot of mirage for the bullet trace to be visible to a marksman.

Therefore, **bullet trace is best seen from directly behind the marksman** where the observer's line-of-sight aligns the most with the bullet's trajectory. (See Image 66, Pg. 63.) (See Image 67, Pg. 63.) Bullet trace is often impossible to see from even a small angle away from the bullet's trajectory because the amount of distorted air that light travels through decreases as the line-of-sight deviates from the trajectory of the bullet

There is no trace in many climate conditions. For example, wind erases trace. In fact, wind can sometimes push trace making it less reliable. High humidity and direct sunlight tend to create a more visible trace. Trace is easiest to see when the sun is not in front to minimize the amount of direct sunlight that can blind an observer. If the background is land and not air, trace is easier to see against a stationary, solid-color backdrop. When the marksman controls the targets, using stationary, solid-color targets helps minimize distractions and makes traces easier to detect.

Because trace is subtle, high-quality optical equipment is necessary; in fact, **marksmen often buy expensive optics specifically to see trace**. Similarly, shooting faster, larger, and generally less aerodynamic bullets causes a larger surrounding pressure wave and thereby a stronger bullet trace. In contrast, aerodynamic bullets are designed to minimize air resistance, and thereby the pressure wave. Therefore, marksmen must be aware that upgrading to a bullet with a higher ballistic coefficient may make trace more difficult to see.

If a spotter is looking for trace, they sit behind the marksman's rifle barrel by a few meters and align themself to the orientation of the rifle barrel as closely as they can. (See Image 66, Pg. 63.) That is, the spotter can be as far behind the marksman as they want. While sometimes a spotter may lay on top of a marksman to align themself as closely as possible to the marksman, this is only for concealment; simply sitting a short distance behind the marksman is sufficient to see trace and easily communicate. (See Image 67, Pg. 63.)

The best way to view trace is with a **low-magnification setting** to see a larger portion of trajectory of the bullet (usually between one third and two thirds of the total trajectory). To maximize the amount of trace seen, an observer can focus on a bullet's breakpoint. Subjectively, this is the point in the bullet's trajectory where the bullet begins to experience significant bullet-drop (i.e., just before the apex of the bullet's arc). Objectively, the breakpoint is the endpoint of the longest line that intersects the tunnel of air surrounding the bullet's trajectory. Focusing on the breakpoint ensures that the observer sees a large fraction of the trajectory where trace is clearly visible as close to the target as possible.

Seeing **the breakpoint and the target in one sight picture** is especially useful for determining whether a bullet is deviating too far to the right or left. In contrast, bullet impact is most useful for determining whether a bullet is impacting forwards or backwards. (See Determining Bullet Impact, Pg. 64.)

After positioning their optic, the marksman or spotter then relaxes their eyes and focuses their attention above the target, where a bullet would arc in the air. Bullet trace appears and disappears quickly, so the easiest way to see it is to have the trace appear where the marksman is already looking. Using a camera to record the trace is also helpful. Having a recording also helps an observer to determine the exact location of the trace of future bullets. It also avoids the problem of flinching and losing sight when the bullet is fired.

The fourth and final visual indicator is **vapor trails**. Although sometimes confused with bullet trace, a vapor trail is a rare and separate phenomenon. (Even rarer is a frost trail, which is a frozen vapor trail.) A vapor trail forms when a bullet displaces air, creating a low-pressure pocket behind it. In this

Spotter Position to see Bullet Trace

Image 66: 3rd Special Forces Group Soldiers engage targets with the M2010 rifle. To intersect the most with a bullet's trajectory, and therefore see the thickest trace, the spotter must be **vertically in line with the rifle**. Nellis Air Force Base, NV, 27 Aug 2019.

Image 67: Two infantrymen from the 101st Airborne Division conduct sniper training. In tactical settings, it is not uncommon to have the spotter sit or **lay on the marksman** to allow for non-verbal communication. Mielec, Poland, 09 Aug 2022.

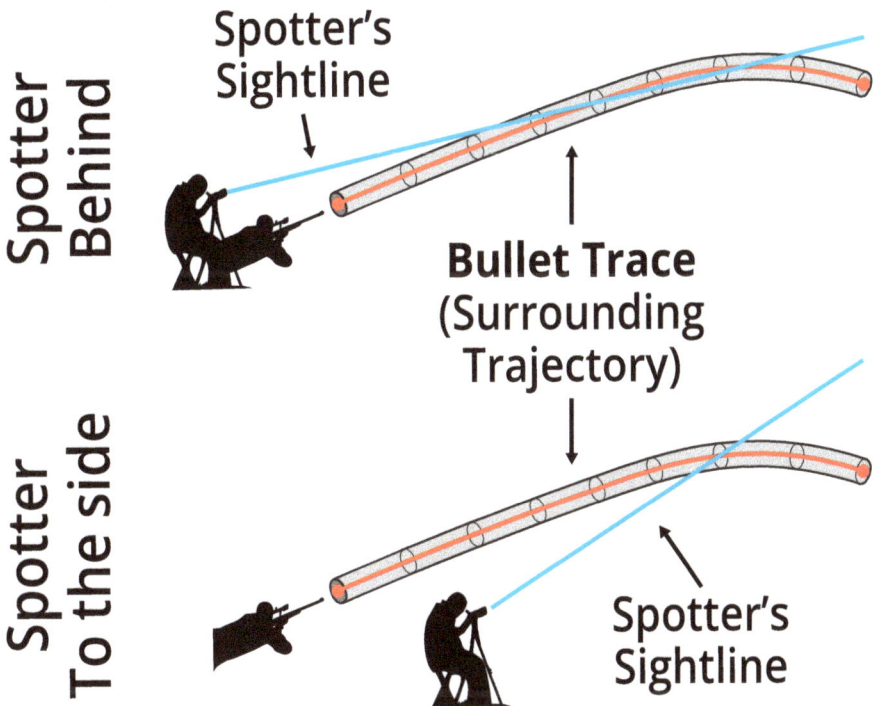

Spotter Behind

Spotter's Sightline

Bullet Trace (Surrounding Trajectory)

Spotter To the side

Spotter's Sightline

Image 68: Bullet trace appears as a rough cone that travels behind a bullet, which is caused by the bullet's supersonic pressure wave. (See Image 95, Pg. 95.) The cone follows the bullet's trajectory, forming a bent cylindrical path over time. The spotter **must intersect their sightline with that bent cylinder** to see the thickest section of the bullet trace possible. Therefore, they position themself behind the marksman's boreline and aim their sightline at the breakpoint of the trajectory. Even a position slightly to the side significantly reduces the intersection.

pocket, water vapor condenses into a **visible white cloud** because lower-pressure air cannot hold as much moisture. Consequently, vapor trails require very high humidity to develop. The resulting cloud can appear as a long line or as a dot trailing the bullet, since the water vapor can be quickly reabsorbed once the pressure returns to normal. Additionally, strong winds can prevent a vapor trail from forming altogether. Overall, the climate conditions necessary for vapor trails are very specific, and these trails can disappear as quickly as they appear.

Unlike glint or trace, a vapor trail following the bullet can be seen from any direction. In fact, since vapor trails appear as white dots or streaks, they do not require special skill, positioning, or training to see. That said, often the white dot is too small or moving too fast to be particularly useful to a marksman. They are also so rare (or at least climate-specific) that it is difficult to practice with them. Many marksmen never see a vapor trail because their bullets are too aerodynamic, small, or slow to form a low enough pressure behind them. Or they simply may never shoot in cold, high humidity weather.

Of course, there is debatably a fifth way of determining a bullet's trajectory, and that is if the bullet **hits something along the way**. When shooting in an area where overhead structures such as bridges or trees are present, there is always a risk of unintentionally hitting something. Therefore, marksmen must always consider and avoid the possibility that the highest point in a bullet's trajectory (i.e., the "maximum ordinate") intersects with some object. This can be an issue when using a scope at a high magnification. In that case, a marksman may only see their target and forget to look for potential obstacles above the sight picture, along the bullet's trajectory.

5.b Determining Bullet Impact

For long-range marksmen, missing some shots is inevitable. Therefore, a marksman must develop the skill of knowing where they missed so they can implement corrections quickly and accurately.

The best method for determining the location of a bullet's impact while shooting is to pay attention to the **area surrounding the target**. A bullet typically shows clear signs of where it impacts, which can be seen in the surrounding area. For example, when a bullet hits dirt, the impact typically throws a **characteristic cloud of dirt into the air.** (See Image 69, Pg. 65.) For any dust cloud, the lowest point of that cloud is the most likely impact point, since the cloud is always above the impact site.

Sometimes, dirt clouds can be very difficult to see. Therefore, one trick that many marksmen use is to upgrade to more powerful ammunition. In

Image 69: The dust cloud to the right is caused by a bullet impacting the berm. Because the berm is sloped, dust explodes upwards. Therefore, a marksman knows that **the bullet impacted at the base of the dust cloud**. After the dust enters the air, a marksman can use the dust to determine the prevailing winds. North West Islands, South Korea, 10 Oct 2017.

fact, one of the primary reasons that long-range marksmen use such powerful ammunition is not to get the bullet to the target (less powerful bullets can often travel sufficiently far and cost less), but instead to create larger dirt clouds when the bullet misses. Thereby, the marksman can more easily correct their shots. Another solution to better see dirt clouds is to set up cameras at the target site to see impacts where they happen.

It is important to note that visible signs of dirt displacement do not always indicate a missed shot. For example, when shooting at steel targets, the bullet fragments upon contact. Individual fragments then impact the ground and can cause dirt to kick up in a line across the target's base.

It can be especially difficult to determine whether a shot missed high or low. This is because depth perception does not work well at long distances. Many marksmen at longer ranges **set up their targets on the ground**, and do not elevate their targets like a closer-range marksman would. Thereby, if there is no sign of an impact, the impact could have likely occurred from a bullet falling over and behind the target, hiding the impact from observation. In sum, for a target set up on the ground, any impact below the target must be in front of it, and the bullet went too low.

In contrast, if the target is mounted high off the ground, a bullet may arc over the target and still allow the marksman to see the impact below the target's stilts. With the arc, the bullet would have landed behind the target.

However, because the bullet would be seen below the target, it would create the illusion that the bullet impacted in front of the target instead.

Because ground impacts are not always reliable or visible, marksmen must also look for additional clues. At close range, one sign of a missed shot would be a lack of a hole in a paper target. However, at longer ranges, marksmen use steel targets because they are durable and reusable, and also because they make a clear ring when they are hit. The **absence of that ring** would indicate a missed shot. That said, a lack of something (i.e., a hole or a ring) is not very specific to where the miss was.

When hunters and military shoot living targets, determining whether or not a shot hit or missed can be equally difficult. The response of a deer to a gunshot varies based on several factors, including the location of the impact on the body, the caliber and type of the bullet, the range and the angle of the shot, and the overall health status of the deer.

Observing the **immediate reactions of the deer** can provide insight into where the bullet has struck, thus guiding the subsequent follow-up actions to ensure the deer is dispatched humanely and swiftly. For example, when a deer collapses immediately and exhibits jerky or twitchy movements, it suggests that the bullet hit the key neurological areas of the head, high neck, or spine. If a deer falls immediately but attempts to sit up, it could be an indication that the bullet has damaged the spinal column at the saddle or upper haunch level. If the deer moves a short distance before collapsing, possibly in an uncoordinated manner or lunging forward from its back legs, they could have a chest, lungs, or heart shot. If the deer kicks out its back legs, stands with its body hunched, or walks off slowly with a stilted gait, it might signify a shot to the stomach, intestines, or liver. In this situation, the deer, if left undisturbed, might lie down after a brief period. Finally, a deer that initially collapses but immediately gets up and runs off, possibly indicates that the bullet hit the mid haunch, back leg, front leg, chest, outer neck, or head. These areas, while still painful, might not cause immediate incapacitation, allowing the deer to regain its footing and attempt to flee.

Enemy targets may exhibit all of the above symptoms as well. In addition, clothes may show visible blood splatter and staining. People are often near clean surfaces, so blood surrounding the target is another indication of a successful shot. Humans are also not as prone to running, and may collapse in an unnatural way as if they have lost their structural integrity.

Because there are so many different, unpredictable responses, marksmen must ensure that they **perform a proper follow-through** by reengaging their sight picture. Reengaging allows a marksman to remember and internalize where that target was when the shot was taken. In contrast, the most common response to shooting a deer is to immediately want to chase after the deer, immediately forgetting the location at which the deer was shot! This makes finding the start of the blood trail much more difficult.

5.c Making Adjustments

After a miss, marksmen must readjust their point-of-aim according to a series of steps. The first step is to ensure that the miss was not due to faulty equipment. This is rare, but important. The most common equipment fault is **a scope that is loose or has lost its zero** due to an impact or recoil. Minor equipment failures are extremely difficult to diagnose while shooting, but the marksman must always be aware of that possibility instead of mindlessly trying to adjust their point-of-aim for hours.

Second, the marksman must ensure that they themself are not creating errors. For example, when using a ballistic calculator, it is very easy to input the wrong information. Therefore, checking and double checking inputs can prevent incorrect outputs.

Regarding technique, the marksman must ensure that they are shooting consistently from shot to shot. Specifically, marksmen must know their body mechanics: stance, grip, breath control, and trigger squeeze. To become more consistent, marksmen get in the habit of calling their shots specifically to remember their technique at the moment they shot and to reflect on any possible errors or deviations they may have made at that time.

Third, the marksman must **update the information available to them**. This update mostly involves rereading the wind and reverifying the distance to the target to improve the windage hold and elevation hold respectively. Similarly, climate data changes throughout the day, so marksmen must regularly update their ballistic calculator every hour or so.

Even if the marksman is confident in their windage and elevation holds, if a shot is off to the left or right, something was incorrect; and if it wasn't the equipment, and it wasn't the data, it was likely the information. The **marksman listens to the bullet, the bullet doesn't listen to the marksman**.

If an error was found and an adjustment must be made, the marksman **must correct their information precisely**. While sometimes using educated guesses for an adjustment is the only option, even then, using and recording actual numbers allows a marksman to shoot better with future shots. For example, a marksman can better reference previous shots when they were recorded as "2.2 mils" as opposed to "kind of leftish."

When shooting at a target, any adjustments must be greater than the width of the target. Requiring a minimum width for adjustments is useful because **marksmen as a group habitually underestimate how much they miss by.** By setting the minimum adjustment to the target's width, marksmen are far less likely to miss twice on the same side. Marksmen are also never forced to overcorrect and miss on the other side of the target, since a distance equal to the target's width by definition cannot pass over the target. For example, if a shot impacts 50 cm (20 in) to the left of a 100 cm (39 in) target, then adjusting only 30 cm (12 in) (i.e., a small adjustment) misses again. Instead, the marksman must adjust by at least 100 cm (39 in) (i.e., the width of the target).

A larger adjustment also allows a marksman to **bracket their estimates**. For example, if a shot is overcorrected and misses to the opposite side of the target, then the marksman knows that the proper hold is somewhere between their first and second shot. For example, if a windage hold of 2.0 mils (7.0 MOA) misses to the right and 2.4 mils (8.4 MOA) misses to the left, the marksman can be confident that a windage call of 2.2 mils (7.7 MOA) would hit the target in the center. In contrast, missing twice to the same side due to undercorrection doesn't help the marksman much.

If all else fails, the marksman might consider that a random factor, such as a sudden gust of wind or a corroded cartridge, caused the miss. In this case, there's no immediate solution other than to try again. If another error occurs, it clearly indicates that something within the marksman's control is wrong. However, if no further errors happen, the issue is resolved.

Nonetheless, attempting to shoot again without making any changes is a last resort because it essentially means giving up. Sometimes, recognizing that nothing can be done is the most efficient course of action. However, each time this thought arises, the marksman must ensure they are putting maximum effort and skill into every shot. While dismissing errors as unavoidable might feel reassuring in the short term, it ultimately undermines skill development and progress.

5.d Trajectory Validation for Ballistic Calculators (Truing)

Ballistic calculators predict bullet trajectories using mathematical models that place a curve onto the bullet's zero point to represent its prediction of a bullet's trajectory. These calculators make their predictions by assuming average, ideal conditions. While these predicted trajectories can be highly accurate,

Truing

Image 70: This reticle is a Mildot reticle, so each dot is 1 mil (3.5 MOA) apart. The target is very close to 2.5 mils (8.75 MOA) tall. It is an E-Type target, all of which are 101.6 cm (40 in) tall or about 1 m tall. Therefore, it can be determined that the target is about 400 m (437 yd) away from the marksman. The red line (i.e., the truing line) is 0.5 mils or 20.3 cm (1.75 MOA or 8 in) tall. If the **truing line is hit using the trajectory predicted by the ballistic calculator**, then that predicted trajectory must be accurate within 0.5 mils. That is still a relatively large amount, and most precision long-range marksmen would use a narrower line or a farther placement. Camp Pendleton, CA, 30 Jan 2014.

Image 71: A zero point is where the bullet's point-of-aim matches its point of impact. Because only two points (the marksman's position and the zero point) are not enough to define a single curve, the bullet could theoretically follow infinite trajectories between them. **"Truing" adds a third reference point** to solve this problem. With three points, a ballistic calculator can produce a much more accurate and predictive trajectory curve. In fact, a marksman can repeat the truing process at multiple distances, adding as many points as needed to achieve the most precise curve possible.

real-world factors, such as non-ideal weather or different ammunition, can lead to **discrepancies that shift the actual trajectory forward or backward** from the calculator's predicted trajectory. For example, if a properly zeroed bullet travels faster than the model predicts, it would drop less and hit higher beyond the zero distance (i.e., a forward shifted trajectory). Conversely, if the bullet travels slower than expected, it would land lower than the calculator's projection (i.e., a backward shifted trajectory).

To ensure accurate modeling, marksmen must "true" their system. **"Truing" is to shift a ballistic calculator's predicted trajectory so that it matches the actual trajectory of the bullets a marksman is firing.** This matching is accomplished by giving the ballistic calculator an additional point of data: the truing distance.

The truing distance is essentially a second zero distance. For both truing and zeroing, the first step is to determine how far the marksman's point-of-aim is from the actual point-of-impact. For zeroing, this difference is dialed into the scope so that the point-of-aim intersects the point-of-impact. In contrast for truing, the vertical angular difference is input into the ballistic calculator as the true elevation hold without adjusting the scope.

Thereby, the ballistic calculator has three X-coordinates (i.e., the marksman's location, the zero distance, and the truing distance) and three Y-coordinates (i.e., zero bullet-drop, zero bullet-drop, and the bullet-drop at the truing distance) on which to base its curve. The program can generate a much more accurate predicted curve when it is basing the curve on two verified coordinates instead of just one (i.e., the zero point alone).

The truing process begins by setting up a truing target (a.k.a., a "truing bar"). Ideal truing targets are steel rectangles that are short, wide, and embedded in something. Ideal truing targets are steel because steel makes a loud ring when shot. They are short because they are meant to calibrate the elevation hold within a narrow vertical band. They are wide because the windage hold is not being calibrated, and so where the bullet impacts horizontally is irrelevant. They are embedded because the marksman is expected to miss (if they never missed, truing would be redundant), so being surrounded by a different material helps the marksman to identify whether their shot was high or low.

A good option is to embed two stacked railroad tracks in a dirt berm. When the bullet hits a track, it rings; but when it hits the berm it makes a dust cloud. (See Image 69, Pg. 65.) A much cheaper, single-use option is to paint a line onto a large cardboard target. (See Image 70, Pg. 69.) Then, each bullet hole can be tracked as a hit or a miss. However, without a metallic ring, truing on cardboard can become more difficult at farther distances. Therefore, the marksman may want to set up a remote camera that can live-transmit the area surrounding the truing target.

Because the ideal height of a truing target is an angular height (e.g., mils and MOA) and not linear height (e.g., cm and in), a truing bar at 1000 m or yd must be larger than a truing bar at 400 m or yd. A common truing target height is around ½ mil (1.7 MOA). Although that height may seem tall for a precision rifle, the truing target must be **at least as tall as the inherent dispersion or precision of the rifle system**. As a counterexample, if a target were only 5 cm (2 in) tall (⅛ mils, ½ MOA at 400 m or yd), and a rifle had 15 cm (6 in) of dispersion (⅜ mils, 1.5 MOA at 400 m or yd), then over half of perfectly aimed rounds would miss the target no matter what. On the other hand, the height of the truing target is the maximum accuracy of a trajectory that is derived from it; for example, a trajectory that is trued on a ½ mil (1.7 MOA) target can only be predictive up to a ½ mil (1.7 MOA) accuracy.

When deciding on how tall to make their truing target, a marksman can reference the advertised precision of their rifle and ammunition. For example, a truing target for an off-the-shelf hunting rifle would be much taller than a truing target for an expensive rifle with match-grade ammunition, which may advertise a dispersion of 0.2 mils (0.7 MOA). A rifle system can be trued on a target that is equal to its advertised precision.

Once the target is set up, **the marksman records the elevation hold** that is required to hit the target. To do this, the marksman first uses the elevation hold that their ballistic calculator predicts. If that shot hits, the predicted trajectory is fine. If it misses, the marksman adjusts their elevation hold up or down for their following shots. Once the correct elevation hold is found, the marksman inputs that hold into their calculator so that the calculator can update its predicted trajectory to be either shorter or longer.

Whether the shot is high or low on the target is usually caused by ammunition that respectively travels faster or slower compared to average ammunition. Therefore, **some marksmen use a chronograph** to directly measure their bullets' velocities and feed the average velocities into their ballistic calculators instead of elevation holds. Traditionally, this has been an inferior method because using velocity alone may not account for other environmental causes of high or low shots. However, as ballistic calculators get better at integrating data and designing models for specific bullets, inputting actual velocity is more and more becoming a viable way to true a ballistic calculator's trajectory.

For example, older ballistic calculators had a limited set of trajectories to start truing from. Because these models were highly generalized, they didn't predict the trajectory of any particular kind of bullet particularly well. Many only had two basic trajectory models named "G1" and "G7" (clarification: these are drag curves, and trajectories are based on drag curves). However, modern calculators use hundreds of custom curve models, which are predictive models customized to each type of round that has ever been widely manufactured.

Were a marksman to use the older G1 or G7 models and also want to true those models with a chronograph, the resulting trajectories would not be accurate everywhere since the G1 and G7 models wouldn't have the same drag curve (i.e., aerodynamic drag as a function of bullet velocity). This means that while the predicted trajectory could be accurate for a part of its range, it would always be inaccurate for other ranges.

Some calculators can accept **multiple holds or velocities** at different distances so that a marksman can correct (i.e., true) their model every few hundred meters. In this way the marksman can essentially transform any model into a custom curve model for their ammunition. This can be somewhat redundant to the custom curve models already designed by the ballistic calculator manufactures. However, truing at multiple distances becomes more important at extremely far distances (e.g., over 2 km or 1.5 mi) because errors in predicted values accrue exponentially with distance. And truing at multiple distances is the only way to precisely fit a generalized trajectory (e.g., G1 or G7) to a particular rifle system.

6. Bullet Spin

Bullets are cylinders in the back with a pointy cone facing into the wind. This shape minimizes the air resistance the bullet faces, as the tip slices through an incoming airflow. However, this shape also makes a bullet prone to randomly flipping over (i.e., "tumbling") because the shape has a center-of-mass that is behind its center-of-pressure (i.e., where the air acts on the bullet). This tumbling would make the bullet face more air resistance, and therefore slow down faster, than if it could maintain the tip facing forward.

There are two ways to prevent tumbling. First, many projectiles move the center-of-pressure behind the center-of-mass by adding fins to the rear of the projectile. Such projectiles include: arrows, missiles, and airplanes. The other way **to prevent tumbling is to spin a projectile very fast**. Bullets use this method. Interior grooves in a rifle barrel, called "rifling" forcefully spin

Barrel Twist

Image 72: This 105 mm L7 tank gun barrel (not a rifle barrel) was cut to show the **rifling** inside. Typical rifle barrels use far fewer grooves.

Image 73: **Bullets expand into the rifling** after they are fired (bottom) to both grip the rifling and to plug the barrel to stop gas from slipping past.

bullets as they travel forward through the bore. (See Image 72, Pg. 73.) Specifically, a bullet expands to fill into the rifling grooves, forcing it to follow the orientation of the grooves as it travels. (See Image 73, Pg. 73.)

Why exactly spinning causes the tip of a bullet to continue to point forward as it travels is explained in this section. It is not at all intuitive that spinning a bullet would somehow cause its tip to stay forward. Then, this section explains the side effects of spinning a bullet that cause the bullet to deviate from its initial trajectory, called "aerodynamic jump" and "spin drift."

6.a Barrel Twist

The first step in understanding spin is to understand how it is created. Inside every rifle barrel are grooves that span the length of the barrel while slightly rotating. (See Image 72, Pg. 73.) That is, the grooves are helical. The grooves (cutout section) and the lands (raised section) are together called "rifling" and are the origin of the word "rifle."

When a bullet is shot through a rifle barrel, the extreme force of the expanding gas causes the bullet to deform into the grooves, forming a barrier between the rifle chamber and the muzzle. Because the bullet expands into the grooves, the bullet follows the rotation of the grooves as it moves forward, causing the bullet to rotate as well. Therefore, the bullet makes exactly one spin for each length of distance that the rifling takes to make one rotation (a.k.a., one twist) in the bore. The bullet then continues to rotate at this rate of **one spin per each twist-distance** even after it has exited the bore. For example, if a rifling were to have one rotation in 25 cm (10 in), then a bullet shot through that rifling would continue to make one rotation every 25 cm (10 in).

In other words, bullets spin fast because they travel fast. And the twist rate locks in the relationship between spinning speed and traveling speed at the muzzle. (Notably, velocity degrades faster than spin, so the twist rate lowers as the bullet travels forward.) For example, if a bullet were to travel at 1000 m/s (1093 yd/s) with a 25 cm (10 in) twist, it would spin at 4000 rotations per second at the muzzle (1000 m ÷ 25 cm/rotation = 4000 rotations).

Twist rate is often expressed as a ratio in inches or millimeters, such as 1:10 (1 twist in 10 in) or 1:254 (1 twist in 254 mm). The unit of measure is usually inferred from the number, as twist rates for common firearms only exist in a small range. For example, almost all rifles sold in the United States are between 1 twist per 6 in to 1 per 20 in (i.e., 1:6 to 1:20). In contrast, metric twist rates have a larger ratio because millimeters are smaller than inches.

The twist rate has a nominal effect on muzzle velocity because some of the energy required to spin is taken from the energy propelling the bullet forward; however, the loss is less than 1% of the total. Therefore, changing the twist rate (and only the twist rate) does not significantly affect bullet velocity.

6.b Spin Stabilization (Precession)

Spinning is effective at maintaining an object's orientation because of a weird quirk of physics: when any object spins and an external force is applied to it, **that force acts at a right angle to the direction from which it was applied**, and not in the direct line of the applied force. (See Image 74, Pg. 75.) This is known as "gyroscopic precession," or "precession" for short.

A common example of this phenomenon is a spinning top. When a top is still, it falls over. But when it is spun, it stays upright. A bullet, like a top, resists changes to its orientation because of gyroscopic precession. When any force is applied to the side of a spinning bullet like trying to push a top over while it's spinning (in physics terms, the "overturning moment or torque"), it doesn't move in the direction one would expect (directly away from the push). Instead, the spinning bullet moves at a right angle to the applied force.

Much as tops wobble on uneven ground, bullets wobble in the air because they travel through air pockets of different densities. Therefore, bullets only point exactly forward on average, and do not necessarily point forward at any particular instant. To stay pointed forward on average, a statically stable bullet rotates (in physics terms, "precesses") in a circle around the exact direction of airflow. This precession is similar to how the handle of a spinning top slowly circles the vector of gravity. (See Image 80, Pg. 78.)

This kind of deflection is known as "**static stability**." Static stability causes the force trying to overturn the bullet (i.e., air resistance) to be deflected sideways so that the bullet maintains its near-forward orientation. The effectiveness of static stability is directly related to how fast the bullet is

Precession

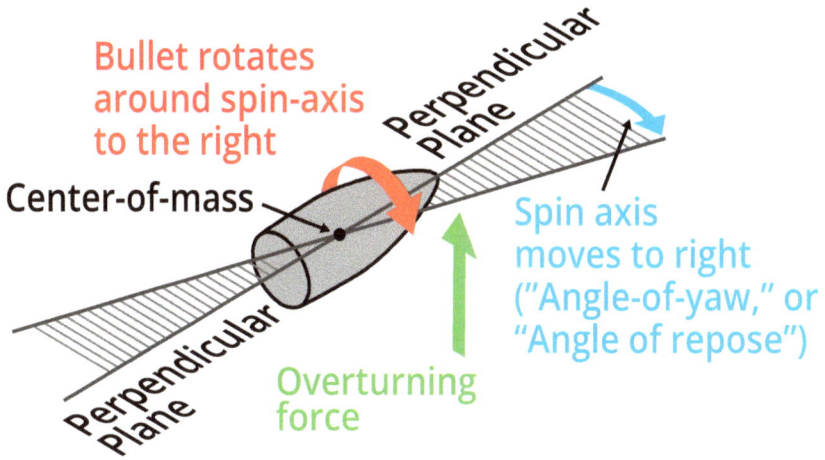

Bullet rotates around spin-axis to the right

Perpendicular Plane

Center-of-mass

Spin axis moves to right ("Angle-of-yaw," or "Angle of repose")

Perpendicular Plane

Overturning force

Image 74: Precession is a phenomena whereby when an overturning force (i.e., a force that causes a rotation perpendicular to the spin-axis) is applied to a spinning object, **the spin-axis of that object rotates 90 degrees** relative to the direction of applied force.

Yaw and pitch for a bullet relative to the direction-of-travel

1:15.24 cm (1:6 in) twist

1:22.86 cm (1:9 in) twist

Image 75: The spin-axis of a flying bullet deviates from the bullet's direction-of-travel by a small vertical angle, called the "pitch angle," and a small horizontal angle, called the "yaw angle." This is caused by an overturning force (air resistance) being diverted through precession from changing the pitch to changing the yaw. **The greater the angular velocity of the bullet's rotation (i.e., faster spinning caused by a shorter twist), the more overturning force is diverted** to affecting the yaw instead of the pitch. Thereby the pitch cannot reorient as quickly. At the initial stage of flight, the bullet's tip rotates in a dampening spiral caused by the initial yaw angle formed when the bullet leaves the rifling (i.e., lateral throwoff). After approximately 200 m or yd, the angles stabilize due to dynamic stability.

Spin-axis Alignment

Image 76: **Bullets perform just like spinning tops**. That is, both reorient to align with the vector of force (gravity for tops, and air resistance for bullets) by facing resistance to their spinning (surface friction for tops, and air resistance for bullets).

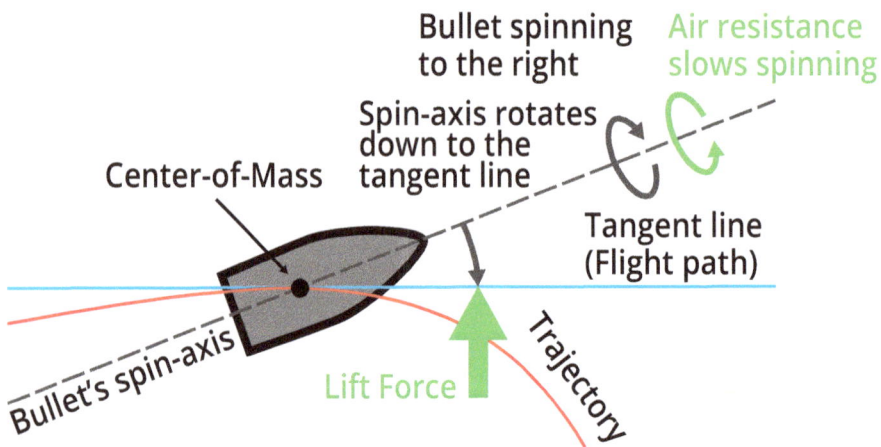

Bullet spinning to the right

Air resistance slows spinning

Spin-axis rotates down to the tangent line

Center-of-Mass

Tangent line (Flight path)

Bullet's spin-axis

Lift Force

Trajectory

Image 77: There's a balancing act between two types of friction on a spinning bullet. One is the lift force (See Image 83, Pg. 83.) that drives the bullet's precession around a circle (See Image 74, Pg. 75.); the other friction is the air friction that slows the bullet's spinning. The friction that slows the bullet's spin is what causes its spin-axis to align with the incoming airflow, whereas the friction resisting its precession causes the bullet to unalign.

spinning. However, the spinning speed cannot be increased indefinitely due to material limitations. For example, if a copper jacket were spun too fast, it would tear away from the rest of the bullet (i.e., mechanical failure due to centrifugal forces).

The **gyroscopic stability factor** (Sg) measures how fast a bullet must spin to maintain its orientation during flight (i.e., its static stability). By definition, an Sg <1 means that a bullet is not stable and cannot maintain any orientation. In contrast, an Sg ≥1 means, by definition, a bullet is spinning fast enough to maintain its orientation.

The specific Sg of a bullet depends on the air's density and the bullet's density, length, and shape. As a rule though, longer bullets require faster spin rates to stay stable than shorter bullets do. (In physics terms, the farther apart the center-of-mass and center-of-pressure are, the faster a bullet has to spin to be stable in a given airflow.) Because bullet length increases the ballistic coefficient (BC) (i.e., how aerodynamic a bullet is) but decreases the static stability, more aerodynamic bullets usually need more spin to remain stable.

That in turn means long, narrow bullets must be fired with a higher twist rate or a faster velocity. Increasing the twist rate directly increases the spin, and is a simple solution to keeping long, narrow bullets stable. However, increasing the velocity both increases the amount of spin (for a given twist rate) and also increases the amount of spin required for stability. In sum, a 3% increase in velocity may only result in a 1% increase in spin rate.

When determining the optimal twist rate or velocity for any given bullet, marksmen must be careful to measure or determine the Sg at sea level (or at least at a lower altitude than they intend to fire). Because thicker air decreases Sg, what may be a sufficient Sg at a higher altitude may become insufficient when shooting at a lower altitude. (See Image 100, Pg. 99.) For example, because the stability factor scales inversely with air density, and air density at the top of Mount Everest is 34.8% of the air density at sea level, a bullet with an Sg of 2.87 (1 ÷ 0.348) on top of the mountain would only have an Sg of 1 at sea level. Oddly enough, because of this fact, **marksmen at lower altitudes must use rifles with lower twist ratios**.

Determining the optimal twist rate to achieve a desired spin factor for a specific bullet and climate involves complex formulas. Today, manufacturers typically perform these calculations due to their complexity. These calculations take into account various factors, including bullet mass distribution, shape, spin dynamics over time, spin rate, behavior in transonic and subsonic velocity ranges, velocity, and air density. Additionally, manufacturers have access to empirical data that is not available to the general public. They are

Helical Spinning

**Straight
(False)**

**Helix or Spiral
(Stable True)**

**Tumbling
(Unstable True)**

Image 78: In animation and movies, bullets are usually depicted as traveling in a straight line. Actually, **bullets revolve about the trajectory as they travel**.

Image 79: This image depicts the angle at which an actual example bullet pointed for a given segment of its travel. (See Coning and Nutation, Pg. 118.)

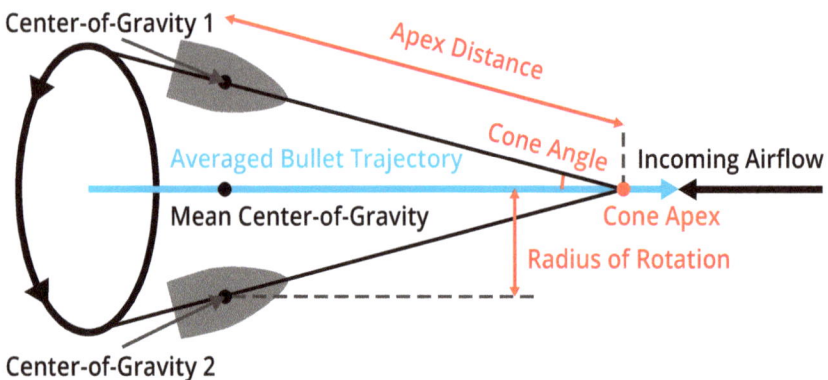

Image 80: A bullet is often depicted as spinning its tip more than its tail. The more likely configuration is that the tail rotates about a larger radius than the tip.

also better equipped to determine optimal twist rates for non-standard bullets, such as polymer-tipped or tungsten-cored varieties. As a result, conducting a simple online search usually provides more accurate information on suitable twist rates for different types of ammunition than any layman could hope to achieve on their own.

If a marksman already owns a rifle, they would need to research ammunition before purchasing it to ensure that it is compatible with their rifle's twist rate. If a marksman is buying a rifle before knowing what ammunition they would use in it, they are better off buying the lowest twist rate available so they can use the widest variety of ammunition. In contrast, if a marksman does know the ammunition they would use, they can look up the manufacturer-recommended twist rate for that ammunition. If the marksman plans to use multiple kinds of ammunition, they must choose the lowest of the various recommended twist rates.

To determine the twist rate of a rifle, a marksman begins by attaching enough cotton to the end of a cleaning rod to engage their rifle's rifling. The cleaning rod is then inserted a short distance into the bore. Using a marker, the marksman draws a line across both the rifle's crown and the cleaning rod, ensuring the lines are aligned. The rod is pushed downward until it completes one full rotation, causing the marked line on the rod to realign with the mark on the crown. The marksman marks the rod again at this point and withdraws it from the bore. The distance between the two markings on the cleaning rod represents the length of one full twist in the rifle's rifling.

6.c Bullet Orienting to Vector (Weathervaning, Tractability)

A statically stable object maintains the orientation of its spin-axis despite the application of an external force. A second kind of stability, **dynamic stability,** is the ability of an object to reorient its spin-axis to align with incoming force. The way that dynamic stability works is that the tip of a bullet precesses (i.e., rotates around the spin-axis) in increasingly small circles. In contrast, if a bullet is statically stable but not dynamically stable, then the bullet would precess in increasingly large circles until the Sg falls below 1 and the bullet tumbles. In other words, for a dynamically stable bullet, the angular difference (i.e., the angles of yaw and pitch) between the direction of incoming airflow (i.e., the bullet's direction-of-movement plus wind) and the orientation of its spin-axis becomes smaller with time.

Tops demonstrate dynamic stability by adjusting themselves to stand perfectly vertical, regardless of the initial direction of their spin-axis. Similarly, bullets reorient themselves to align with incoming airflow because they experience air resistance as an external force, much like how tops are affected by gravity.

The mechanism by which dynamic stability works is that the slowing of the spinning itself causes the bullet to reorient. (See *How Do Gyroscopes Lift Themselves Up?* from The Action Lab on Youtube.) Bullet's can continually lose spin while increasing their static stability at the same time because they lose velocity faster than they lose spin. A slower bullet faces less air resistance, and so can remain stable with less spin. In fact, a bullet's static stability at the point-of-impact can be more than 5, depending on how much a bullet has slowed down.

As a bullet travels in an arced trajectory, it goes up and then comes down. Therefore, the airflow it faces (i.e., the air resistance) comes from above and then from below, since airflow and air resistance occur in the opposite direction of a bullet's movement. A dynamically stable bullet always points its tip into incoming airflow because the spin-axis of a bullet goes through the bullet's tip. In that way, **a dynamically stable bullet is a weathervane**, pointing into incoming airflow. Because all modern bullets exit the muzzle with dynamic stability, bullets rotate their spin-axis forward as they travel through their trajectory.

Because the reorientation of the spin-axis is due to incoming airflow, crosswinds also affect where the spin-axis points. (See Aerodynamic (Muzzle) Jump, Pg. 86.) However, the effect of crosswinds on the spin-axis is minimal because wind speeds are typically much lower than bullet speeds. For example, if a 16 km/h (9.9 mi/h) crosswind were to combine with a perpendicular 1600 km/h (994.19 mi/h) incoming airflow (i.e., 100 times as fast), it would only alter the incoming airflow direction by 0.573°. In contrast, long-range bullets can regularly follow trajectories that arc almost 90°. Therefore, the most significant cause of reorientation is due to gravity changing the trajectories of bullets and thereby the air resistance they face.

6.d Optimal Spin Rates

While bullets can be dynamically stable with an Sg = 1 in theory, they are **far more likely to be stable with an Sg's of at least 1.5** at the muzzle. (See Image 96, Pg. 98.) This 1.5 number is somewhat arbitrary, and generally ensures that even if bullets exit the muzzle with a bit of misalignment, their dynamic stability can overcome it. (See Image 97, Pg. 98.)

Tractability

Image 81: This artillery shell (essentially a very large bullet) impacted the tree at a downward angle, exemplifying the **forward turn** that bullets make as they travel.

Tractable (Actual)   Overstabilized (False)

Image 82: Because a dynamically stable bullet always reorients into the incoming airflow, **it always reorients to its trajectory**. The myth of overstabilization comes from the concepts of gyroscopes, which resist reorientation. And bullets are gyroscopes. However, gyroscopes do not eliminate overturning force; they instead redirect it at a 90-degree angle into spin drift (See Spin Drift, Pg. 89.) or aerodynamic jump (See Aerodynamic (Muzzle) Jump, Pg. 86.). Spinning a bullet faster causes more of both, and makes bullets slower to turn forward (See Image 75, Pg. 75.), causing a negligibly higher drag coefficient that is accounted for with truing. (See Trajectory Validation for Ballistic Calculators (Truing), Pg. 68.)

Bullets are rarely intentionally spun faster than an Sg = 1.5 because additional spin cannot significantly get the tip pointed forward any faster and therefore cannot significantly increase the aerodynamics of the bullets. On the other hand, attempting a faster-than-necessary spin requires extra energy that could have been applied to the bullet's velocity instead. That said, bullets are regularly shot with Sg's > 1.5 when marksmen shoot one kind of ammunition out a rifle with a twist rate optimized for another kind of ammunition.

That said, **not much extra energy is required to spin a bullet faster**. As a practical example, a 5.56 mm (0.223 in) caliber rifle bullet experiences twice as many spins in a 1:152 mm (1:6 in) twist bore than in a 1:304 mm

(1:12 in) bore. When traveling 30 cm (12 in), the bullet's outer surface would travel 306.9 mm (12.08 in) with a 1:152 mm (1:6 in) twist, versus 305.4 mm (12.02 in) with a 1:152 mm (1:6 in) twist. That's a difference of about 0.5%. The radial midpoint of mass is about ⅓ of the radius below the surface; therefore, the average mass of the bullet travels less than 0.5% farther inside the bore when the twist rate is halved. That travel distance equates to less than 0.5% of the energy that is transferred from velocity to spin. For a bullet traveling at 800 m/s (2624 ft/s), doubling the twist rate causes a loss in muzzle velocity of about 2 m/s (6 ft/s). Empirical data roughly matches these numbers.

Excessive spin also carries two additional risks. First, a bullet's materials may experience such high stress due to the unnecessary rotational energy that the bullet catastrophically fails and bursts midair. Second, if mass were unevenly distributed in the bullet, then the center-of-mass would not be in the center of the bullet's shape. This discrepancy would cause a bullet to unstably rotate, or wobble, just like the buzz of a phone is caused by a motor spinning a lopsided weight. Increasing the spin would directly increase the wobble. That said, modern match-grade bullets are so well made and have such a small mass distribution error, that wobble is not a major concern.

Importantly, **over-stabilization is also not a concern**. The term "over-stabilization" refers to the idea that spinning a bullet too fast would cause it to maintain its the original orientation of its spin-axis, rather than reorienting the spin-axis to align with the incoming airflow and air resistance. In simpler terms, an over-stabilized bullet would spin so rapidly that its spin-axis would remain fixed in space. The farther this mythical overstabilized bullet were to travel along its curved trajectory, the more its tip would point away from the direction of travel, increasing air resistance and reducing efficiency.

But a bullet can never become fixed in space. This is because the reorientation comes from the loss of spin, and bullets that are spun faster have more spin to lose in the first place.

That said, **bullets that are spun faster are indeed slower to reorient** to the incoming airflow, causing increased drag. (See Image 75, Pg. 75.) This is because precession more effectively redirects the reorientation of the spin-axis 90 degrees, and not because reorientation is not occurring. (See Image 74, Pg. 75.)

But this **slower reorientation does not have a negative effect on accuracy nor precision** because the effect is predictable and repeatable, and so is accounted for by zeroing, ballistic calculators, and truing. In fact, while some experts claim that an Sg greater than 1.7 has is over-stabilized, other

Lift Force

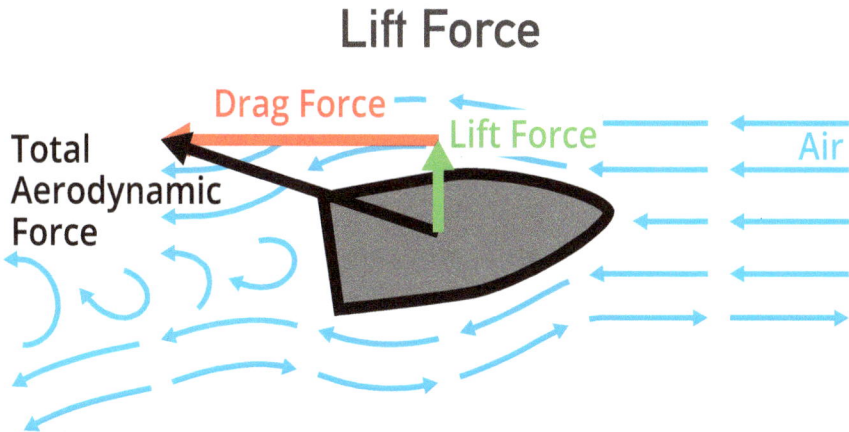

Image 83: When an object **impacts air at an angle**, one side of the object contacts more resistance than the other side. This has two effects. First, the air directly pushes against that side more. Second, the air slows down more on the side it contacts more. Bernoulli's principle states that an decrease in the speed of a fluid occurs simultaneously with an increase in pressure, also pushing that side more. The combination of these two effects (direct and indirect pressure) cause an object to accelerate into the direction that the object is pointing.

experts regularly shoot bullets with an Sg of 2.3 or more with no decrease in precision.

The misconception about over-stabilization likely stems from analogizing bullets to gyroscopes, which are specifically designed to resist reorientation. A spinning top is a better analogy because they experience significant friction on their tips which reorients their spin-axis. Tops, like bullets, would have their materials fly apart due to the rotational energy of spinning long before the spinning caused destabilization.

6.e Vector Orienting to Bullet (Lift Force)

When an object moves through the air, it faces air resistance, a force that opposes its motion. Air resistance consists of two main components: drag and lift. Drag acts in the direction opposite to the object's movement, slowing it down. **Lift acts perpendicular to the airflow,** causing the object to change its path. (See Image 83, Pg. 83.) For example, aircraft use engines or jets to produce thrust, which counteracts drag. By overcoming drag, the aircraft allows lift to continuously act on it, pushing it upward to balance gravity and keep the plane in the air.

In simple terms, lift is generated when air exerts more force on one side of an object than the other. For bullets, lift occurs when they are not pointed exactly forward. The reasons why air pushes more on one side are highly complex. In addition to standard aerodynamic principles like Bernoulli's principle, supersonic fluid dynamics also play a role. However, this book does not explore those details, as the simple explanation is enough.

When experiencing lift, a bullet changes its trajectory away from the side experiencing more force until the air resistance is equalized between both sides of the bullet. In this way, **bullets are pushed into the direction they are pointing**. Put another way, the lift force pushes bullets into the path of least resistance in the air. Because a bullet maintains its narrowest orientation in only two ways (i.e., tip-forward or tail-forward), and a spin-stabilized bullet is already tip-forward, its direction-of-travel adjusts until it matches whichever direction the tip points.

The degree to which lift deflects a bullet depends on how effectively the bullet interacts with the air to create a difference in force on each of its sides. This force difference is influenced by several factors. Higher air density provides more air for interaction, increasing the force difference and enhancing lift. A greater speed differential between the bullet and the surrounding air causes each air molecule to impart more force, resulting in increased lift. Additionally, long, narrow bullets with a higher surface-area-to-volume ratio interact more with the air, further amplifying lift. In contrast, bullets in a vacuum, those traveling at the same speed as the surrounding air, and spherical objects do not experience lift.

6.f Precession-Induced 90-Degree Turning

As previously mentioned, precession is when the spin-axis of a spinning object rotates 90 degrees from the direction of an applied rotational-force (i.e., a torque). (See Image 74, Pg. 75.) In ballistics, two main factors can cause such rotating forces on a bullet's spin axis: crosswinds and gravity.

A crosswind is a wind that blows perpendicular to the bullet's path. By blowing into the side of a bullet, a crosswind changes the airflow around the bullet, affecting the thicker tail (more surface area) more than the thinner tip (less surface area). As a result, the bullet experiences a sideways rotation, altering its spin-axis. (See Image 84, Pg. 85.)

Gravity has a more subtle effect. As a bullet travels, gravity causes it to follow a curved trajectory. From the bullet's perspective, the air appears to

Spin-axis Rotation and Precession

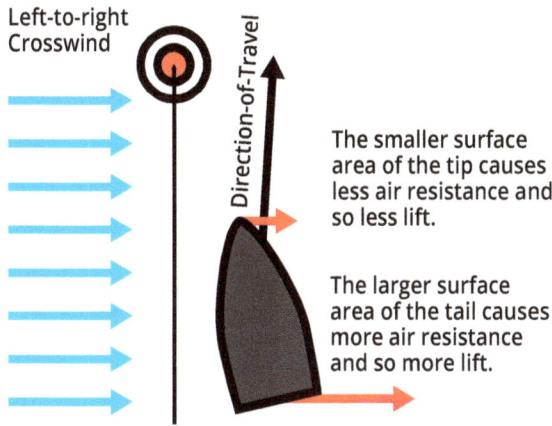

Image 84: The rear of a bullet has more surface area than the front and so the rear experiences **more force for the same pressure**, turning the bullet. Thereafter, precession redirects a portion of that force 90 degrees, pointing the bullet either more up or more down.

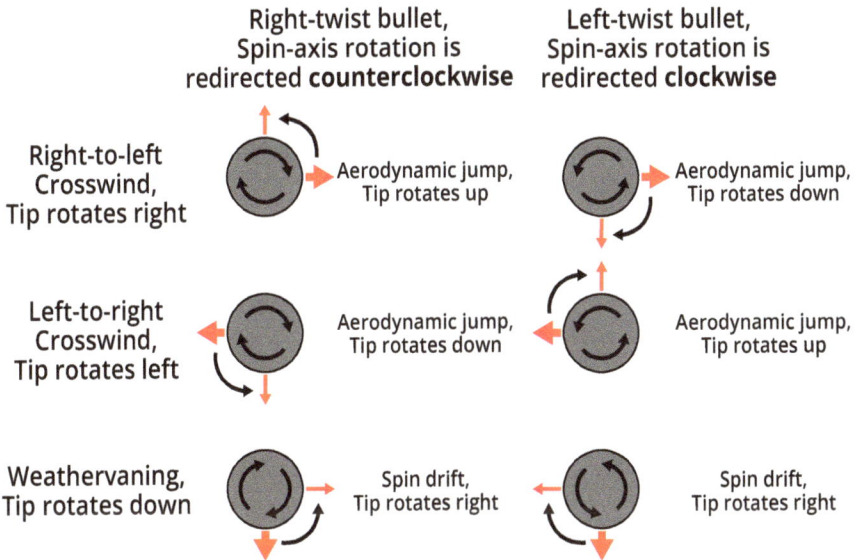

Image 85: This view is of the rear of a bullet. When a changing airflow applies a torque to a spinning bullet, that **torque is partially applied at a 90-degree angle**, resulting in either aerodynamic jump or spin drift. A right-twist bullet applies the torque 90 degrees counterclockwise (to the tip when viewed from the rear), while a left-twist bullet applies the torque 90 degrees clockwise.

accelerate upwards. This upward airflow increases the air pressure on the underside of the thicker tail compared to the tip, causing the spin axis to rotate forward.

When a crosswind or gravity applies a torque to a bullet, gyroscopic forces deflect part of that force to a 90-degree angle. Whether that 90 degrees is clockwise or counterclockwise depends on which direction the bullet is spinning. For a right-spinning bullet shot from a right-twist barrel, the deflection is to the right, and vice versa for a left-spinning bullet. (See Image 85, Pg. 85.)

The strength of the 90-degree torque is directly proportional to the strength of both the gyroscopic effect and the initial torque applied. The gyroscopic effect is determined by the bullet's spin rate (i.e., the gyroscopic stability factor). A faster-spinning bullet converts a larger percentage of applied torque 90 degrees and takes longer to realign with the incoming airflow. The aerodynamic force acting on the bullet depends on the wind strength or the steepness of the bullet's trajectory. The distance to the target can be relevant, but only insofar as steeper trajectories occur at farther distances.

So far, this explanation has described what makes a bullet point 90 degrees to the direction of an uneven application of pressure. However, it has yet to describe why a bullet that points in a particular direction would travel in that direction. **Bullets travel in the direction that they point because of lift.** (See Vector Orienting to Bullet (Lift Force), Pg. 83.)

An astute observer may notice that rotating the spin-axis 90 degrees would itself activate a second precession 90 degrees again (180 degrees in total). And this does in fact happen; however, precession only ever redirects a portion of an applied torque. Therefore, that second-order torque is small and masked by the stronger initial torque. In theory, there are infinite higher-order precessions of decreasing strength. However, because each successive iteration is only a fraction of the previous one, the net forces are only the initial reorientation of the spin-axis and the secondary 90-degree precession.

6.g Aerodynamic (Muzzle) Jump

Aerodynamic jump (a.k.a., muzzle jump) is the name for a precession-induced turn caused by a crosswind. It is called a "jump" because it occurs most strongly when the bullet leaves the muzzle and initially contacts and reorients to the wind.

Essentially, aerodynamic jump places a bullet on a higher or lower trajectory from the start (i.e., when it hits a wind). Therefore, a bullet experiences double the jump for double the crosswind. For example, a bullet

Aerodynamic Jump for a Right Twist

Image 86: At a given distance, aerodynamic jump occurs on a line. **A faster crosswind pushes and rotates a bullet more**; thereby precession redirects more of that rotation 90 degrees, allowing lift to push the bullet up or down more.

Image 87: **The slope of that line is proportional to the static stability**. Bullets with more spin, and thereby a higher gyroscopic (static) stability, redirect a larger portion of the crosswind's rotation 90 degrees, and so create a steeper slope.

Image 88: As a bullet travels farther and slows down, the crosswind becomes a greater proportion of the bullet's total air resistance, shifting the vector of incoming airflow. The **largest jump occurs as the bullet exits the muzzle**, when the crosswind first hits and spins the bullet.

Image 89: This diagram combines the previous two. The largest jump occurs when the bullet exits the muzzle and meets the crosswind. Thereafter the jump is proportional to the velocity of the bullet. And **a shorter twist causes greater static stability, which in turn causes more jump**.

Aerodynamic Jump 2

Aerodynamic jump in 16 k/h (10 m/h) wind

Image 90: This graph is generalized; the effect of aerodynamic jump is not truly a straight line. The purpose of this graph is rather to show that the effect of aerodynamic jump at even 1000 m or yd is only plus or minus a few cm or in.

that experiences 0.1 mils of jump with a 10 km/h or mi/h crosswind would experience 0.2 mils of jump in a 20 km/h or mi/h crosswind. Therefore, **aerodynamic jump for a weapon system can generally be defined at a mil deviation per 1 km/h or mi/h of effective crosswind.** Accounting for an aerodynamic jump can be especially important when truing a rifle at a far target in a strong crosswind.

For a right-spinning bullet, a right-to-left crosswind rotates a bullet down, and a left-to-right crosswind rotates a bullet up. (See Image 85, Pg. 85.) Therefore, a right-spinning bullet shot in different winds would form a straight line of bullet holes in a target from the top left through the center to the bottom right. (See Image 86, Pg. 87.) A mirrored line would appear for a left-spinning bullet or a vertical wind.

The angle of the aerodynamic-jump line depends on how fast the bullet is spinning (i.e., the percent of rotation translated 90 degrees), and how much lift the bullet experiences. (See Image 87, Pg. 87.) For example, shooting at 100 m or yd may create: an 18-degree line from a 35-cm (14-in) twist; a 19-degree line from a 30-cm (12-in) twist; or a 22-degree line from a 25-cm (10-in) twist. (See Image 87, Pg. 87.) This characteristic line is very familiar to precision marksmen at 100 to 200 m or yd when all other factors are accounted for.

The angle of the aerodynamic-jump line also depends on the distance to the target. (See Image 88, Pg. 87.) This is because the angle is determined by the ratio of the crosswind speed to the bullet's velocity, which reflects the direction of the incoming airflow. As bullets travel, they slow down. Consequently, a constant crosswind becomes a larger proportion of the incoming airflow, causing the bullet to rotate more. Increased torque results in additional aerodynamic jump. In other words, halving the bullet's velocity has the same effect as doubling the crosswind speed. Unfortunately, the relationship between the aerodynamic-jump line angle and distance to the target is not simple because bullet velocity does not decrease linearly.

This book uses the term "aerodynamic jump," but other sources might use different names for the same phenomenon. Additionally, some concepts like lateral throwoff are sometimes included under the umbrella of "aerodynamic jump," which can be confusing. Lateral throwoff is a separate phenomenon because it can occur in any direction, making its effect random rather than linear. Moreover, any non-random (predictable) lateral throwoff is virtually undetectable since it is already accounted for when zeroing the rifle.

In contrast, aerodynamic jump is significant because it can be consistently predicted using historical data and crosswind speed. Unlike lateral throwoff, which is random and varies in direction, aerodynamic jump follows a predictable pattern based on specific conditions, making it a more reliable factor to consider. (See Image 90, Pg. 88.)

6.h Spin Drift

Gravity causes bullets to experience a perpetually increasing updraft as gravity accelerates bullets to the ground. In other words, from the marksman's perspective, the air appears stable while the bullet's arc is curved; however, from the bullet's perspective, it remains stable itself while the incoming airflow takes on an increasingly upward velocity. The bullet always points slightly above the vector of incoming airflow as gravity constantly accelerates a bullet downwards, causing the direction of incoming airflow to constantly rotate. The result is a constant tiny overturning torque that tries to rotate the bullet's nose upward. For this reason, the act of the updraft caused by gravity pushing against the bullet's tip is sometimes referred to as the "overturning force."

Gyroscopic precession translates that torque on the spin-axis 90 degrees into a right or left rotation. (See Image 85, Pg. 85.) The resulting angle between the direction-of-movement and the spin-axis is called the "yaw-of-repose" or "angle-of-yaw." (See Image 74, Pg. 75.) A right-spinning and a left-spinning bullet would respectively have a right-facing and a left-facing angle-of-yaw.

Spin Drift

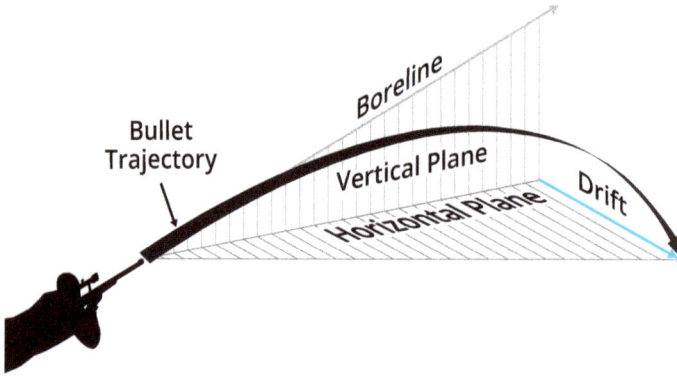

Image 91: Spin drift causes bullets to spin to the right for a right-twisted bullet and to the left for a left-twisted bullet. The drift is the induced **horizontal offset**.

Predicted spin drift for 5 bullets according to 3 programs at 3 distances

Spin drift in in (cm)

Target distance in yd
1000 yd = 914 m, 1500 yd = 1372 m, 2000 yd = 1829 m)

Bullets: 1) Sierra 168 Intl. · 2) 175 M118LR · 3) Sierra 220 SMK
4) Berger 215 Hybrid · 5) Hornady .338 285 BTHP

Prediction software: ■ B&R   ■ Hornady 4DOF   ■ Litz

Image 92: The amount of spin drift predicted by a ballistic calculator **depends on the formulas programmed** into that calculator. This graph shows predictions from three different softwares for five different bullets at three different distances. Clearly, the programs agree for the most part. At even 1000 m or yd, spin drift is still a minor effect, so the absolute differences between the programs only amount to a few cm or in. However as distance increases, the differences between the predictions become more pronounced and significant.

In sum, **spin drift uses the same principles as aerodynamic jump**, but with the shift in airflow being caused by falling down as opposed to by a crosswind. And just as with aerodynamic jump, once a bullet points in a new direction, lift causes it to go in that direction. (See Vector Orienting to Bullet (Lift Force), Pg. 83.) (See Image 90, Pg. 88.)

The amount of spin drift increases exponentially with distance. This exponential relationship occurs because spin drift is caused by gravity, which continuously accelerates bullets downward. As a result, the bullet's spin axis turns more and more over time, leading to an exponentially greater degree of spin drift the farther the bullet travels.

This complicated exponential relationship makes the calculation of spin drift mathematically complicated, and **many ballistic calculators are notoriously bad at calculating spin drift**. (See Image 91, Pg. 90.) In fact, many marksmen continue to turn off the spin drift calculation in their software because of that historical inaccuracy. Instead, these marksmen estimate that: spin drift is nominal below 500 m or yd; it equals between 1% and 2% of the elevation hold for slow and fast spinning bullets respectively at 1000 m or yd; and it increases exponentially thereafter. However, time has led to many technological advances and modern ballistic calculators are better at estimating spin drift than humans are.

The calculation of spin drift is made even more complicated by the fact that spin drift is influenced by many factors specific to each rifle system. For example, two identical cartridges fired from different rifles with different twist rates would experience different spin drift. And two bullets with different shapes fired from the same weapon system can also experience wildly different spin drift. At 1000 m or yd, a bullet can experience anywhere between 3 cm to 34 cm (1 in to 14 in) of spin drift, with most bullets experiencing between 20 cm and 30 cm (8 in and 12 in).

Another confounding factor is rifle cant. If a rifle were tilted to the right, a fired bullet would drift rightward. This cant-induced drift can be confused with spin drift because both result in significant sideways movement of a bullet at long ranges. However, the two types of drift can be distinguished. While both cant-induced drift and spin drift increase with respect to gravity, only spin drift also increases with respect to a change in a bullet's velocity. That means that only cant-induced drift appears at short distances. Therefore, when shooting at long ranges, marksmen must first perform a vertical target test (as explained in *Long Range Shooting: an Illustrated Manual*) at short range to ensure their scope is not canted.

The method for empirically measuring spin drift is rather ingenious. It involves using two identical rifles, with the only difference being that one has a left twist and the other a right twist. Both rifles fire the same ammunition simultaneously at a target. Since the twist direction is the only variable, any differences in where the bullets hit the target are solely due to spin drift. After firing many shots, the average distance between each pair of bullet impacts is calculated. This average distance is then divided by two to determine the spin drift for that specific weapon system under the given conditions. This approach effectively isolates and measures the effect of spin drift by accounting for all other variables.

7. More Air Effects

As a bullet travels through the air, several atmospheric factors can influence its trajectory. Crosswinds, which blow perpendicular to a bullet's trajectory, have an immediate and strong effect. Other factors, such as headwinds, tailwinds (winds that blow parallel to a bullet's path), and the transition from supersonic to subsonic speeds (i.e., the transonic range), take longer to affect a bullet. However, the effects of these slower-acting influences grow exponentially with distance, making them increasingly important to consider at very long ranges.

7.a Headwinds and Tailwinds (Parallel Winds)

Both headwinds and tailwinds blow parallel to a bullet's path. A headwind blows against the bullet's direction of travel, slowing it down. A tailwind blows from behind, pushing it forward. However, because bullets fly faster than any wind can blow, a bullet never truly experiences a "tailwind" from its own perspective. Instead, what feels like a tailwind to the marksman is simply a slower headwind to the bullet.

The faster that a parallel wind blows against the front of a bullet, the **more air resistance that the bullet experiences**. Therefore, stronger headwinds cause more air resistance. Conversely, tailwinds reduce air resistance, allowing a bullet to fly farther (given the same amount of initial energy). Notably, parallel winds only affect forward motion; a bullet falls to the Earth at roughly the same speed regardless of any headwind or tailwind.

Headwinds and tailwinds have a relatively small impact on a bullet's trajectory because these winds are minor compared to the bullet's forward

Headwinds and Tailwinds

Vertical change in point-of-impact (i.e., bullet-drop) versus distance a bullet travels due to a 25 k/h (10 m/h) wind incoming from 8 directions for a right-twist bullet

Image 93: Headwinds (blue, bottom, 0°) and tailwinds (red, top, 180°) do not affect the point-of-impact of a bullet until well after 1,000 m or yd. This effect can be seen in the diagram, where the red and blue lines do not diverge from the center until about 1100 m or yd for this particular ammunition. However, **the effect is exponential**, so it becomes vital to account for their effect thereafter. Crosswinds (0° and 270°) have a vertical effect due to aerodynamic jump (See Aerodynamic (Muzzle) Jump, Pg. 86.). The diagonal winds are a combination of the headwind-tailwind effect and the aerodynamic jump effect.

velocity. For instance, if a bullet travels at 1000 m/s against a headwind of 5 m/h, its combined airflow becomes 1005 m/s, meaning the wind is only 0.5% of the total. Even for a slower bullet traveling at 500 m/s, the same 5 m/s headwind still makes up just 1% of its airflow, and therefore just 1% of its total air resistance. Increased air resistance causes a bullet to drop faster than gravity alone would. Hence, headwinds and tailwinds only become significant when that 1-2% change in bullet-drop due to air resistance alone becomes large enough to matter.

Another way to frame the small effect of a parallel wind versus a crosswind is that **bullets are specifically designed to be aerodynamic** into an incoming airflow. That is, the shapes of bullets are designed to negate parallel winds.

To find a headwind or tailwind, a marksman must first measure the actual wind. Most of the time, winds are not perfectly parallel nor perfectly perpendicular; they instead approach from a diagonal angle with a crosswind component and a headwind-tailwind component. The crosswind component is frequently referred to as the wind value. In the same way, the parallel wind component can be labeled as either the headwind value or the tailwind value. Whereas the crosswind value is the cosine of the angle of the actual wind from the 12 o'clock direction, the parallel wind is instead the sine of that angle. Notably, because wind values are determined trigonometrically, the crosswind value and the parallel wind value sum to a number greater than one.

Unlike crosswinds, parallel winds are never calculated manually. Marksmen simply enter the wind's direction and velocity into a ballistic calculator, which automatically factors in any headwind or tailwind. Although a marksman could theoretically calculate parallel wind effects by hand, doing so for shots beyond 1 km or 1 mi would take so long that the wind would likely have shifted anyway.

7.b Transonic Range (Supersonic to Subsonic Shift)

Sound travels through air at a specific speed, commonly referred to as "Mach 1." The number after "Mach" indicates how many times the speed of sound an object is traveling, so Mach 2 is twice the speed of sound, and Mach 0.5 is half. Exactly what Mach 1 translates to in km/h or mi/h depends on environmental conditions. For example, sound travels faster in warmer air because air molecules move and transmit energy more quickly. It also moves faster at sea level, where air pressure is higher, since the molecules are closer together and interact more readily than at higher altitudes. Mach 1 at sea level and at 20°C (68°F) is **approximately 343 m/s (1125 ft/s)**, or about 1235 k/h (767 mi/h).

An object traveling slower than Mach 1 is called "subsonic," and an object traveling at or faster than Mach 1 is called "supersonic." The interactions of a subsonic object and a supersonic object with air are fundamentally different. This is because supersonic objects experience a strong form of drag called "wave drag." To understand wave drag, first, the fundamental cause of sound must be explained.

All sounds are made by the vibrations of molecules. For example, when a person hits a drum or a cymbal, the object deforms. Specifically, when an object is struck and depresses, a pocket of low air pressure is created, and

Transconic Range

Transonic Range

Bullets experience this graph backwards as they slow down

Drag Coefficient

0.50
0.45
0.40
0.35
0.30
0.25
0.20
0.15
0.10

0.5   1   1.5   2   2.5   3

Velocity in Mach
(Mach is the multiple of the speed of sound)

Image 94: This graph shows an example **relationship between the velocity of a bullet and its drag coefficient**. A higher drag coefficient means that an object experiences more resistance for a given airflow (e.g., a cube), and a lower coefficient means less resistance (e.g., a plane). The coefficient depends not only on an object's shape but also on its velocity. In both the subsonic and the supersonic range, the relationship shown here is relatively linear. However around Mach = 1, there is a sudden drop. This S-shaped section is called the "transonic range" and spans from Mach = 0.8 to Mach = 1.2 (because the range is defined by the complexity of the curve, the range is approximate). **The transonic range can be substantially different for different bullets** because of their different shapes, and so generalizing one curve to all bullets causes inaccurate results.

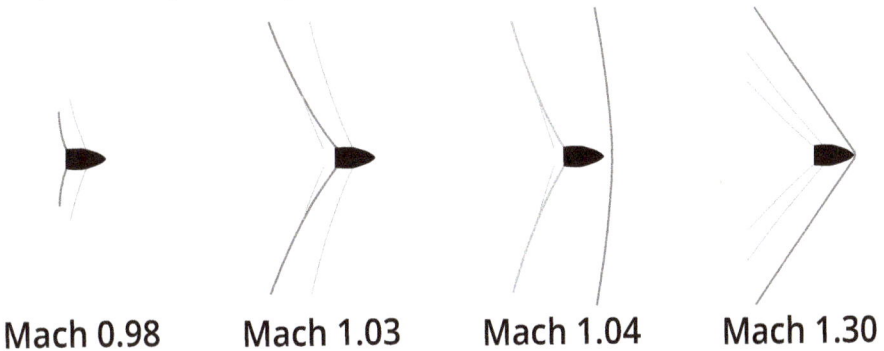

Mach 0.98   Mach 1.03   Mach 1.04   Mach 1.30

Image 95: The darker line represents **the pressure wave of supersonic air**. Due to complicated fluid dynamics, air surrounding an object moves at different speeds in different locations around the object. That means that in the transonic range, some air is supersonic while other air is subsonic.

when the object rebounds, a pocket of high air pressure is created. When objects depress and rebound in quick succession, that is called "vibration." When a bullet travels through the air, it creates an area of high pressure in front of it and a pocket of low pressure behind it.

These pockets or areas of high and low pressure travel outwards, away from their origin, and have the secondary effect of vibrating whatever they come into contact with. For example, when these pressure pockets vibrate an eardrum, a person hears those vibrations as "sound." Therefore, the speed of sound is the speed at which compressions and decompressions can travel through air (or any other material).

An object is considered supersonic when it travels faster than the speed at which pressure waves (or pockets of air pressure) can move outward. At these high speeds, **the pressure waves build up into shock waves** rather than dispersing. Because the pressure cannot spread out as it does at lower, subsonic speeds, it becomes more concentrated, increasing the drag on the object. This extra drag is known as "wave drag." It is clearly observed when a cone-shaped shock wave forms around a supersonic object. The sudden release of this shock wave creates a "sonic boom," which signifies a sharp jump in drag.

Although all long-range ammunition starts out traveling faster than the speed of sound, bullets can sometimes slow down to subsonic speeds before reaching their target. The transition from supersonic to subsonic does not happen abruptly right at Mach 1. Instead, it occurs gradually because **the air surrounding the bullet moves at different speeds and densities at various locations on the bullet**. For example, the air striking the bullet's nose might slow to subsonic speeds, while air flowing around the sides can still accelerate to supersonic speeds. This mixed-speed zone is called the "transonic range," and it generally spans bullet velocities between Mach 0.8 and Mach 1.2. (See Image 94, Pg. 95.)

The transonic range is often associated with a decrease in the accuracy and stability of bullets; however, this perception is only partly true. Air does not interact with bullets in a novel way that it does not in the subsonic or supersonic ranges. Instead, as the bullet moves from supersonic to subsonic speeds, the airflow around it becomes more complex, causing rapid shifts in the drag coefficient and other aerodynamic properties. Also, external forces have more time to act on a bullet per unit distance traveled because transonic bullets travel slower than supersonic bullets. Therefore, while the trajectories of transonic bullets are more difficult to predict, their trajectories are not inherently more random or chaotic. In physics, "chaos" means a small change in initial conditions can lead to a large change in outcome, and so the opposite of chaos is repeatability. The transonic range demonstrates this repeatability: **when two identical bullets are fired under the same conditions, they generally follow the same transonic trajectory**.

In fact, bullets experience a slower loss of velocity once they transition into the subsonic range, where they are traveling below the speed of sound. This occurs because the rate at which a bullet decelerates is tied directly to its drag, and drag itself is largely a function of velocity. When a bullet is moving at higher, supersonic speeds, it encounters a greater amount of drag; as a result, it slows down more rapidly. However, once the bullet's velocity drops below the speed of sound, the drag forces become less intense, causing the bullet to decelerate at a slower rate. Therefore, bullets in the subsonic range maintain their speeds more effectively over a given distance compared to bullets traveling faster. For example, a bullet may lose more absolute velocity in the first 100 m or yd of supersonic flight than in 500 m or yd of subsonic flight.

The reputation of transonic flight as being less accurate likely stems from the inability to generalize the transonic curve of one type of bullet to another kind of bullet. More specifically, when drag (Y-axis) is plotted against bullet velocity (X-axis), the resulting "drag curve" (See Image 94, Pg. 95.) shows how much air resistance a bullet experiences at any given speed. Even small differences in bullet geometry can greatly alter this curve, particularly in the chaotic transonic range. Therefore, any standardized drag model, such as G7 and G1 are relatively inaccurate and unhelpful in the transonic range.

A solution to this lack of generalizability is **custom drag models**, where each kind of bullet starts with a drag curve that is based on that bullet's particular geometry. However, even custom drag models are not perfect. They do account for how varying climate conditions can also affect a drag curve. For example, because the speed of sound increases when temperature increases, bullets enter and exit the transonic range sooner in their trajectories (i.e., at a higher velocity) in hotter climates than in colder ones. Therefore, truing is still necessary. (See Determining Bullet Trajectory, Pg. 60.)

A marksman can improve their performance in the transonic range by decreasing the effect of air resistance in the first place. Bullets with a high ballistic coefficient (i.e., low drag) are long and narrow, allowing them to slice through the air more efficiently than blunt bullets and have a smaller sonic boom. Further, because air is less compressed for a narrow bullet and it is less disruptive than a blunt bullet, a narrow bullet design has a

Static versus Dynamic Stability Factors

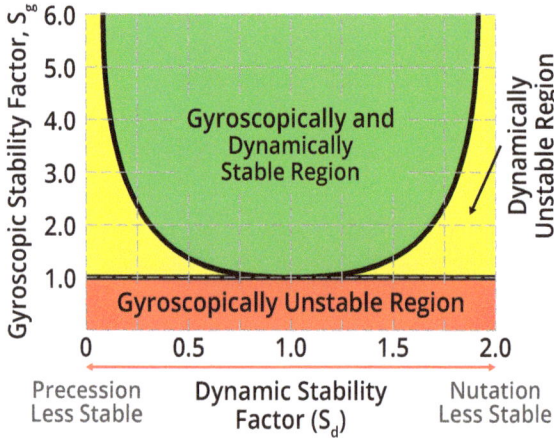

Image 96: **Gyroscopic (static) stability** is the ability of a bullet to have its tip point forward on average. By definition, a tip points forward on average over time when the static stability factor is above 1. Bullets that spin faster at a given velocity have higher static stability than those that spin slower.

Dynamic stability is the ability of the tip of a bullet to align to the incoming airflow. In contrast, a dynamically unstable (but still statically stable) bullet's tip rotates in larger and larger circles around the axis of travel as it moves forward. The dynamic stability factor determines whether a bullet is dynamically stable or not. A change in bullet spin does not have a large effect on dynamic stability.

This graph shows **three regions**. In the green, middle region, bullets are both statically and dynamically stable. Therefore, they travel with a constant, minimal drag. In the red, bottom region, bullets are not statically stable and so they tumble. In the yellow, side region, bullets are not tumbling, but their orientation is constantly changing. Drag depends on orientation. (Bullets are only aerodynamic if they face directly forward.) Therefore, **a changing orientation complexly increases drag and eventually leads to tumbling.**

Image 97: The above relationship between stability factors is why most marksmen use twist rates that generate at least a static stability factor of at least 1.5. A marginally stable bullet with Sg < 1.5 can **easily become dynamically unstable**.

Falling Velocity's Effect on Stabilities

Image 98: The tapering line is the bullet's falling velocity. Going transonic causes **a sharp decrease** in the DS factor. Once the velocity line crosses the frontier line, the bullet loses dynamic stability.

Image 99: The frontier line is a parabola, so increasing static stability does allow the DS factor to decrease more before dynamic stability is lost. But increasing static stability has **exponentially decreasing returns**.

Image 100: Decreasing air density lowers the transonic threshold speed.

Image 101: Different bullets can have **vastly different stability curves**.

narrower transonic range. Therefore, narrower bullets retain their velocity for a longer time.

However, once narrower bullets do become unstable, they tumble far sooner. Longer bullets catch more air on their sides (per unit weight) if their tips deviate from pointing forward. In more technical terms, narrower bullets are more susceptible to dynamic instability because the center-of-pressure moves forward through the transonic range, increasing the overturning torque applied to the bullet. Dynamic instability means that a bullet's tip still points forward on average; however at any instant, the tip is slowly diverging from the bullet's direction of travel (i.e., angle-of-yaw increases). Dynamically unstable bullets eventually tumble, which massively increases air resistance and decreases predictability. Therefore, **long bullets better retain velocity but short bullets better retain stability**.

For example, when comparing a short 250-grain match bullet with a longer, 300-grain bullet, the 300-grain bullet performs better in the supersonic range. Moreover, the 300-grain bullet experiences a slower loss of velocity and slows to the transonic range at a longer distance. However, the shorter 250-grain bullet has a higher stability and endures the transonic range better than the longer, heavier bullet. Another example is that a boat tail of greater than 10 degrees makes a bullet more aerodynamic, but also narrower and less stable in the transonic range. This tradeoff between better supersonic performance and better transonic performance is a common consideration when choosing the optimal bullet for long-range shooting.

Another method to counter the extreme drag of the transonic range is to simply spin a bullet faster. Faster gyroscopic stabilization can overcome the lower dynamic stability inherent to longer bullets traveling in the transonic region. However, spinning bullets has diminishing returns. Increasing the twist rate by 10% may potentially sustain stability for perhaps an extra 5% of distance or less. (See Image 99, Pg. 99.) In fact, increasing the rifle twist rate for extremely lengthy bullets may yield no results at all.

Again, the exact relationship between velocity and stability varies greatly depending on the bullet because the effects of the transonic range are so sensitive to minute change. (See Image 101, Pg. 99.) If transonic information for a specific rifle system cannot be obtained from a manufacturer nor from a ballistic calculator, **trial-and-error** usually yields the best results, even if it is tedious to do.

8. Acceleration on a Rotating Body

The Coriolis effect is a pseudo-force that causes bullets to drift to the right in the Northern Hemisphere and to the left in the Southern Hemisphere. This effect becomes more pronounced when a marksman is farther from the equator and closer to either pole.

It is considered a pseudo-force rather than a true force because the apparent drift results from two interacting factors: first, the Earth's surface and everything on it are constantly rotating; and second, a bullet takes time to reach its target. Due to these factors, the target that a bullet was fired towards moves from its initial position by the time the bullet arrives. This movement makes the bullet appear deflected from where it was originally aimed at. In reality, the bullet is not being forced to drift; instead, the target moves out of position during the bullet's flight.

Although a simplified mathematical explanation is provided below, marksmen do not perform complex Coriolis calculations on the shooting range. Instead, they simply plug geographic data into a ballistic calculator to get the correct outputs. Calculators are necessary because the Coriolis effect is so small and complex. For example, depending on the rifle and ammunition, the Coriolis effect only causes about a 10-cm (4-in) deflection at a distance of 1000 m or yd. (That said, as the distance to the target increases, the effect grows exponentially since it scales linearly with both the distance to the target and the bullet's flight time.)

8.a Latitudinal (North-South) Coriolis Effect

Latitude is a geographic coordinate that indicates a point's North-South position on Earth's surface. When a bullet is fired toward either the North or South Pole, it experiences an eastward deflection. Conversely, when a bullet is fired toward the equator, it deflects to the West. This phenomenon means that bullets in the Northern Hemisphere are deflected to the right, while those in the Southern Hemisphere are deflected to the left.

The reason why bullets deflect when fired across latitudes is because the bullet becomes physically closer to Earth's spin-axis. The distance from the spin-axis is important because every part of the Earth completes one rotation in the same amount of time: one day. To achieve this uniform rotation period, the equator must spin much faster than the poles. This is necessary because objects located farther from the spin axis must travel a larger circular path compared to those closer to the axis. Specifically, the rotational speed at the equator is about 464 m/s (1522 ft/s), whereas the rotational speed at the poles is nearly zero.

A bullet fired from one latitude to another latitude retains the rotational speed of its original latitude circle. In other words, due to the conservation of angular momentum, the bullet continues to rotate around the Earth at the same speed regardless of when the Earth's rotation becomes faster (closer to the equator) or slower (closer to the poles). (See Image 101, Pg. 99.)

This is why shooting towards the poles (i.e., into a smaller rotation circle) deflects to the East: because the bullet would rotate faster than the surface underneath it. Similarly, shooting towards the equator (i.e., into a larger rotation circle) deflects to the West because the bullet would rotate slower than the ground underneath it. (Moving East or West is by definition to have a different rotational speed compared to the Earth's surface.)

When imagining a change in latitude, it is important to remember that bullets are fired tangentially to the Earth's surface. (See Image 103, Pg. 105.) So for example, if a marksman in Portland, Oregon (located at ~45° latitude) were to fire a bullet North or South, it would travel at about a 45° angle towards the North Pole or the equator respectively. Therefore, the Coriolis effect is more powerful as marksmen get closer to the poles, and the surface of the Earth is more perpendicular to the Earth's spin-axis.

Coriolis drift is calculated with the formula:

▸ Coriolis Drift = $\omega \times \sin(\text{Latitude}) \times (D^2 \div V)$. (Technically, this is only valid when vertical motion is negligible, and calculators may use more complicated three-dimensional formulas, but this is sufficient for demonstrative purposes.)

▸ ω represents the Earth's rotation rate, which is 0.00007292 radians/second (rad/s). There are 2π (~6.28) radians in a circle. (A radian is a unit of angular measure defined by the ratio of an arc length to its radius. Since both the arc length and the radius are measured in the same unit, their units cancel each other out. As a result, a radian is considered a dimensionless quantity.)

▸ Latitude is measured in degrees from the equator, with positive values indicating positions in the Northern Hemisphere and negative values in the Southern Hemisphere. By applying the sine function to the latitude, the component of Earth's rotation that affects an object at a given latitude can be determined. This mathematical relationship allows for the calculation of how Earth's rotational speed influences objects differently depending on their North-South position on the planet.

Latitudinal Coriolis Effect

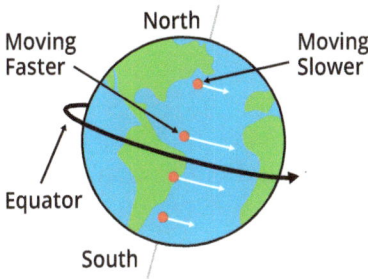

Objects closer to the equator move slower than objects closer to the equator

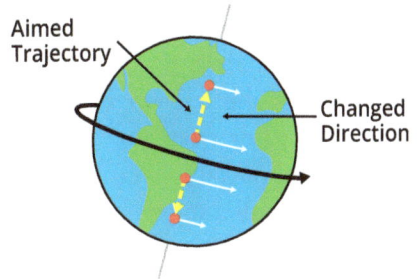

A Bullet is fired at a target at a slower-moving location.

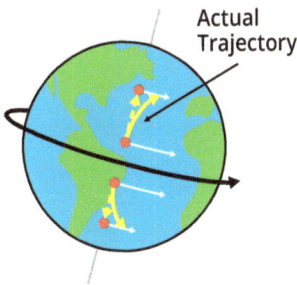

$$\text{Ground Velocity} = \text{Difference in Latitude velocities}$$

The object retains its initial, faster eastward velocity when it arrives at its final destination, farther from the equator.

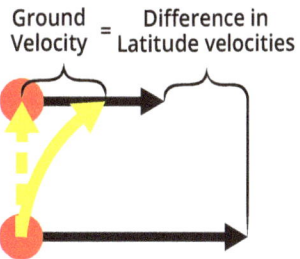

Its final ground speed is equal to difference between its intial rotational velocity minus its final rotational velocity.

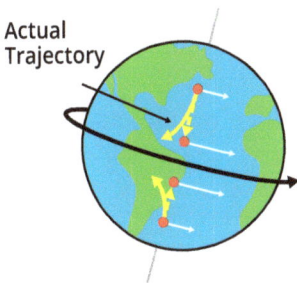

$$\text{Ground Velocity} = \text{Difference in Latitude velocities}$$

The object retains its initial, slower eastward velocity when it arrives at its final destination, closer to the equator.

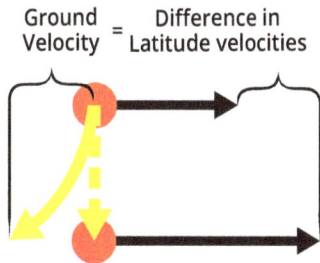

Its final ground speed is equal to difference between its intial rotational velocity minus its final rotational velocity.

Image 102: When a bullet is fired in the North-South direction, the Coriolis effect occurs because bullets retain the rotational velocity they have at their original location when they arrive at their new location at a different latitude.

▸ D represents the distance that the bullet travels. Both the distance and the bullet's velocity must use the same linear units (here, meters). Because the distance to the target is a squared variable, the effect quickly increases at longer distances.

▸ V is the bullet's average velocity from start to finish. (The Coriolis effect, which arises from the conservation of angular momentum, causes the bullet to deviate from its initial path. Importantly, the actual trajectory that the bullet follows does not impact the calculation of this deviation.)

▸ (A simplification of (D² / V) is (distance × flight time); however, using (D² / V) better illustrates why the Coriolis effect is only significant at longer ranges.)

Consider a bullet fired at Anchorage, Alaska (61.22° N) to a target 1000 m due North, at a speed of 1000 m/s.

▸ Total Coriolis drift = ω × sin(Latitude) × (D² ÷ V)

▸ 0.00007292 rad/s × sin (61.22°) × (1000 m)² ÷ 1000 m/s

▸ 0.00007292 rad/s × 0.876475 × 1000 m × 1 s

▸ 0.06391 m ≈ 6.4 cm.

The bullet would experience about 6.4 cm of Coriolis effect.

8.b Longitudinal (East-West) Coriolis Effect

A longitude is a geographic coordinate that specifies the East-West position of a point on Earth's surface. When a bullet is fired eastward, it deflects toward the equator. Conversely, when fired westward, it deflects toward the North Pole in the Northern Hemisphere or toward the South Pole in the Southern Hemisphere. In other words, bullets in the Northern Hemisphere are deflected to the right, while those in the Southern Hemisphere are deflected to the left.

In fact, the equation for longitudinal Coriolis effect and latitudinal Coriolis effect are identical:

▸ Coriolis Drift = ω × sin(Latitude) × (D² ÷ V).

This can be reinforced with a thought experiment. At the poles, East-West shooting merges into North-South shooting because there is neither an East nor a West at the poles.

The East-West Coriolis effect arises for a more subtle reason than its North-South counterpart. When a bullet is fired at a target situated on the same latitude as its origin (i.e., traveling across longitudes), it does not remain on that latitude for the entire flight (unless it is at the equator). Because

Longitudinal Coriolis Effect 1

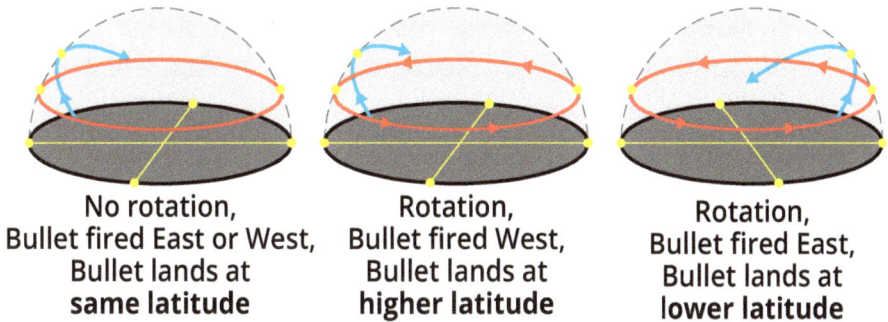

No rotation,
Bullet fired East or West,
Bullet lands at
same latitude

Rotation,
Bullet fired West,
Bullet lands at
higher latitude

Rotation,
Bullet fired East,
Bullet lands at
lower latitude

Image 103: Bullet trajectories follow the curve of the Earth, and not lines of longitude. Therefore, for a marksman to hit a target on the same latitude as themselves, their bullet must travel through latitudes closer to the poles before falling back down to the original latitude. **Firing a bullet westward reduces the centrifugal force applied to the bullet**, preventing it from fully returning to its original latitude. In contrast, firing a bullet eastward increases the amount of centrifugal force experienced, propelling the bullet closer to the equator. This diagram shows the same effect as the following diagram, but in a more three-dimensional way. (See Image 104, Pg. 106.)

---

bullets must counteract gravity to travel, they follow an arc that is tangential to Earth's surface at their location. As a result, in the Northern Hemisphere, a bullet's path shifts into latitudes closer to the North Pole before returning to the latitude of the target roughly halfway through its flight; in the Southern Hemisphere, it moves toward latitudes closer to the South Pole and then falls back to the target's latitude. (See Image 103, Pg. 105.)

When a bullet is fired eastward, the Earth's rotation adds to its angular momentum, effectively pushing it outward and increasing the centrifugal force it experiences. As a result, the bullet tends to overshoot its intended latitude, dropping through additional latitudes and ultimately ending up closer to the equator.

Similarly, **firing westward opposes and cancels out the rotation of the Earth**, causing the bullet to not fall as far as it needs to return to its original latitude. Therefore, it contacts the Earth earlier and remains at a higher latitude. (See Image 104, Pg. 106.)

These two examples assume that the stopping point, target, or destination is a location on the ground that is visible to the marksman. This assumption is important because if a westward-fired bullet is fired fast enough, it could enter an orbit and continue to a lower latitude than the original target is on (albeit at a lower velocity than an identically fired eastward-bound bullet).

Longitudinal Coriolis Effect 2

Image 104: Bullets are fired perpendicular to the ground. In contrast, lines of latitude are perpendicular to the Earth's spin-axis. Therefore, a fired bullet must traverse latitudes to go from the muzzle to the target. In the Northern Hemisphere, such a bullet briefly moves to a higher latitude before dropping back down.

On a rotating Earth, a bullet already has velocity before being fired. Whether the bullet is fired westward or eastward affects how its velocity combines with the Earth's rotation. A bullet that is fired westward is already moving away from the target's location, and a bullet fired eastward is already moving towards the target's location. This causes a westward bullet to have a slower velocity relative to the ground; therefore, it remains at a higher latitude instead of falling down to its original latitude. In contrast, an eastward-fired bullet has the bullet's velocity and the rotational velocity added together, so it overshoots its original latitude to end at a lower latitude than it was at originally.

Because the Earth rotates to the East, a bullet in the Northern Hemisphere always impacts to the right of the target, and a bullet in the Southern Hemisphere impacts to the left.

There are some other minor causes of the Coriolis effect. For example, the Earth is not a perfect sphere, and is actually 43 km (27 m) wider around the equator than around the poles. However, that is negligible for the Coriolis effect as it applies to bullets.

8.c Altitude (Up-Down) Eotvos Effect

Previously, the North-South and East-West Coriolis effects were discussed. They both describe how objects behave as they move on the Earth's surface. However, there is a third Coriolis effect, called the Eotvos effect, that describes how objects move in the third dimension: towards and away from the Earth's surface.

In sum, **the Eotvos effect causes a bullet fired to the East to land higher on a target than the initial point-of-aim, and a bullet fired to the West to land lower than the initial point-of-aim.** The Eotvos effect occurs in ballistics because the Earth rotates eastward during a bullet's flight time. (If a bullet were fired westward and instantly arrived, it would impact exactly on the point-of-aim since the target would have no time to advance into the bullet.)

As a bullet flies eastward, its target rotates away from the bullet; and as a bullet flies westward, the target rotates towards the bullet. It helps to imagine the bullet as flying in a straight line like a laser. Rotating into a straight line leads to a lower point-of-impact, whereas rotating away from the straight light raises the point-of-impact.

Another way to understand this phenomenon is by considering the bullet's perspective in terms of rotational speed. The bullet, when at rest, rotates in sync with Earth's rotation and thus experiences centrifugal force. Firing the bullet eastward increases its rotational speed and the associated centrifugal force, causing the bullet to gain additional centrifugal force. Conversely, firing it westward reduces its effective rotational speed and centrifugal force, allowing gravity to pull the bullet downward more forcefully. (See Image 105, Pg. 109.)

In contrast to the two other Coriolis effects, the Eotvos effect is strongest at the equator and is zero at the poles. However, much like the other Coriolis effects, the Eotvos effect is not very strong. There is a centrifugal force equal to only 0.34% of the force of gravity at the equator. Therefore, even eliminating the centrifugal force entirely doesn't have a large effect.

The formula for the Eotvos effect is:

- Eotvos effect = $((Distance\text{-}to\text{-}target) \times (Time\text{-}of\text{-}flight)) \times$
  $(\omega \times \cos(Latitude) \times \sin(Azimuth))$.

Because time-of-flight can be difficult to find, the formula can be rearranged.

- $(Distance\text{-}to\text{-}target) \times (Time\text{-}of\text{-}flight) = ((V \times 2 \times (Bullet\text{-}Drop)) \div g)$
- Eotvos effect = $((V \times 2 \times (Bullet\text{-}Drop)) \div g) \times$
  $(\omega \times \cos(Latitude) \times \sin(Azimuth))$.

- V is the bullet's average velocity.
- *(Bullet-Drop)* is the expected distance a bullet vertically drops at the target distance. This number can be found by firing North or South, where there is no Eotvos effect.
- g is the acceleration due to gravity (9.8 m/s^2).
- ω represents the Earth's rotation rate (0.00007292 radians/second).
- *(Latitude)* is degrees from the equator.
- *(Azimuth)* is the direction in which the bullet is fired, clockwise from true North.

So for example, if a bullet were to travel 1000 m in 1 s, at 0 degrees latitude (i.e., the equator), directly West (270 degrees), the formula with units omitted would be:

- Eotvos effect = $(1000 \times 1) \times$
  $(0.00007292 \times \cos(0) \times \sin(270))$
  $\approx$ -0.073 m $\approx$ 7 cm below expected.

Using the second formula, assuming a drop of 4.9 m (16 ft) and a velocity of 1000 m/s at 0 degrees latitude (i.e., the equator) and directly West (270 degrees), the formula with units omitted would be:

- Eotvos effect = $((2 \times 4.9 \times 1000) \div 9.8) \times$
  $(0.00007292 \times \cos(0) \times \sin(270))$
  $\approx$ -0.073 m $\approx$ 7 cm below expected.

For this example, a bullet would drop $\sim$7 cm (2.8 in) more than would be expected without the Eotvos effect. Using the same numbers except shooting East would result in 7 fewer centimeters of drop, since sin(90) is positive whereas sin(270) is negative. Therefore for these numbers, there is a $\sim$14 cm (5.5 in) differential in drop when shooting East compared to West.

Altitude Eotvos Effect

Image 105: The origin of this graph represents the baseline centrifugal force an object experiences on Earth's surface when at rest. Adding eastward or westward velocity to an object changes the amount of centrifugal force that object experiences.

The Eotvos effect is closely tied to the longitudinal Coriolis effect. As a bullet travels westward, the velocity of the shot **effectively cancels out** a portion of Earth's eastward rotation. At the equator, where Earth spins at 465 m/s (1526 ft/s), firing a bullet westward at that same speed would reduce the bullet's angular velocity to zero, thereby eliminating its centrifugal force. If a bullet is fired even faster in the westward direction, a counter-rotation begins, reintroducing a different centrifugal force. Because the Eotvos effect depends on Earth's rotational speed, the effect diminishes with increasing latitude, where local rotational speed is lower.

De Minimis Contents

9. Local Gravity   111

10. Thermal Effects   113

11. Magnus Force   116

12. Coning and Nutation   118

13. Additional Effects   118

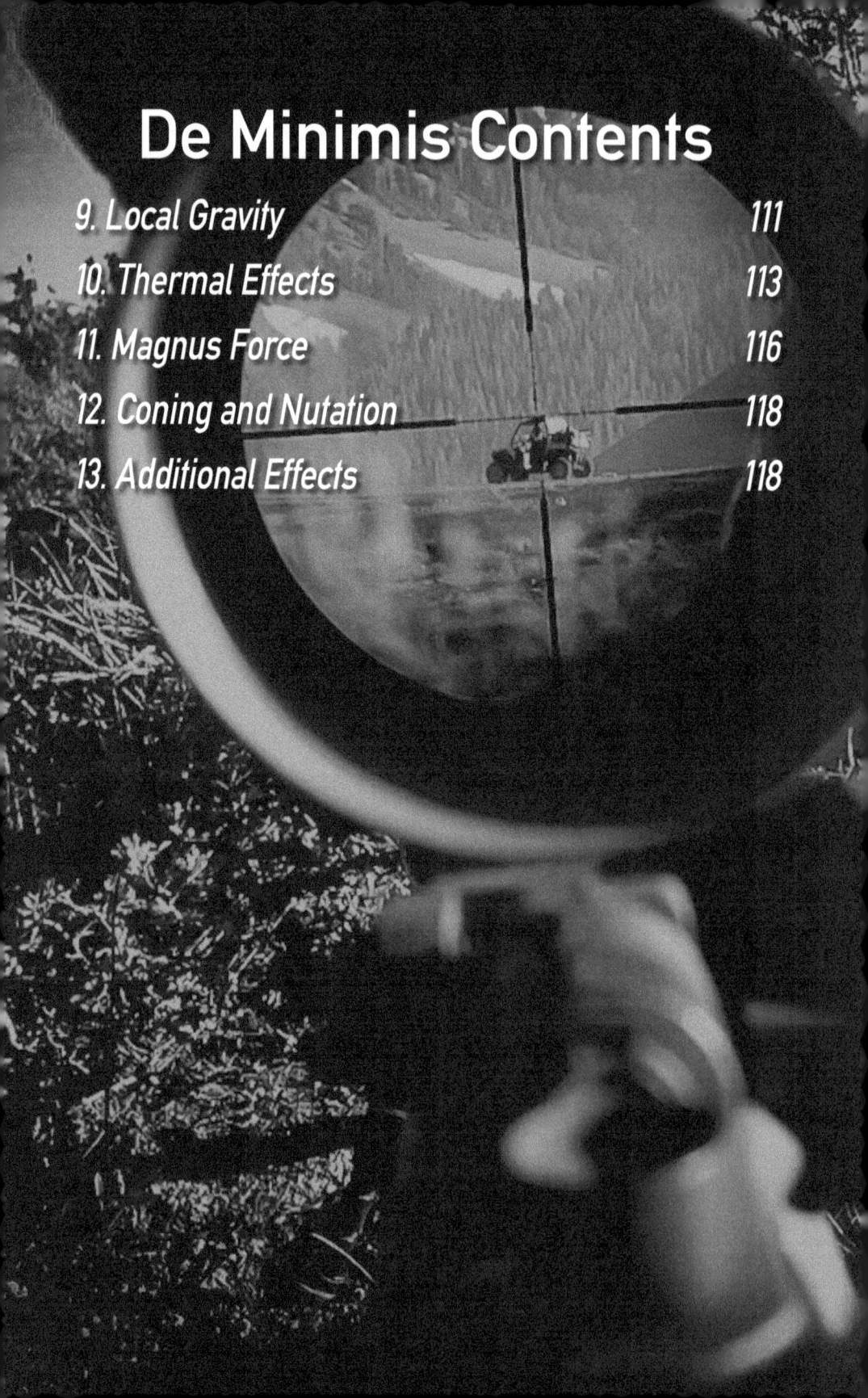

De Minimis Effects

De minimis non curat praetor. ("A praetor is unconcerned with trifles.")
—A legal principle dating back to at least the 15th century

"De minimis" means too trivial or minor to merit consideration. Many ballistic effects exist that simply do not affect the trajectory of bullets enough that they are worth calculating with modern technology. These de-minimis effects can be ignored because humans cannot physically hold rifles still enough or with enough precision to adjust for these effects.

Even if a bullet were fired from a robotic setup, the theoretical marginal benefit would be extremely slight to the point that it would be an **inefficient use of time** to code a program or input the data. This time inefficiency exists in large part because many of these effects are accounted for by other processes. For example, almost all effects of local gravity are accounted for by simply re-zeroing and re-truing in the local area in which the marksman plans to shoot. In fact, these effects are so small that attempting to account for them could introduce additional errors if the program overcorrects.

Finally, even if a robot did exist that could accurately account for these effects, the result would not be to definitively change any bullet's trajectory from a miss to a hit. A bullet's trajectory at the distance these effects matter passes through kilometers or miles of randomized wind that would unpredictably alter the trajectory anyway. Therefore, accounting for these effects would only increase the probability of a hit (by a microscopic amount).

In sum, perfectly accounting for these effects would result in a bullet being ever so slightly more likely to be on target in the same way that buying a lottery ticket technically makes one more likely to win the lottery. That is, **it would be technically true but a waste of time.**

9. Local Gravity

Gravity is not the same everywhere on Earth. The acceleration due to Earth's gravity on its surface varies from 9.76 to 9.83 m/s^2 (32.02 to 32.25 ft/s^2), which amounts to a range of about 1%. This difference exists for a few reasons. For example, gravity is about 0.3% stronger at the poles than at the equator due to the influence of centrifugal force. (The equator rotates at about 460 m/s (1509 ft/s), while the poles rotate at 0 m/s (0 ft/s.)) That is, as the Earth rotates, it throws everything on its surface into space, counteracting a small percentage of gravity.

Gravity is also less at the equator because the Earth is not a perfect sphere. The same centrifugal force squishes the Earth out at the equator, and so objects at the equator are actually farther from the Earth's center-of-mass than objects at the poles. These mass effects can get very complicated because the ellipsoid shape of the Earth adds mass at the equator and removes mass from the poles. In brief, the shape of the Earth can cause objects at the poles to be 0.6% heavier at the equator.

Further, local terrain can affect gravity in a variety of ways. For example, gravity differs depending on the local altitude. For example, the difference in gravity from the top of Mount Everest to sea level is about 0.2%. Gravity also varies depending on the density of local material by about 0.1%. This fact was first proven by the Schiehallion experiment, an 18th-century experiment that was performed to determine the mean density of the Earth. The experiment measured how much the gravity of a nearby mountain could deflect a tool sideways. In sum, gravity is more intense near expansive areas of rock, such as mountain ranges. (See Image 106, Pg. 113.)

In order of the largest effect to smallest effect, with a spherical, homogeneous earth as a hypothetical base gravity, the variations from that base gravity and their scale would be:

- 10^{-1}% flattening, centrifugal acceleration
- 10^{-2}% mountains, valleys, ocean ridges, subduction
- 10^{-3}% density variations in crust and mantle
- 10^{-4}% salt domes, sediment basins, ores
- 10^{-5}% tides, atmospheric pressure
- 10^{-6}% temporal variations: oceans, hydrology
- 10^{-7}% ocean topography, polar motion
- 10^{-8}% general relativity

Gravity data for the United States is readily available from the National Geodetic Survey, a United States federal agency, through their Gravity for the Redefinition of the American Vertical Datum (GRAV-D) program. This data is useful, for example, in determining where water flows because the exact gravity of an area can better predict where water flows than altitude changes can predict alone.

The reason that local gravity is not more important is because, while across the globe gravity may differ by about 1%, rarely are marksmen zeroing targets at one extreme (e.g., the Hudson Bay) and then shooting at another extreme (e.g., Mount Everest). Realistically, gravity only varies by 0.1% or

Image 106: This "gravity anomaly" map shows where Earth's gravity field **differs from a simplified model** that assumes the Earth is perfectly smooth and featureless. Areas colored yellow, orange, red, and brown are areas where the actual gravity field is larger than the featureless-Earth model predicts (e.g., the Himalayan Mountains in Central Asia). In contrast, the other shades indicate places where the gravity field is less (e.g., the area around Hudson Bay in Canada). This map was derived from NASA's Gravity Recovery and Climate Experiment (GRACE) Gravity Model 05. The color scale spans from −80 to +80 milligals (mGal), where one milligal is one thousandth of Earth's average surface gravity.

less at common shooting locations within a nation. Furthermore, marksmen typically zero and true in the area they are planning to shoot, eliminating even more of the difference.

10. Thermal Effects

Temperature is the measure of the average amount of heat in an area. Heat is how much energy atoms have. As atoms in a material increase or decrease their heat (i.e., energy), their physical properties change. For example, most materials expand as they heat up and contract as they cool down. These phenomena are called "thermal expansion" and "thermal contraction" respectively. Another change is in elasticity: as materials heat up, they usually stretch and deform more easily. In contrast, as materials cool down, they become more rigid. This phenomenon is called "thermoelasticity."

Thermal expansion is a significant problem (i.e., not de minimis) when an object contains multiple materials, such as a steel barrel attached to a stock (i.e., not freefloating), or wood in general (wood is composed of cellulose, lignin, and hemicellulose, and contains water). The effect of thermal expansion on rifle components has been well-documented. There was a famous instance of thermal expansion causing significant problems in 2015, in which the U.S. Government sued the company L-3 (EOTech's former parent company) for civil fraud. The U.S. accused L-3 of covering up defects in their rifle sights that the company knew about as early as 2006. One such defect was "thermal drift," which shifted the aiming points of the sights by as much as 30 cm (12 in) at 274 m (300 yd) as the sights heated or cooled.

It is exactly because of how well-known thermal expansion is, that manufacturers put in great effort to eliminate its negative effects. For example, almost all extreme long-range rifles do not use wood, and many favor metal chassis (a.k.a., stocks). These rifles also almost all have free floating barrels that prevent the barrel from touching the forestock.

The **de minimis effects are those that occur in single-material components**. For example, thermal expansion has long been suspected of causing issues in barrels. The hypothesis is that changing the physical properties of the barrel's metal alters how bullets travel through the bore and thereby affects a bullet's trajectory. This is borne out in the attention given to "cold-bore" shots, which is the first shot taken with a rifle. The classic thinking is that the first shot is especially inaccurate because the rifle barrel is the coldest, and therefore the smallest and most rigid, it will ever be during a shooting session.

However, there is no compelling evidence that the "cold-bore" inaccuracy exists. Of course the theory makes sense, and the thermal expansion does exist; rather, the lack of evidence points to the fact that the effect is so small as to be immeasurable. The lack of a measurable "cold-bore" effect could be the case for a few reasons.

First, **a single-metal object does not twist** due to a uniform temperature change. That is, if a free-floating barrel points one way, it always points that way, no matter the temperature. Similarly, a gas-powered semiautomatic usually has a gas tube made of the same material as the barrel, so it expands and contracts at the same rate. A metal barrel could deform if heating were unevenly distributed, but because the thermal conductivity of steel is approximately 2000 times that of air, heat quickly transfers to other parts of the rifle to evenly heat it.

Second, **the expansion itself is very small**. Carbon steel expands about 0.12% per 100°C (180°F). However, even a 100°C (180°F) change is a significant increase in temperature and rarely occurs in a barrel without

Thermal Expansion

continuous firing. If a 66 cm (26 in) barrel were heated up by 100°C (180°F), it would increase by a length of about 0.8 mm (0.03 in). And if the bore were 1 cm (0.4 in) in diameter, it would increase by 0.012 mm (0.0005 in). Notably, bore tolerance in mass-market rifles is already around 0.024 mm (0.0009 in). Such a tolerance is acceptable and performs well because bullets are designed to expand into rifling grooves as they travel through a bore.

That said, some marksmen do report a lower muzzle velocity from just a 0.012 mm difference in bore diameter due to the increased volume inside the bore for the gas to fill. The bore volume for a 100°C (180°F) increase amounts to 0.36%, or ~⅕ cm^3 (0.01 in^3). However even in this case, even though the gas must expand more into more volume, it has more time to do so since the bullet would take longer to leave the bore. Therefore, any decrease in velocity would be smaller than the increase in bore volume. Plus, to account for any increase in volume due to thermal expansion, a marksman would have to record and input the barrel's temperature as it actively changed, making any input by definition momentary and inaccurate.

Related to thermal expansion, some professionals claim that **heating a barrel during shooting relieves preexisting stresses** in the metal that are induced in the manufacturing process. These internal stresses do exist, but are already relieved by any reputable manufacturer by tempering the steel. Therefore, stress relief is a theoretical but unlikely possible source of heat-based inaccuracy for the first few shooting sessions with a new barrel.

Thermoelasticity is the relationship between the elastic properties of a material and its temperature. As materials heat up, materials stretch and deform more easily. The change in elasticity of steel is small at normal temperatures. Young's modulus (E) is a measurement that indicates how easily a material can stretch and deform. For steel, the Young's modulus decreases by roughly 2.16% for every 100°C (180°F) increase in temperature, within the range of 0°C to 300°C (32°F to 572°F). Even a 100°C (180°F) change is a significant increase in temperature, and firing a single bullet through a long-range rifle would not heat a barrel nearly that much. Therefore, there is no significant change to the Young's modulus from firing a single bullet through a long-range rifle.

A change in elasticity combined with a change in size could in theory affect a barrel's harmonics. "Barrel harmonics" is how the barrel shifts and moves as the shockwave of energy propagates through it from back to front. However, there is no reason to believe that a hot barrel would be significantly different from the harmonics of a cold one. Where harmonics do exist, they are very repeatable from one shot to the next, meaning they are accounted for during the zeroing process.

To cool down a rifle, a marksman can wait. But they can also slowly pour water down the barrel. As long as the barrel is below the tempering temperature of steel (up to approximately 150°C (302°F)), the micro-structure of the steel would remain unaffected no matter how quickly the steel cools.

11. Magnus Force

The Magnus force, or Magnus effect, occurs when fluid flows across an object **perpendicular to the spinning object's spin-axis**. For example, a crosswind blowing past a bullet perpendicular to its spin-axis generates a Magnus force. In contrast, a headwind or tailwind both align with the axis of a bullet's rotation, so do not induce a Magnus force.

The Magnus force is caused by the different forces of resistance on either side of the rotating body as a fluid, in this case air, passes over it. Fluid experiences **more resistance as it moves faster over a surface**. Therefore, if air blows into a rotating object or an object moves through static air, the air experiences more resistance on the side that is rotating towards the airflow's origin than that side that is rotating away from the airflow's origin. (Whether the object, the air, or both are moving is irrelevant; all that matters is that there is a speed differential between the two.)

More resistance causes slower movement. Slower movement causes more pressure due to Bernoulli's principle. Therefore, when an object rotates perpendicular to an incoming airflow, the object experiences a force that is 90 degrees from the air's direction of flow.

For example, for a right-twist bullet (i.e., spinning clockwise when viewed from the rear) in a right-to-left crosswind, there is a downward Magnus force. A left-twist bullet in the same crosswind experiences an upward force.

Magnus Force for Left-to-Right Crosswind
(Flip the red arrows for Right-to-Left Crosswind)

Right-Twist Bullet   Left-Twist Bullet

Magnus Force — Less resistance, Faster air, Lower pressure

More resistance, Slower air, Higher pressure

Airflow

More resistance, Slower air, High pressure

Less resistance, Faster air, Lower pressure

Magnus Force

Image 109: The Magnus force occurs when a **spinning object** moves through and spins perpendicular to a **moving fluid**. This image is of the back of a bullet, where the airflow is a crosswind. The surface of the bullet that is spinning with the air causes less resistance than the surface spinning against the direction of the air. More resistance slows the air down more. Slower air causes more pressure according to Bernoulli's principle. Therefore, spinning in a fluid creates a force that is 90 degrees perpendicular to the direction the fluid is flowing. For bullets, this effect is negligible.

Reversing the direction of the crosswind also flips the direction of the Magnus force. In this way, **the Magnus force works in the opposite direction of aerodynamic jump**.

The magnitude of the Magnus force is determined by many factors. Because a major factor of resistance is friction, bullets that have rougher surfaces experience more Magnus force. The second factor is surface area for the same reason: a larger surface area has more friction than a smaller area. Third is the difference in relative speed that air experiences when blowing over either side of the bullet. The higher the relative speed differential (e.g., the faster the crosswind), the greater the Magnus force.

In practical terms, **the Magnus force on a bullet is insignificant** for a few reasons. First, the Magnus force depends on the surface of the bullet gripping the air. At supersonic speeds, the surface of the bullet (including the rifling marks) is covered by a boundary layer of air, reducing direct contact with the sideways airflow. Second, as the bullet travels in an airflow, it weathervanes into the direction of incoming air, progressively reducing the amount of air that passes over the bullet perpendicular to its spin-axis.

It is purported that the Magnus force can be measured in theory whenever winds exceed 15 km/h; however, since the force works directly opposite of a stronger force that scales in the same way (i.e., aerodynamic jump), there is no good way to measure it separately.

That said, the Magnus force can significantly affect a bullet's stability as it spins. Due to manufacturing imperfections, the bullet's spin-axis does not perfectly align with its center-of-mass. Additionally, as the bullet moves through air pockets with slightly different densities, its spin-axis also misaligns with the center-of-pressure. These factors cause the bullet to wobble in the air, resulting in complex Magnus-force effects. This destabilizing influence makes theoretical ballistic coefficients less accurate compared to those obtained from empirical measurements. However, these complex stability issues are not something that an individual marksman can effectively account for.

12. Coning and Nutation

In addition to one large precession (See Spin Stabilization (Precession), Pg. 74.), bullets also have secondary and simultaneous secondary rotations, called nutations. (The combined pattern of precession and nutations are sometimes referred to as "epicyclic swerve.") (See Image 111, Pg. 119.)

Nutation appears as a wobbling motion of the bullet's spin-axis around its precession-axis during flight. (See Image 110, Pg. 119.) Because nutations are **secondary rotations**, when they are averaged together, the precession is unaffected. Therefore, nutations are impractical to measure individually. And because they can be averaged together without affecting the precession, they have a nominal net effect on a bullet's trajectory.

13. Additional Effects

There are a few other forces that are ignored when calculating bullet trajectory because they are simply too small to detect or are only theoretical. For example, all texts assume that bullets are perfectly rigid to make calculations

Nutations

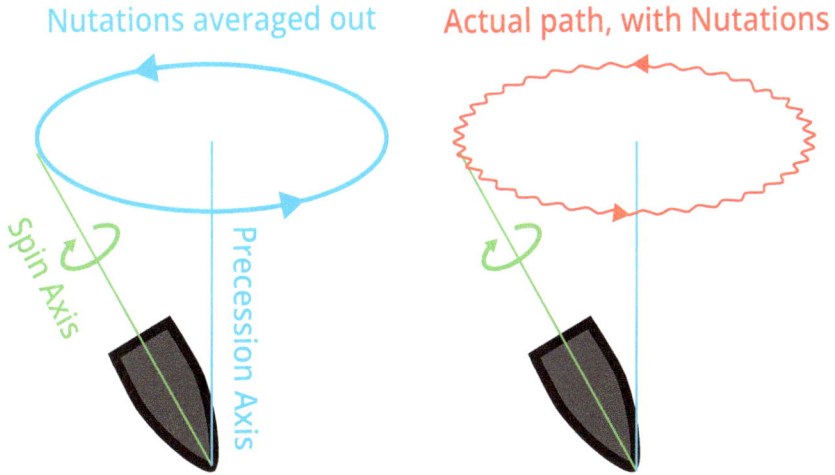

Image 110: Spin-stabilized objects actually have **two methods of rotation**. The first is the large circle that the spin-axis follows, called the precession. However, along the precession circle, the spin-axis also travels in many more smaller, **secondary circles, called nutations**. Nutations can vary greatly in size. Gyroscopes often have very small nutations as they precess in a circle. In contrast, bullets can have nutations that are very large in proportion to their primary circle of precession. Dynamic stability can be compromised if either the precession or the nutation become unstable. (See Image 96, Pg. 98.)

Image 111: This is a two-dimensional view of where an actual bullet's spin-axis pointed over time as the bullet traveled forward a set distance. Although the spin-axis followed a generally circular path, there were many smaller, secondary rotations (i.e., nutations) that complicated the curve.

simple; however, every material has its own elastic properties. In theory, bullets flex as they fly, affecting their ballistic properties ever so slightly.

Different material properties even apply to the air itself. While air density effects are considered for air resistance, different air density also affects other forces. For example, the Magnus force depends on the viscosity of the air. Also, as the bullet flies, the center-of-pressure shifts along the bullet, and how that happens is also influenced by the properties of the air. For another example, all objects in a fluid experience buoyancy forces, which are also affected by the properties of the air.

Another insignificant approximation is when continuous paths are entered into formulas as points. For example, when calculating the Coriolis effect, only the initial and final coordinates are used in the equation. However, as the bullet curves, it changes latitude, and a new instant Coriolis force would have to be calculated for complete accuracy. In theory, a true Coriolis force would require calculus to calculate; however, using just one coordinate is accurate enough for the purposes of long-range shooting.

The state of the rifle itself changes between different rounds. A common example is powder combustion residue, which is left over from each shot. Also, each shot degrades the chamber to some microscopic effect, which necessarily alters a bullet's path to an insignificant amount.

Some effects are theoretical, if not apocryphal. For example, the Poisson Effect supposes that the tip of the projectile above the direction of travel causes an air cushion to build up underneath it. Therefore, there is an increase of friction between this cushion and the bullet so that the bullet tends to spin off the cushion and move sideways.

In the future, scientists, engineers, and programmers may eventually not only develop computer programs that can account for these effects, but also build a robot that can take advantage of that program. However, as it stands now, **humans can accurately shoot for kilometers and miles while only using the knowledge carefully described in previous chapters.**

Best of luck shooting!

Appendices

14. Glossary   123

15. Credits   127

Appendices

The shepherd drives the wolf from the sheep's throat, for which the sheep thanks the shepherd as his liberator, while the wolf denounces him for the same act as the destroyer of liberty.

Plainly, the sheep and the wolf are not agreed upon a definition of liberty.
—*Abraham Lincoln, 16th commander-in-chief of the United States Army*

14. Glossary

Aerodynamic Jump	The minor vertical shift in a bullet's trajectory caused by a crosswind rotating a bullet sideways.
Airflow	The movement of air across and over a bullet or object, influencing drag and flight stability.
Ambushing	A shooting technique where the marksman holds the sights still and fires the moment a moving target crosses the point-of-aim.
Angle-of-Yaw	The angle between a projectile's longitudinal axis and its actual flight path.
Angular Momentum	Angular momentum is momentum going in a circle. It's conserved, which means the total amount of momentum must be the same unless acted on by a force.
Average Velocity	The mean speed of a projectile over a specified distance.
Barrel Mirage	Heat waves rising from a warm barrel that create a visual distortion, potentially affecting the marksman's aim.
Barrel Twist	The rate at which a rifle's rifling spins a bullet, typically measured in centimeters or inches per turn.
Berm	An elevated earthen backstop designed to safely contain bullets and reduce ricochets on a shooting range.
Blasting Cap	A small, sensitive explosive component used to initiate a larger explosive charge.
Bullet Spin	The rotational motion imparted on a projectile by rifling to enhance its stability in flight.
Bullet Trace	The visible disturbance due to pockets of different air pressure left in a bullet's path, used by marksmen to spot impacts.
Bullet Wake	The turbulent airflow behind a projectile that influences drag and downrange stability.
Buoyancy Forces	The minor upward lift on a projectile caused by density differences in the surrounding air.
Chronograph	A device that measures a projectile's velocity by detecting its passage across multiple sensors.
Clip-Ons	Auxiliary optic or imaging attachments that mount directly onto a rifle's primary scope for specialized viewing.

Cone Cells	Photoreceptor cells in the human eye responsible for high-acuity color vision critical for precise aiming.
Coning	The slight circular or oscillatory motion of a spinning bullet's nose around its primary axis of flight. The spin-axis forms a cone shape as it rotates around the mean center-of-mass.
Coriolis Effects	The apparent drift of a bullet's trajectory due to Earth's rotation.
De Minimis	A term indicating that an effect or amount is so minor as to be considered negligible.
Detonating (Det) Cord	A flexible explosive line that rapidly transmits a detonation wave between connected charges.
Drag Coefficient	A dimensionless value indicating how smoothly or turbulently a projectile passes through air.
Dynamic Instability	The tendency of a spinning projectile to not align the spin-axis to the applied force. A dynamically unstable bullet does not align its spin-axis to the incoming airflow.
Eotvos Effect	A change in the amount of centrifugal force an object experiences when it is fired East or West.
Epicyclic Swerve	The small, cyclic oscillation of a spinning bullet's nose around its primary flight path. See nutation.
E-type NATO Target	A standard, life-size silhouette target commonly used in military training to simulate an upright human figure.
Fragmentation Bullets	Projectiles engineered to break apart on impact, creating multiple wound channels for increased lethality.
Free Floating Barrel	A rifle barrel design that ensures no contact with the stock or fore-end, reducing external pressures for improved accuracy.
Frost Trail	A transient condensation trail formed behind a projectile under specific cold and humid atmospheric conditions.
Gaskets	Sealing components used in firearms or explosive devices to prevent unwanted gas or fluid leakage.
Gimbal Stabilizer	A pivoting support mechanism that keeps mounted equipment (e.g., cameras or scopes) steady despite external movement.
Glint	A noticeable flash of light reflecting off a surface.
Grains	A unit of weight (1/7000th of a pound) commonly used to measure bullets and powder charges in ammunition.
Grooves	The helical channels cut into a barrel's rifling that grip the bullet and impart stabilizing spin.
Hashmark	The reference lines or markings on a scope reticle used for precise aiming, windage, or elevation adjustments.
Haunch	The upper part of a deer's thigh.
Headwinds	Winds blowing directly against a projectile's flight path, decreasing its velocity and requiring additional elevation compensation.
Helical Spinning	The corkscrew-like motion of a bullet traveling through the air due to its rifling-induced rotation.
Hollow-Point Bullets	Projectiles with a concavity in the nose designed to expand on impact for increased energy transfer and wound channel size.
Housings	Protective or structural enclosures (such as scope bodies) that safeguard internal components from environmental damage.
Incoming Airflow	The volume and direction of air meeting a projectile in flight, influencing its drag and overall trajectory.

Jerk	To pull a trigger too fast so that the rifle shakes. Also, the rate of change of acceleration (third derivative of velocity), often felt as a sudden jolt in recoil dynamics.
Lateral Throwoff	A sideways deviation in bullet flight caused by uneven barrel harmonics or slight imbalances in bullet spin.
Laze	To use a laser rangefinder to measure target distance, enabling precise ballistic calculations.
Lead	The forward aim point applied to track and hit a moving target, compensating for its direction and speed.
Loadouts	Configurations of ammunition and equipment that a marksman carries, tailored to specific mission or shooting requirements.
Mach	A measure of speed relative to the speed of sound (Mach 1), used to describe supersonic, transonic, or subsonic projectile travel.
Magnus Force	A lift force acting on spinning projectiles due to asymmetric airflow around their rotating surfaces.
Mean Center-of-Gravity	The averaged or effective point where the center-of-gravity for an object would be for a moving object if the mass at all locations were averaged together.
Mil	To measure angular distance with a reticle.
Milligals	A unit of acceleration (1/1000 of a gal) used in precise gravitational measurements relevant to ballistic calculations.
Movement Value	A fraction by which the actual movement speed is multiplied to attain the equivalent sideways movement speed, perpendicular to the bullet's trajectory.
Nutation	The small, conical wobble of a spinning bullet's nose around its circle of precession.
Over-stabilization	A myth that a projectile's spin can be so high that it resists aerodynamic forces, potentially compromising trajectory and terminal performance.
Overturning Force	An aerodynamic force acting on a spinning projectile's nose, attempting to tilt it away from its flight path. Most commonly, this is the lift force.
Parallax	An optical effect where the reticle and target appear to move relative to each other if the marksman's eye position shifts.
Pitch	The angular rotation of a projectile or firearm around its transverse axis, influencing elevation and point of impact.
Precession	A principle of gyroscopes (e.g., a spinning bullet) that states when a force is applied to a spinning object, the force occurs in part 90 degrees later in the direction of rotation.
Rifling	The helical grooves inside a firearm's barrel that impart spin to a bullet, enhancing its stability in flight.
Right-Twist	A rifling pattern in a barrel that rotates clockwise away from the marksman. A projectile spins clockwise when viewed from the back as it travels through the barrel.
Rod Cells	Photoreceptor cells in the human eye responsible for low-light and peripheral vision, aiding target detection in dim conditions.
Spall	Fragmented material that flakes or breaks off a surface (e.g., armor or bullet) upon impact or extreme stress.
Spider Strap	A multi-armed securing system used to stabilize equipment or a person. It is named as the different arms loosely resemble a spider's web.

Spin Drift	A phenomena whereby the precession of a spinning bullet transfers gravitational energy into a lateral force that pushes a bullet to the side of its twist. A right-twisted bullet drifts to the right and a left-twisted bullet to the left.
Spin-Stabilized	Describes a projectile that maintains stable flight primarily through precession.
Spotter	An observer who assists a marksman by estimating range, calling windage, and correcting shots for more precise impacts. The spotter is usually more experienced than their paired marksman.
Spotting Scope	A high-powered optic used to observe targets, bullet trace, and impacts at extended distances.
Static Stability	A gyroscope's (e.g., a bullet's) ability to resist a change in orientation and maintain its orientation in flight despite the forces being applied to it.
Supine Position	A shooting stance where the marksman lies on their back with the firearm angled forward toward the target.
Tailwinds	Winds blowing in the same direction as a projectile's flight, effectively increasing its speed and flattening its trajectory.
Thermal Expansion	The tendency of materials, including firearm barrels, to expand when heated, potentially affecting accuracy.
Trace	The mirage that follows a supersonic bullet as it changes the air density along its trajectory as it travels.
Tracers	Projectiles with an added incendiary compound that burns brightly, helping marksmen observe the bullet's path.
Tracking	Consistently adjusting sights to follow a moving target or observe bullet impact.
Tractability	The ability of a spin-stabilized bullet to reorient its spin-axis to the direction of incoming airflow.
Transonic Range	The range of speeds where some of the air surrounding a bullet travels at supersonic speeds while other air travels at subsonic speeds.
Truing	The process of adjusting a rifle's ballistic data or scope settings based on observed impact points at various distances.
Truing Bar	A rigid reference rod or device used to verify that an actual trajectory matches a ballistic calculator's predicted trajectory.
Tumbling	The end-over-end flipping of a projectile in flight.
Twist Rate	The distance in a barrel (in inches) required for the rifling to complete one full rotation of a bullet.
Vapor Trail	The visible condensation path (i.e., a cloud) behind a fast-moving bullet that is created under specific atmospheric conditions.
Wave Drag	The aerodynamic resistance created by shock waves forming around a supersonic projectile.
Weathervaning	A projectile's natural tendency to turn its nose into the oncoming airflow, reducing drag and improving stability.
Windage Hold	The horizontal sight adjustment (measured in angular units) applied to compensate for crosswind drift.
Wobble	A minor oscillation of a spinning bullet's nose that dampens over distance, affecting flight precision.
Yaw	The angular deviation between a bullet's longitudinal axis and its actual trajectory.

Yaw-of-Repose	A small, steady angular offset adopted by a spinning bullet due to aerodynamic and gyroscopic forces.
Young's Modulus	A measure of a material's stiffness, indicating its resistance to elastic deformation under stress (such as in barrel flex).

15. Credits

The explicit and implied contents herein do not imply or constitute endorsement by the U.S. DOD or any of its branches.

Front Cover: U.S.M.C. CPL Manuel A. Estrada
Back Cover 1: Louisiana Army N.G. SSG Greg Stevens
Back Cover 2: U.S. Army SPC Uriel Ramirez
Back Cover 3: U.S. Army SGT Patrik Orcutt
Back Cover 4: U.S.M.C. CPL Amelia Kang
TOC Image 1: U.S. Army SSG Austin Berner
TOC Image 2: U.S.M.C SGT Esdras Ruano
TOC Image 3: U.S.M.C LCPL Juan Carpanzano
CT Image 1: U.S.M.C CPL Brandon Salas
CT Image 2: U.S.M.C CPL Khalil Ross
CT Image 3: U.S. Army SPC Samuel Hyer
CT Image 4: U.S. Army SGT Patrik Orcutt
Image 1: U.S.M.C PFC Eric Keenan
Image 5: Scott Bauer
Image 8: U.S. Army 1LT Benjamin Haulenbeek
Image 9: Anne Barth
Image 10: U.S. Army 1LT Benjamin Haulenbeek
Image 11: U.S. Army Justin Connaher
Image 12: U.S. Army SGT Patrik Orcutt
Image 13: U.S.M.C CPL Kirstin Merrimarahajara
Image 14: U.S.M.C LCPL Paley Fenner
Image 15: U.S.M.C CPL Ricky Gomez
Image 16: U.S. Army N.G. SGT Remi Milslagle
Image 17: U.S. Navy Seaman Christopher Williamson
Image 18: U.S. Army SGT Patrik Orcutt
Image 19: U.S. Army K. Kassens
Image 21: Warren LeMay
Image 23: U.S.M.C CPL Dean Gurule
Image 24: U.S. Army K. Kassens
Image 25: U.S. Army SSG Russell Klika
Image 26: U.S. Army SGT Connor Mendez
Image 27: U.S. Army SPC Taylor Shaffer
Image 28: U.S. Army SPC Taylor Shaffer
Image 29: U.S. Army SSG Sidney Sale
Image 30: U.S. Air Force Justin Connaher
Image 31: U.S. Army SPC Zachary Bouvier
Image 32: U.S.M.C SSGT Jestin Costa
Image 33: U.S. Army K. Kassens
Image 34: U.S. Army SPC Taylor Shaffer
Image 35: U.S. Army SSG William Waller
Image 36: U.S. Army SGT Patrik Orcutt
Image 37: U.S.M.C SGT Tia Nagle
Image 38: U.S. Army K. Kassens
Image 39: U.S.M.C CPL Matthew A. Callaha
Image 40: U.S. Army SGT Connor Mendez
Image 41: U.S.M.C SGT Tia Nagle
Image 43: U.S.M.C SGT Tia Nagle
Image 44: U.S.M.C SGT Tia Nagle
Image 45: U.S. Army PVT Austin Anyzeski
Image 46: U.S. Army SPC Caroline Schofer
Image 47: U.S.M.C CPL Sean Potter
Image 48: U.S. Air Force SRA Brennen Lege
Image 49: Courtesy photo
Image 50: Courtesy photo
Image 51: U.S. Air Force SSGT Trevor McBride
Image 52: U.S.M.C CPL Israel Chincio
Image 53: U.S.M.C CPL Diana Jimenez
Image 54: U.S. Army SPC Michael J MacLeod
Image 55: U.S.M.C SGT Danny Gonzalez
Image 56: U.S.M.C SGT Isaac Ibarra

Image 57: U.S. Army SPC Kimberly Gonzalez
Image 58: JIP
Image 59: Kh1604
Image 60: U.S. Army PFC Lloyd Justine Villanueva
Image 61: U.S. Army SGT Karen Sampson
Image 62: U.S.M.C CPL Manuel A. Estrada
Image 63: U.S.M.C CPL Manuel A. Estrada
Image 65: U.S. Air Force CPT Kippun Sumner
Image 66: U.S. Army SGT Steven Lewis
Image 67: U.S. Army SSG Malcolm Cohens-Ashley
Image 69: U.S.M.C CPL Aaron Patterson
Image 70: U.S.M.C CPL Demetrius Morgan
Image 76: Albarubescens
Image 72: Pollyanna1919
Image 73: Vic2015
Image 87: Robert L. McCoy
Image 88: Bryan Litz
Image 89: Bryan Litz
Image 92: James A. Boatright and Gustavo F. Ruiz
Image 93: Hornady
Image 98: Bryan Litz
Image 99: Bryan Litz
Image 100: Bryan Litz
Image 101: Bryan Litz
Image 106: NASA and German Aerospace Center (DLR)
Image 107: Jacek Halicki
Image 108: Jan Polák
Image 111: James A. Boatright and Gustavo F. Ruiz
LRSU Front Cover: U.S. Air Force TSGT Gregory Brook
LRSU Back Cover 1: U.S. Army K. Kassens
LRSU Back Cover 2: U.S. Army CPT Thomas Cieslak
LRSU Back Cover 3: U.S.M.C. CPL Ricky S. Gomez
LRSU Back Cover 4: U.S. Army SSG Oscar Gollaz

Please leave a review. Thanks!

Positive reviews from awesome people like you help other people to benefit from the valuable instructions in this manual. Could you take 60 seconds to share your thoughts?

Thank you in advance for helping the community!

If you liked this book, consider buying, *Long Range Shooting.*

Or consider both books together, *Long Range Shooting United.*

www.ingramcontent.com/pod-product-compliance
Lightning Source LLC
Chambersburg PA
CBHW041913190326
41458CB00023B/6251